Padé Approximants

Part I: Basic Theory

GIAN-CARLO ROTA, *Editor*
ENCYCLOPEDIA OF MATHEMATICS AND ITS APPLICATIONS

GIAN-CARLO ROTA, *Editor*
ENCYCLOPEDIA OF MATHEMATICS AND ITS APPLICATIONS

Other volumes in preparation

ENCYCLOPEDIA
OF MATHEMATICS
and Its Applications

GIAN-CARLO ROTA, Editor
Department of Mathematics
Massachusetts Institute of Technology
Cambridge, Massachusetts

Editorial Board

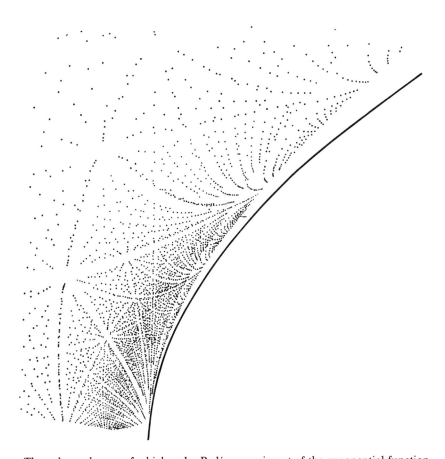

The poles and zeros of a high order Padé approximant of the exponential function delineate remarkable trajectories in the complex plane. The illustration, reproduced by kind permission of Professors E. B. Saff and R. S. Varga, conspicuously demonstrates the interplay and common features of many such trajectories. See pages 229 ff. for further details.

GIAN-CARLO ROTA, *Editor*

ENCYCLOPEDIA OF MATHEMATICS AND ITS APPLICATIONS

Volume 13

Section: Mathematics of Physics
Peter A. Carruthers, *Section Editor*

Padé Approximants
Part I: Basic Theory

George A. Baker, Jr.
Los Alamos National Laboratory
Los Alamos, New Mexico

Peter Graves-Morris
University of Kent
Canterbury, Kent, England

Foreword by
Peter A. Carruthers
Los Alamos National Laboratory

1981

Addison-Wesley Publishing Company
Advanced Book Program
Reading, Massachusetts

London · Amsterdam · Don Mills, Ontario · Sydney · Tokyo

Library of Congress Cataloging in Publication Data

Baker, George A. (George Allen), Jr., 1932–
  Padé approximants.

  (Encyclopedia of mathematics and its applica-
tions; 13–14. Section, Mathematics of physics)
    Bibliography: p.
    Includes index.
    Contents: pt. 1. Basic theory—pt. 2. Extensions
and applications.
    1. Padé approximant. I. Graves-Morris, P. R. (Peter Russell), 1941–
II. Title. III. Series: Encyclopedia of mathe-
    matics and its applications; v. 13–14.
IV. Series: Encyclopedia of mathematics and its applications.
Section, Mathematics of physics.
QC20.7.P3B35   530.1′514   81-3546
ISBN 0-201-13512-4 (v. 1)   AACR2
ISBN 0-201-13513-2 (v. 2)

American Mathematical Society (MOS) Subject Classification Scheme (1980): 41A21

To Our Wives
Elizabeth Baker *and* **Lucia Graves-Morris**
and to Our Families

Contents of Part I

Contents of Part II

Editor's Statement

A large body of mathematics consists of facts that can be presented and described much like any other natural phenomenon. These facts, at times explicitly brought out as theorems, at other times concealed within a proof, make up most of the applications of mathematics, and are the most likely to survive changes of style and of interest.

This ENCYCLOPEDIA will attempt to present the factual body of all mathematics. Clarity of exposition, accessibility to the non-specialist, and a thorough bibliography are required of each author. Volumes will appear in no particular order, but will be organized into sections, each one comprising a recognizable branch of present-day mathematics. Numbers of volumes and sections will be reconsidered as times and needs change.

It is hoped that this enterprise will make mathematics more widely used where it is needed, and more accessible in fields in which it can be applied but where it has not yet penetrated because of insufficient information.

GIAN-CARLO ROTA

Foreword

New insights into mathematical problems and physical applications continue to arise from the study of power series representation of functions. The present work is devoted to an intensive description of the Padé approximant technique*. The value of this scheme of approximation in a wide variety of physical problems has been increasingly recognized in recent years. This two-part presentation is a fine example of the interplay between physics and mathematics, each stimulating the other to new concepts and techniques.

One could not imagine better qualified authors for a contemporary major set of volumes on Padé approximants. Baker and Graves-Morris are widely known for their original contributions both to the mathematics and serious physical applications. The result is a lucid and explicit treatment of the subject which does not compromise mathematical accuracy yet is easily accessible to the modern theorist.

We may mention that this work is an example of a healthy trend developing in recent years in which modern mathematical developments are increasingly providing the language in which the most advanced physical theories are expressed. In the present case the renaissance in statistical mechanics and field theory studies in recent years has required such developments as Wilson's renormalization group method and Padé approximants. We may also mention the serious studies of continuous groups and their representations inspired by efforts to unite the weak, electromagnetic, strong, and gravitational forces. These same theories seem to be best formulated as non-Abelian gauge field theories, whose content and consequences involve the concepts of differential geometry and topology.

Stated briefly, the Padé approximant represents a function by the ratio of two polynomials. The coefficients of the powers occurring in the polynomials are, however, determined by the coefficients in the Taylor series expansion of the function. Thus, given a power series expansion

$$f(z) = c_0 + c_1 z + c_2 z^2 \dots$$

we can by the methods described in the text, make an optimal choice of the

---

*Among other related techniques we mention continued fractions treated in volume 11 of this Encyclopedia (Jones and Thron).

coefficients a_i, b_i in the Padé approximant

$$\frac{a_0 + a_1 z + \cdots + a_L z^L}{b_0 + b_1 z + \cdots + b_M z^M}$$

Exploitation of this simple idea and its extensions has led to many insights and by now has become practically a major industry. I shall not spoil the story by revealing more of the plot.

Inspection of the table of contents reveals an intensive development of the mathematical texture inherent in the subject. Many excellent examples illustrate the concepts. Some recent results appear here for the first time in monograph form. Among these are included the developments of reliable algorithms (I.2.1, 2.4, 4.5, and II.1.1), Saff-Varga theorems on Padé approximants to the exponential (I.5.7), the theory of convergence in capacity of Pommerenke (I.6.6), Canterbury (two variable) approximants (II.1.4) and results from $\lambda\phi^4$ Euclidean field theory derived using Padé approximants. In addition the approach to Laguerre's method in (II.3.7) is new. The treatment of applications is well done and has sufficient depth to be useful to the research scientist. Part II, Chapter 2, describes the connection with integral equations and quantum mechanics. The connection with numerical analysis is made in Part II, Chapter 3. The authors close with a frontier topic, the application of Padé approximants to problems in quantum field theory. Finally, an extensive bibliography documents the subject and provides references to the treatment of further related problems.

This two-volume presentation is a fine example of a creative review because it weaves together the vital ideas of the subject of Padé approximants and sets the stage for vigorous new developments in theory and applications. It should fill this role for some time to come.

PETER A. CARRUTHERS
General Editor, Section on Mathematics of Physics

Preface

These two volumes are intended to serve as a basic text on one approach to the problem of assigning a value to a power series. We have attempted to present the basic results and methods in as transparent a form as possible, in line with the general objectives of the Encyclopedia. The general topic of Padé Approximants, which is, among other things, a highly practical method of definition and of construction of the value of a power series, seems to have begun independently at least twice. Padé's claim for credit is based on his thesis (1892), in which he developed the approximants and organized them in a table. He paid particular attention to the exponential function. He was presumably unaware of the prior work of Jacobi (1846), who gave the determinantal representation in his paper on the simplification of Cauchy's solution to the problem of rational interpolation. Also, Padé's work was preceded by that of Frobenius (1881), who derived identities between the neighboring rational fractions of Jacobi. It is interesting to note that Anderson seems to have stumbled upon some Padé Approximants for the logarithmic function in 1740. A photograph of H. Padé is to be found in *The Padé Approximant Method and Its Application to Mechanics*, edited by H. Cabannes. A copy of his autographed thesis is to be found in the Cornell University Library.

This work has been distilled from an extensive literature, and *The Essentials of Padé Approximants*, written by one of us, has been an essential reference. We use the abbreviation EPA for this book, and refer to it often for a different or fuller treatment of some of the more advanced topics. While each book is entirely self-contained, our notation is normally compatible with EPA, and to a large extent the books complement each other. An important exception is that the Padé table in EPA is reflected through its main diagonal in our present notation. The proceedings of the Canterbury Summer School and International Conference, edited by the other of us, contain diverse contributions which initiated in print the multidisciplinary view of the subject—a view we hope we have transmitted herein. The many publications which have contributed substantially to our text are listed in the bibliography. We are grateful to our numerous colleagues at Brookhaven, Canterbury, Cornell, Los Alamos, and Saclay in freely discussing so many topics which have made possible the breadth of our treatment. Especially, we thank Roy Chisholm, John Gammel, and Daniel Bessis for many conversations, and the C.E.A. at Saclay, where part of this book was written, for hospitality.

Our hardest task in writing this book was to choose a presentation which is both correct and readily comprehensible. A fully precise system based on rigorous analysis and set-theoretic language would have ensured total obscurity of the more practical techniques. Conversely, omission of all the conditions under which the theorems hold good would be absurdly misleading. We have chosen a level of presentation suitable for the topic in hand. For example, the connectivity of sets is mentioned where it is important, and otherwise it is omitted. The meaning of the order notation is clear in context. Both applications in physics and techniques recently developed are treated in a practical fashion.

Equations are referenced by a default option. Equation (I.6.5.3) is Equation (5.3) of Part I, Chapter 6; the Part and Chapter are dropped by default if they are the same as the source of the reference.

Finally, a spirit of evangelism may be detected in the text. When a review of rational approximation in 1963 can claim that Padé approximants cannot approximate on the entire range $(0, \infty)$ and be believed, a revision of view is overdue.

GEORGE A. BAKER, JR.
PETER GRAVES-MORRIS

Padé Approximants

Part I: Basic Theory

CHAPTER 1

Introduction and Definitions

1.1 Introduction and Notational Conventions

Suppose that we are given a power series $\sum_{i=0}^{\infty} c_i z^i$, representing a function $f(z)$, so that

$$f(z) = \sum_{i=0}^{\infty} c_i z^i. \tag{1.1}$$

This expansion is the fundamental starting point of any analysis using Padé approximants. Throughout this work we reserve the notation c_i, $i=0,1,2,\ldots$, for the given set of coefficients, and $f(z)$ is the associated function. A Padé approximant is a rational fraction

$$[L/M] = \frac{a_0 + a_1 z + \cdots + a_L z^L}{b_0 + b_1 z + \cdots + b_M z^M} \tag{1.2}$$

which has a Maclaurin expansion which agrees with (1.1) as far as possible. We give a more complete and precise definition of Padé approximants in Section 1.4. Notice that in (1.1) there are $L+1$ numerator coefficients and $M+1$ denominator coefficients. There is a more or less irrelevant common factor between them, and for definiteness we take $b_0 = 1$. This choice turns out to be an essential part of the precise definition, and (1.2) is our conventional notation with this choice for b_0. So there are $L+1$ independent numerator coefficients and M independent denominator coefficients, making $L+M+1$ unknown coefficients in all. This number suggests that normally the $[L/M]$ ought to fit the power series (1.1) through the orders

ENCYCLOPEDIA OF MATHEMATICS and Its Applications, Gian-Carlo Rota (ed.). Vol. 13: George A. Baker, Jr., and Peter R. Graves-Morris, Padé Approximants: Basic Theory, Part I   ISBN 0-201-13512-4

$1, z, z^2, \ldots, z^{L+M}$. In the notation of formal power series,

$$\sum_{i=0}^{\infty} c_i z^i = \frac{a_0 + a_1 z + \cdots + a_L z^L}{b_0 + b_1 z + \cdots + b_M z^M} + O(z^{L+M+1}). \qquad (1.3)$$

Example.

$$f(z) = 1 - \tfrac{1}{2}z + \tfrac{1}{3}z^2 + \cdots,$$

$$[1/0] = 1 - \tfrac{1}{2}z = f(z) + O(z^2),$$

$$[0/1] = \frac{1}{1 + \tfrac{1}{2}z} = f(z) + O(z^2),$$

$$[1/1] = \frac{1 + \tfrac{1}{6}z}{1 + \tfrac{2}{3}z} = f(z) + O(z^3).$$

Returning to (1.3) and cross-multiplying, we find that

$$\left(b_0 + b_1 z + \cdots + b_M z^M\right)\left(c_0 + c_1 z + \cdots\right)$$
$$= a_0 + a_1 z + \cdots + a_L z^L + O(z^{L+M+1}) \quad (1.4)$$

Equating the coefficients of $z^{L+1}, z^{L+2}, \ldots, z^{L+M}$, we find

$$b_M c_{L-M+1} + b_{M-1} c_{L-M+2} + \cdots + b_0 c_{L+1} = 0,$$
$$b_M c_{L-M+2} + b_{M-1} c_{L-M+3} + \cdots + b_0 c_{L+2} = 0,$$
$$\vdots \qquad (1.5)$$
$$b_M c_L \qquad + b_{M-1} c_{L+1} \qquad + \cdots + b_0 c_{L+M} = 0.$$

If $j < 0$, we define $c_j = 0$ for consistency. Since $b_0 = 1$, Equations (1.5) become a set of M linear equations for the M unknown denominator coefficients:

$$\begin{vmatrix} c_{L-M+1} & c_{L-M+2} & c_{L-M+3} & \cdots & c_L \\ c_{L-M+2} & c_{L-M+3} & c_{L-M+4} & \cdots & c_{L+1} \\ c_{L-M+3} & c_{L-M+4} & c_{L-M+5} & \cdots & c_{L+2} \\ \vdots & \vdots & \vdots & & \vdots \\ c_L & c_{L+1} & c_{L+2} & \cdots & c_{L+M-1} \end{vmatrix} \begin{pmatrix} b_M \\ b_{M-1} \\ b_{M-2} \\ \vdots \\ b_1 \end{pmatrix} = - \begin{pmatrix} c_{L+1} \\ c_{L+2} \\ c_{L+3} \\ \vdots \\ c_{L+M} \end{pmatrix},$$

$$(1.6)$$

from which the b_i may be found. The numerator coefficients, $a_0, a_1, \ldots, a_L$, follow immediately from (1.4) by equating the coefficients of $1, z, z^2, \ldots, z^L$:

$$a_0 = c_0,$$

$$a_1 = c_1 + b_1 c_0,$$

$$a_2 = c_2 + b_1 c_1 + b_2 c_0,$$

$$\vdots$$ (1.7)

$$a_L = c_L + \sum_{i=1}^{\min(L, M)} b_i c_{L-i}.$$

Thus (1.6) and (1.7) normally determine the Padé numerator and denominator and are called the Padé equations; we have constructed an $[L/M]$ Padé approximant which agrees with $\sum_{i=0}^{\infty} c_i z^i$ through order z^{L+M}. Because the starting point of these manipulations is the given power series, we do not ever need to know about the existence of any function $f(z)$ with $\sum_{i=0}^{\infty} c_i z^i$ as its Maclaurin series, as in (1.1). Of course, we expect that a well-chosen sequence of Padé approximants will normally approximate a function $f(z)$ with the Maclaurin expansion $\sum_{i=0}^{\infty} c_i z^i$, but it is important to distinguish between problems of convergence of Padé approximants and problems of construction of Padé approximants. Given the power series, (1.6) and (1.7) show how the Padé approximants are constructed.

Every power series has a circle of convergence $|z| = R$. If $|z| < R$, the series converges, and if $|z| > R$, it does not. If $R = \infty$, the power series represents an analytic function (functions analytic everywhere we often call *entire*) and the series may be summed directly for any value of z to yield the function $f(z)$. If $R = 0$, the power series is undoubtedly formal. It contains information about $f(z)$, but just how this information is to be used is not immediately clear. However, if a sequence of Padé approximants of the formal power series converges to a function $g(z)$ for $z \in \mathcal{D}$, then we may reasonably conclude that $g(z)$ is a function with the given power series. In certain circumstances (see Chapter 5) we make such statements precise and prove them. Nevertheless, in this book we will not be hampered by a lack of rigorous justification of any technique, and empirical convergence is regarded as entirely satisfactory within its limitations. If the given power series converges to the same function for $|z| < R$ with $0 < R < \infty$, then a sequence of Padé approximants may converge for $z \in \mathcal{D}$ where $\mathcal{D}$ is a domain larger than $|z| < R$. We will then have extended our domain of convergence. This is frequently a practical approach to what amounts to analytic continuation. The method of expansion and reexpansion due to Weierstrass is more suited to principle than practice. As an example of how well Padé approximants may work in their natural context, we consider an example.

Example.

$$f(z) = \sqrt{\frac{1 + \frac{1}{2}z}{1 + 2z}} = 1 - \tfrac{3}{4}z + \tfrac{39}{32}z^2 - \cdots.$$

To calculate [1/1], Equation (1.6) becomes

$$\left(-\tfrac{3}{4}\right)b_1 = -\tfrac{39}{32},$$

and so $b_1 = \tfrac{13}{8}$. Equation (1.7) gives $a_0 = 1$ and $a_1 = \tfrac{7}{8}$, with the check

$$\left(1 + \tfrac{13}{8}z\right)\left(1 - \tfrac{3}{4}z + \tfrac{39}{32}z^2\right) = 1 + \tfrac{7}{8}z + O(z^3).$$

Hence

$$[1/1] = \frac{1 + \tfrac{7}{8}z}{1 + \tfrac{13}{8}z},$$

and in Figure 1 we compare this with $f(z)$ for $z \geqslant 0$. In particular, $f(\infty) = 0.5$ and $[1/1](\infty) = \tfrac{7}{13} = 0.54\ldots$, giving 8% accuracy at infinity. This example shows remarkable accuracy for a function with a radius of convergence of $\tfrac{1}{2}$, using just three terms of the series.

There is one feature of the calculation of Padé approximations to be emphasized at the start—these calculations require more numerical accuracy than one might at first expect. The Padé approximant exploits the differences of the coefficients to do its long-range extrapolation, and so the differences must all be accurate. We consider the problem of deciding how much numerical accuracy is needed to calculate an $[L/M]$ Padé approximant in Section 2.4.

Thus far, we have assumed that Padé approximants are calculated directly from (1.6) and (1.7) without implying any particular method. If Cramer's rule is used, we may calculate $b_0 : b_1 : \cdots : b_M$ from (1.6) and hence the denominator of (1.2). Aside from a common factor, the result is

$$Q^{[L/M]}(z) = \begin{vmatrix} c_{L-M+1} & c_{L-M+2} & \cdots & c_L & c_{L+1} \\ c_{L-M+2} & c_{L-M+3} & \cdots & c_{L+1} & c_{L+2} \\ \vdots & \vdots & & \vdots & \vdots \\ c_{L-1} & c_L & \cdots & c_{L+M-2} & c_{L+M-1} \\ c_L & c_{L+1} & \cdots & c_{L+M-1} & c_{L+M} \\ z^M & z^{M-1} & \cdots & z & 1 \end{vmatrix}. \quad (1.8)$$

We take (1.8) to define $Q^{[L/M]}(z)$ and use this convention throughout.

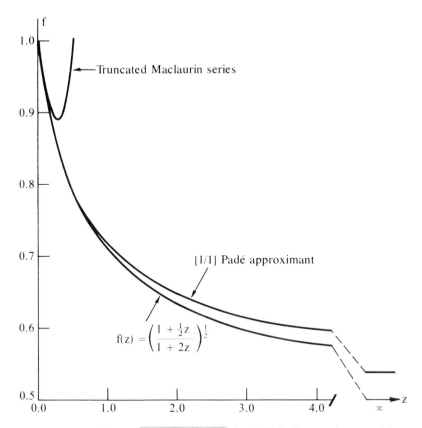

Figure 1. Values of $f(z)=\sqrt{(1+z/2)/(1+2z)}$, its $[1/1]$ Padé approximant, and its truncated Maclaurin series, $1-3z/4+39z^2/32$.

Again, recall that $c_j=0$ if $j<0$. Now consider

$$Q^{[L/M]}(z)\sum_{i=0}^{\infty}c_iz^i=\begin{vmatrix}c_{L-M+1} & c_{L-M+2} & \cdots & c_{L+1}\\ c_{L-M+2} & c_{L-M+3} & \cdots & c_{L+2}\\ \vdots & \vdots & & \vdots\\ c_{L-1} & c_L & \cdots & c_{L+M-1}\\ c_L & c_{L+1} & \cdots & c_{L+M}\\ \displaystyle\sum_{i=0}^{\infty}c_iz^{M+i} & \displaystyle\sum_{i=0}^{\infty}c_iz^{M+i-1} & \cdots & \displaystyle\sum_{i=0}^{\infty}c_iz^i\end{vmatrix}.$$

By subtracting z^{L+1} times the first row from the last, z^{L+2} times the second row from the last, etc., up to z^{L+M} times the penultimate row from the last, we reduce the series in the last row. They become lacunary series, with a gap of M terms missing. Using the initial terms of these series, we define

$$P^{[L/M]}(z) = \begin{vmatrix} c_{L-M+1} & c_{L-M+2} & \cdots & c_{L+1} \\ c_{L-M+2} & c_{L-M+3} & \cdots & c_{L+2} \\ \vdots & \vdots & & \vdots \\ c_{L-1} & c_L & \cdots & c_{L+M-1} \\ c_L & c_{L+1} & \cdots & c_{L+M} \\ \sum_{i=0}^{L-M} c_i z^{M+i} & \sum_{i=0}^{L-M+1} c_i z^{M+i-1} & \cdots & \sum_{i=0}^{L} c_i z^i \end{vmatrix}. \quad (1.9)$$

Again, (1.9) is our notational convention. We now prove our first theorem.

THEOREM 1.1.1. *With the definitions* (1.8) *and* (1.9),

$$Q^{[L/M]}(z) \sum_{i=0}^{\infty} c_i z^i - P^{[L/M]}(z) = O(z^{L+M+1}). \quad (1.10)$$

Proof. We note that $\deg\{P^{[L/M]}\} \leqslant L$, $\deg\{Q^{[L/M]}\} \leqslant M$ and that the remainder is

$$Q^{[L/M]}(z) \sum_{i=0}^{\infty} c_i z^i - P^{[L/M]}(z)$$

$$= \begin{vmatrix} c_{L-M+1} & c_{L-M+2} & \cdots & c_{L+1} \\ c_{L-M+2} & c_{L-M+3} & \cdots & c_{L+2} \\ \vdots & \vdots & & \vdots \\ c_{L-1} & c_L & \cdots & c_{L+M-1} \\ c_L & c_{L+1} & \cdots & c_{L+M} \\ \sum_{i=L+1}^{\infty} c_i z^{M+i} & \sum_{i=L+2}^{\infty} c_i z^{M+i-1} & \cdots & \sum_{i=L+M+1}^{\infty} c_i z^i \end{vmatrix}$$

$$= \sum_{i=1}^{\infty} z^{L+M+i} \begin{vmatrix} c_{L-M+1} & c_{L-M+2} & \cdots & c_{L+1} \\ c_{L-M+2} & c_{L-M+3} & \cdots & c_{L+2} \\ \vdots & \vdots & & \vdots \\ c_{L-1} & c_L & \cdots & c_{L+M-1} \\ c_L & c_{L+1} & \cdots & c_{L+M} \\ c_{L+i} & c_{L+i+1} & \cdots & c_{L+M+i} \end{vmatrix}. \quad (1.11)$$

Equation (1.11) is occasionally a useful form for the error using Padé approximation. Equation (1.10) goes a long way towards satisfying (1.3). To this end, consider

$$Q^{[L/M]}(0) = \begin{vmatrix} c_{L-M+1} & c_{L-M+2} & \cdots & c_L \\ c_{L-M+2} & c_{L+M+3} & \cdots & c_{L+1} \\ \vdots & \vdots & & \vdots \\ c_{L-1} & c_L & \cdots & c_{L+M-2} \\ c_L & c_{L+1} & \cdots & c_{L+M-1} \end{vmatrix}$$

This is called a Hankel determinant, because of the systematic way in which its rows are formed from the given coefficients c_i. Notice that if $Q^{[L/M]}(0) \neq 0$, then the linear equations (1.6) are nonsingular and the solution given by (1.8) is unambiguous. Furthermore, we may divide (1.10) by $Q^{[L/M]}(z)$, yielding

$$\sum_{i=0}^{\infty} c_i z^i - \frac{P^{[L/M]}(z)}{Q^{[L/M]}(z)} = O(z^{L+M+1}).$$

This result has proved our second theorem:

THEOREM 1.1.2 [Jacobi, 1846]. *With the definitions (1.8) and (1.9), the* $[L/M]$ *Padé approximant of* $\sum_{i=0}^{\infty} c_i z^i$ *is given by*

$$[L/M] = \frac{P^{[L/M]}(z)}{Q^{[L/M]}(z)} \tag{1.12}$$

provided $Q^{[L/M]}(0) \neq 0$.

The only difficulties, which we defer to Section 1.4, are those occurring when $Q^{[L/M]}(0) = 0$. We extend the notation $[L/M]$ of (1.12) as $[L/M]_f$ to emphasize approximation of $f(z)$, and as $[L/M](z)$ to emphasize the z-dependence. We will thus have the various forms

$$[L/M] = [L/M]_f = [L/M](z) = [L/M]_f(z)$$

available for convenience. It is common practice to display the approximants in a table, called the Padé table, shown as Table 1.

Table 1. The Padé table.

L \ M	0	1	2	$\cdots$
0	[0/0]	[1/0]	[2/0]	$\cdots$
1	[0/1]	[1/1]	[2/1]	$\cdots$
2	[0/2]	[1/2]	[2/2]	$\cdots$
$\vdots$	$\vdots$	$\vdots$	$\vdots$	$\ddots$

Table 2.   Part of the Padé table of $\exp(z)$ [Padé, 1892].

L M	0	1	2
0	$\dfrac{1}{1}$	$\dfrac{1+z}{1}$	$\dfrac{2+2z+z^2}{2}$
1	$\dfrac{1}{1-z}$	$\dfrac{2+z}{2-z}$	$\dfrac{6+4z+z^2}{6-2z}$
2	$\dfrac{2}{2-2z+z^2}$	$\dfrac{6+2z}{6-4z+z^2}$	$\dfrac{12+6z+z^2}{12-6z+z^2}$

Among other things, we prove in Section 1.2 that part of the Padé table of $\exp(z)$ is given by the entries in Table 2.

1.2 Padé Approximants to the Exponential Function

The coefficients c_i of the Maclaurin expansion of the exponential function are sufficiently simple that explicit forms of the numerator and denominator of the Padé approximants can be found. In this section we will calculate the denominator $Q^{[L/M]}(z)$. The numerator follows by an extremely simple and elegant trick, based on the identity $\exp(-z) = 1/\exp(z)$, and this derivation is discussed in Section 1.5. Padé, in his thesis, elaborated the properties of his approximants with special emphasis on the example of the exponential function: it is a beautiful example of how the approximants work in an ideal situation. Further properties of Padé approximants of $\exp(z)$ are to be found in Sections 4.6, 5.7, II.3.3 and 3.4.

Our task is to calculate

$$Q^{[L/M]}(z)$$

$$= \begin{vmatrix} \dfrac{1}{(L-M+1)!} & \dfrac{1}{(L-M+2)!} & \cdots & \dfrac{1}{L!} & \dfrac{1}{(L+1)!} \\[2mm] \dfrac{1}{(L-M+2)!} & \dfrac{1}{(L-M+3)!} & \cdots & \dfrac{1}{(L+1)!} & \dfrac{1}{(L+2)!} \\[2mm] \vdots & \vdots & & \vdots & \vdots \\[2mm] \dfrac{1}{L!} & \dfrac{1}{(L+1)!} & \cdots & \dfrac{1}{(L+M-1)!} & \dfrac{1}{(L+M)!} \\[2mm] z^M & z^{M-1} & \cdots & z & 1 \end{vmatrix}. \quad (2.1)$$

It is easier to begin with the constant term in (2.1), and so we define $C(L/M) \equiv Q^{[L/M]}(0)$, which is the coefficient of the "1" in the lower right-hand corner of (2.1),

$$C(L/M) = \begin{vmatrix} \dfrac{1}{(L-M+1)!} & \dfrac{1}{(L-M+2)!} & \cdots & \dfrac{1}{L!} \\[2ex] \dfrac{1}{(L-M+2)!} & \dfrac{1}{(L-M+3)!} & \cdots & \dfrac{1}{(L+1)!} \\[2ex] \vdots & \vdots & & \vdots \\[2ex] \dfrac{1}{L!} & \dfrac{1}{(L+1)!} & \cdots & \dfrac{1}{(L+M-1)!} \end{vmatrix}.$$

(2.2)

We assume that $L \geqslant M-1$. If this condition does not hold, the factorial functions must be suitably reinterpreted as gamma functions for the analysis to be valid. We remove the denominators from each row, by defining

$$p = \prod_{i=1}^{M} \frac{1}{(L+i-1)!},$$

and then

$$C(L/M) = p \begin{vmatrix} \dfrac{L!}{(L-M+1)!} & \dfrac{L!}{(L-M+2)!} & \cdots & L & 1 \\[2ex] \dfrac{(L+1)!}{(L-M+2)!} & \dfrac{(L+1)!}{(L-M+3)!} & \cdots & L+1 & 1 \\[2ex] \vdots & \vdots & & \vdots & \vdots \\[2ex] \dfrac{(L+M-1)!}{L!} & \dfrac{(L+M-1)!}{(L+1)!} & \cdots & L+M-1 & 1 \end{vmatrix}.$$

(2.3)

In (2.3), the determinant has M rows. Subtract the $(M-1)$th row from the Mth, then the $(M-2)$th row from the $(M-1)$th, etc. The identity

$$\frac{r!}{s!} - \frac{(r-1)!}{(s-1)!} = (r-s)\frac{(r-1)!}{s!}$$

(2.4)

is used repeatedly. In column 1 of (2.3), $r-s = M-1$; in column 2, $r-s =$

$M-2$; etc., and so one finds that

$$C(L/M)=p(M-1)!\begin{vmatrix} \dfrac{L!}{(L-M+1)!} & \dfrac{L!}{(L-M+2)!} & \cdots & L & 1 \\[2mm] \dfrac{L!}{(L-M+2)!} & \dfrac{L!}{(L-M+3)!} & \cdots & 1 & 0 \\[2mm] \vdots & \vdots & & \vdots & \vdots \\[2mm] \dfrac{(L+M-2)!}{L!} & \dfrac{(L+M-2)!}{(L+1)!} & \cdots & 1 & 0 \end{vmatrix}$$

$$=p(-)^{M-1}(M-1)!\begin{vmatrix} \dfrac{L!}{(L-M+2)!} & \dfrac{L!}{(L-M+3)!} & \cdots & 1 \\[2mm] \dfrac{(L+1)!}{(L-M+3)!} & \dfrac{(L+1)!}{(L-M+4)!} & \cdots & 1 \\[2mm] \vdots & \vdots & & \vdots \\[2mm] \dfrac{(L+M-2)!}{L!} & \dfrac{(L+M-2)!}{(L+1)!} & \cdots & 1 \end{vmatrix}.$$

$$(2.5)$$

This is a $(M-1)\times(M-1)$ determinant with a form identical to (2.3) but with M replaced by $M-1$. Consequently, an obvious inductive argument shows that

$$C(L/M)=p\prod_{i=1}^{M}(-1)^{i-1}(i-1)!$$

$$=(-1)^{M(M-1)/2}\prod_{i=1}^{M}\frac{(i-1)!}{(L+i-1)!}.\qquad (2.6)$$

Thus, for the case $M=1$,

$$C(L/1)=\frac{1}{L!},$$

and for the case $M=2$,

$$C(L/2)=\begin{vmatrix} \dfrac{1}{(L-1)!} & \dfrac{1}{L!} \\[2mm] \dfrac{1}{L!} & \dfrac{1}{(L+1)!} \end{vmatrix}=\frac{-1}{L!(L+1)!}.$$

The sign pattern of (2.6) distinguishes Polýa frequency series, to which we refer in Section 5.7. The row operations we have performed to deduce (2.6) from (2.2) are still permissible with the form (1), except that the situation is more complicated. We consider the coefficient of $(-z)^j$ in $Q^{[L/M]}(z)$, which is

$$(-)^j q_j^{[L/M]}$$

$$= \begin{vmatrix} \dfrac{1}{(L-M+1)!} & \dfrac{1}{(L-M+2)!} & \cdots & \vdots & \dfrac{1}{(L-j+1)!} & \vdots & \cdots & \dfrac{1}{(L+1)!} \\ \dfrac{1}{(L-M+2)!} & \dfrac{1}{(L-M+3)!} & \cdots & \vdots & \dfrac{1}{(L-j+2)!} & \vdots & \cdots & \dfrac{1}{(L+2)!} \\ \vdots & \vdots & & \vdots & \vdots & \vdots & & \vdots \\ \dfrac{1}{L!} & \dfrac{1}{(L+1)!} & \cdots & \vdots & \dfrac{1}{(L+M-j)!} & \vdots & \cdots & \dfrac{1}{(L+M)!} \end{vmatrix},$$

$$(2.7)$$

where the column surrounded by $\vdots \ \vdots$ is deleted. We perform a similar analysis: define

$$p' = \prod_{i=1}^{M} \frac{1}{(L+i)!},$$

and then

$$(-)^j q_j^{[L/M]} = p' \begin{vmatrix} \dfrac{(L+1)!}{(L-M+1)!} & \cdots & \vdots & \dfrac{(L+1)!}{(L-j+1)!} & \vdots & \cdots & 1 \\ \dfrac{(L+2)!}{(L-M+2)!} & \cdots & \vdots & \dfrac{(L+2)!}{(L-j+2)!} & \vdots & \cdots & 1 \\ \vdots & & \vdots & \vdots & \vdots & & \vdots \\ \dfrac{(L+M)!}{L!} & \cdots & \vdots & \dfrac{(L+M)!}{(L+M-j)!} & \vdots & \cdots & 1 \end{vmatrix}.$$

$$(2.8)$$

Subtracting rows, and using the identity (2.4),

$$(-)^j q_j^{[L/M]} = (-)^M p' \frac{M!}{j}$$

$$\times \begin{vmatrix} \dfrac{(L+1)!}{(L-M+2)!} & \cdots & \vdots & \dfrac{(L+1)!}{(L-j+2)!} & \vdots & \cdots & 1 \\[2mm] \dfrac{(L+2)!}{(L-M+3)!} & \cdots & \vdots & \dfrac{(L+2)!}{(L-j+3)!} & \vdots & \cdots & 1 \\[2mm] \vdots & & \vdots & \vdots & \vdots & & \vdots \\[2mm] \dfrac{(L+M-1)!}{L!} & \cdots & \vdots & \dfrac{(L+M-1)!}{(L+M-j)!} & \vdots & \cdots & 1 \end{vmatrix},$$

which again is an $(M-1)\times(M-1)$ determinant with a form similar to (2.8). We make j similar reductions from (2.8) to obtain

$$(-)^j q_j^{[L/M]} = \pm \frac{p'}{j!} \prod_{i=1}^{j} (M-i+1)!$$

$$\times \begin{vmatrix} \dfrac{(L+1)!}{(L-M+j+1)!} & \cdots & \dfrac{(L+1)!}{L!} & \vdots & 1 \\[2mm] \dfrac{(L+2)!}{(L-M+j+2)!} & \cdots & \dfrac{(L+2)!}{(L+1)!} & \vdots & 1 \\[2mm] \vdots & & \vdots & \vdots & \vdots \\[2mm] \dfrac{(L+M-j)!}{L!} & \cdots & \dfrac{(L+M-j)!}{(L+M-j-1)!} & \vdots & 1 \end{vmatrix}.$$

Removing a common factor from each row,

$$(-)^j q_j^{[L/M]} = \pm \frac{p'}{j!} \frac{(L+M-j)!}{L!} \prod_{i=1}^{j} (M-i+1)! \begin{vmatrix} \dfrac{L!}{(L-M+j+1)!} & \cdots & 1 \\[2mm] \dfrac{(L+1)!}{(L-M+j+2)!} & \cdots & 1 \\[2mm] \vdots & & \vdots \\[2mm] \dfrac{(L+M-j-1)!}{L!} & \cdots & 1 \end{vmatrix}$$

The analysis now follows the familiar pattern using the identity (2.4), and we deduce that

$$(-)^j q_j^{[L/M]} = \pm \left\{ \prod_{i=1}^{M} \frac{1}{(L+i)!} \right\} \frac{(L+M-j)!}{L!j!}$$

$$\times \left\{ \prod_{i=1}^{j} (M-i+1)! \right\} \prod_{i=1}^{M-j-1} i!$$

$$= \pm \frac{(L+M-j)!}{L!j!(M-j)!} \prod_{i=1}^{M} \frac{i!}{(L+i)!}. \tag{2.9}$$

The sign of the right-hand side of (2.9) is easily determined to be the same as that of (2.6), because the determinants (2.2) and (2.7) have the same dimension, and are expanded by the same top right-hand elements recursively. Hence

$$(-)^j q_j^{[L/M]} = (-)^{M(M-1)/2} \frac{(L+M-j)!}{L!j!(M-j)!} \prod_{i=1}^{M} \frac{i!}{(L+i)!}. \tag{2.10}$$

Notice that (2.6) emerges as the special case with $j=0$. Consequently we have

$$q_j^{[L/M]} = (-)^j C(L/M) \frac{(L+M-j)!}{(L+M)!} \frac{M!}{(M-j)!} \frac{1}{j!}$$

and

$$Q^{[L/M]}(z) = C(L/M) \sum_{j=0}^{M} \frac{(L+M-j)!}{(L+M)!} \frac{M!}{(M-j)!} \frac{(-z)^j}{j!}$$

$$= C(L/M) \left\{ 1 + \frac{M}{L+M} \frac{(-z)}{1!} + \frac{M(M-1)}{(L+M)(L+M-1)} \frac{(-z)^2}{2!} + \cdots \right\}$$

$$= C(L/M) \left\{ 1 + \frac{-M}{-L-M} \frac{(-z)}{1!} + \frac{-M(-M+1)}{(-L-M)(-L-M+1)} \frac{(-z)^2}{2!} + \cdots \right\}$$

$$= C(L/M) \, _1F_1(-M; -L-M; -z). \tag{2.11}$$

Following the method of Section 1.6, we may deduce from (2.11) that

$$P^{[L/M]}(z) = C(L/M) \, _1F_1(-L, -L-M; z),$$

and hence the $[L/M]$ Padé approximant for $\exp(z)$ is

$$[L/M] = \frac{{}_1F_1(-L, -L-M; z)}{{}_1F_1(-M, -L-M; -z)}. \qquad (2.12)$$

1.3 Sequences and Series; Obstacles

In Section 1.1 we showed how Padé approximants are constructed from given power series, and in Section 1.2 we saw how the Padé approximants of $\exp(z)$ are obtained. In this section, we preview a few of the techniques and problems to be discussed in later chapters. The Padé method is directly applicable for the improvement of convergence of series and sequences. This application is fully discussed in Chapter 3. For the moment, assuming convergence of a sequence of approximants $[L_k/M_k](z)$, $k=1,2,\ldots$ at the point $z=1$,

$$\sum_{k=0}^{\infty} c_k = \lim_{k \to \infty} [L_k/M_k](1). \qquad (3.1)$$

In this sense, ordinary Padé approximants may be used to sum series. Likewise, given a sequence $\{S_n, \ n=0,1,2,\ldots\}$, we define a series from it using forward differences,

$$c_0 = S_0,$$
$$c_{n+1} = S_{n+1} - S_n = \Delta S_n, \qquad n=0,1,2,\ldots,$$

and then Padé approximants may be used to extrapolate sequences. It is common practice to use the diagonal sequence of Padé approximants unless there are reasons to the contrary. We take up these points in Chapter 3. From (1.8),

$$Q^{[L/M]}(z) = \begin{vmatrix} c_{L-M+1} & c_{L-M+2} & \cdots & c_{L+1} \\ c_{L-M+2} & c_{L-M+3} & \cdots & c_{L+2} \\ \vdots & \vdots & & \vdots \\ c_L & c_{L+1} & \cdots & c_{L+M} \\ z^M & z^{M-1} & \cdots & 1 \end{vmatrix}.$$

We reduce this by subtracting z times each column from the previous column, to yield

$$Q^{[L/M]}(z) = \begin{vmatrix} c_{L-M+1} - zc_{L-M+2} & \cdots & c_L - zc_{L+1} \\ c_{L-M+2} - zc_{L-M+3} & \cdots & c_{L+1} - zc_{L+2} \\ \vdots & & \vdots \\ c_L - zc_{L+1} & \cdots & c_{L+M-1} - zc_{L+M} \end{vmatrix} \qquad (3.2)$$

which is a compact and symmetric form. For the numerator, we use (1.9),

$$P^{[L/M]}(z) = \begin{vmatrix} c_{L-M+1} & c_{L-M+2} & \cdots & c_{L+1} \\ c_{L-M+2} & c_{L-M+3} & \cdots & c_{L+2} \\ \vdots & \vdots & & \vdots \\ c_L & c_{L+1} & \cdots & c_{L+M} \\ \sum_{i=0}^{L-M} c_i z^{M+i} & \sum_{i=0}^{L-M+1} c_i z^{M+i-1} & \cdots & \sum_{i=0}^{L} c_i z^i \end{vmatrix}$$

and a similar reduction yields

$$P^{[L/M]}(z) = \begin{vmatrix} c_{L-M+1}-zc_{L-M+2} & \cdots & c_L-zc_{L+1} & c_{L+1} \\ c_{L-M+2}-zc_{L-M+3} & \cdots & c_{L+1}-zc_{L+2} & c_{L+2} \\ \vdots & & \vdots & \vdots \\ c_L-zc_{L+1} & \cdots & c_{L+M-1}-zc_{L+M} & c_{L+M} \\ -c_{L-M+1}z^{L+1} & \cdots & -c_L z^{L+1} & \sum_{i=0}^{L} c_i z^i \end{vmatrix}.$$

Dividing each column, except the last, by z^M, z^{M-1},..., z and adding to the last, we find

$$P^{[L/M]}(z) = \begin{vmatrix} c_{L-M+1}-zc_{L-M+2} & \cdots & c_L-zc_{L+1} & c_{L-M+1}z^{-M} \\ c_{L-M+2}-zc_{L-M+3} & \cdots & c_{L+1}-zc_{L+2} & c_{L-M+2}z^{-M} \\ \vdots & & \vdots & \vdots \\ c_L-zc_{L+1} & \cdots & c_{L+M-1}-zc_{L+M} & c_L z^{-M} \\ -c_{L-M+1}z^{L+1} & \cdots & -c_L z^{L+1} & \sum_{i=0}^{L-M} c_i z^i \end{vmatrix}.$$

$$(3.3)$$

We are now led to define the matrices

$$W(z) =$$

$$\begin{pmatrix} c_{L-M+1}-zc_{L-M+2} & c_{L-M+2}-zc_{L-M+3} & \cdots & c_L-zc_{L+1} \\ c_{L-M+2}-zc_{L-M+3} & c_{L-M+2}-zc_{L-M+4} & \cdots & c_{L+1}-zc_{L+2} \\ \vdots & \vdots & & \vdots \\ c_L-zc_{L+1} & c_{L+1}-zc_{L+2} & \cdots & c_{L+M-1}-zc_{L+M} \end{pmatrix}$$

$$(3.4)$$

and

$$\mathbf{c}^T = (c_{L-M+1}, c_{L-M+2}, \ldots, c_L). \tag{3.5}$$

With these definitions, we recognize that $Q^{[L/M]}(z) = \det W(z)$. We expand (3.3) by its last row and then by its last column, using cofactors of $W(z)$, to deduce

$$[L/M] = \sum_{i=0}^{L-M} c_i z^i + \{\mathbf{c}^T W(z)^{-1} \mathbf{c}\} z^{L-M+1}. \tag{3.6}$$

This equation is called Nuttall's compact form for a Padé approximant. If $L < M$, the polynomial term in (3.6) is understood as zero, because we use the convention that $c_j = 0$ if $j < 0$.

Reductions such as these lead to interesting forms for the Padé approximants when they are used for acceleration of convergence of sequences. Given the sequence $\{S_n, n = 0, 1, \ldots\}$, we define

$$c_0 = S_0, \qquad S_j = 0 \quad \text{for } j < 0,$$
$$c_{k+1} = \Delta S_k = S_{k+1} - S_k \qquad \text{for } k = 0, 1, 2, \ldots \text{and} \tag{3.7}$$
$$\Delta^i S_k = \Delta^{i-1} S_{k+1} - \Delta^{i-1} S_k \qquad \text{for all } k \text{ and } i = 1, 2, 3, \ldots.$$

From (1.8),

$$Q^{[L/M]}(1) = \begin{vmatrix} \Delta S_{L-M} & \Delta S_{L-M+1} & \cdots & \Delta S_L \\ \Delta S_{L-M+1} & \Delta S_{L-M+2} & \cdots & \Delta S_{L+1} \\ \vdots & \vdots & & \vdots \\ \Delta S_{L-1} & \Delta S_L & \cdots & \Delta S_{L+M-1} \\ 1 & 1 & \cdots & 1 \end{vmatrix},$$

and after several elementary operations, we find

$$Q^{[L/M]}(1) = (-1)^M \begin{vmatrix} \Delta^2 S_{L-M} & \Delta^3 S_{L-M} & \cdots & \Delta^{M+1} S_{L-M} \\ \Delta^3 S_{L-M} & \Delta^4 S_{L-M} & \cdots & \Delta^{M+2} S_{L-M} \\ \vdots & \vdots & & \vdots \\ \Delta^{M+1} S_{L-M} & \Delta^{M+2} S_{L-M} & \cdots & \Delta^{2M} S_{L-M} \end{vmatrix},$$

which is an $M \times M$ symmetric determinant with all difference operators acting on S_{L-M}. Similarly we find that the numerator is given by the

$(M+1)\times(M+1)$ symmetric determinant:

$$P^{[L/M]}(1)=(-)^{M}\begin{vmatrix} S_{L-M} & \Delta S_{L-M} & \cdots & \Delta^{M}S_{L-M} \\ \Delta S_{L-M} & \Delta^{2}S_{L-M} & \cdots & \Delta^{M+1}S_{L-M} \\ \vdots & \vdots & & \vdots \\ \Delta^{M}S_{L-M} & \Delta^{M+1}S_{L-M} & \cdots & \Delta^{2M}S_{L-M} \end{vmatrix}.$$

The sequence $\{[M/M],\,M=0,1,2,\ldots\}$ is called the diagonal sequence. These formulas suggest, but in no way compel, the choice of diagonal approximants for acceleration of convergence. An element of the final sequence is given by

$$[L/M](1)=\begin{vmatrix} S_{L-M} & \Delta S_{L-M} & \cdots & \Delta^{M}S_{L-M} \\ \Delta S_{L-M} & \Delta^{2}S_{L-M} & \cdots & \Delta^{M+1}S_{L-M} \\ \vdots & \vdots & & \vdots \\ \Delta^{M}S_{L-M} & \Delta^{M+1}S_{L-M} & \cdots & \Delta^{2M}S_{L-M} \end{vmatrix}$$

$$\div \begin{vmatrix} \Delta^{2}S_{L-M} & \Delta^{3}S_{L-M} & \cdots & \Delta^{M+1}S_{L-M} \\ \Delta^{3}S_{L-M} & \Delta^{4}S_{L-M} & \cdots & \Delta^{M+2}S_{L-M} \\ \vdots & \vdots & & \vdots \\ \Delta^{M+1}S_{L-M} & \Delta^{M+2}S_{L-M} & \cdots & \Delta^{2M}S_{L-M} \end{vmatrix}, \quad (3.8)$$

which is a remarkably elegant result [Shanks, 1955].

Finally in this section we mention obstacles. We will take note of some examples which illustrate why precision is mandatory in the treatment of formal power series.

Using the *real and positive variable* x, $f(x)=\exp(-1/x)$ is smooth and infinitely differentiable throughout $0\leqslant x<\infty$. The problem is that $f(0)= f'(0)=\cdots=f^{(n)}(0)=0$ for all n, and so $f(x)$ has the Maclaurin expansion $f(x)=0$. Of course, this example is contrived, and is based on a function not analytic in a neighborhood of the origin (this statement means a domain $|z|<\delta$, where δ is arbitrarily small but positive). It is clear that $\exp(-1/z)$ is not determined by its Maclaurin series, and so our theorems are always phrased so as to exclude the possibility that we are representing such functions.

Another notorious function is Euler's function. This function is a more constructive example, and in Chapter 5 we show how its Padé approximants converge. It is defined by the series

$$E(z)=1-1!z+2!z^{2}-3!z^{3}+\cdots, \quad (3.9)$$

and we assume that a certain sequence of Padé approximants has been empirically found to converge. The full theory relevant to this example is explained in Sections 5.5 and 5.6. The moot point is whether convergence is to $E(z)$, again begging the question of the extent to which $E(z)$ is defined by (3.9). With the information that (3.9) is an asymptotic series, an entirely satisfactory definition exists for $E(x)$ with $x \geqslant 0$.

To be very pragmatic, take $x = \frac{1}{10}$. The magnitudes of the terms in the series are shown in Figure 1, and in the sense of the previous definitions, a plausible value for $E(0.1)$ is reasonably well determined by truncation at the minimum, namely $\sum_{n=0}^{10} n!(-1/10)^n$. This procedure is much less satisfactory for large values of z, and also it would seem to work badly for z small and negative. The problem turns out to be that with the natural approach, $E(z)$ is defined with a branch cut along $-\infty < z \leqslant 0$, and so $E(z)$ is only determined uniquely in the sense of cut-plane analyticity. To be pedestrian, $E(-0.1)$ is really only determined by a convention about the location of the cut. This point is easy to overlook using Padé approximants, because the approximants "choose" the cut in the natural place in a sense we describe in Section 2.2.

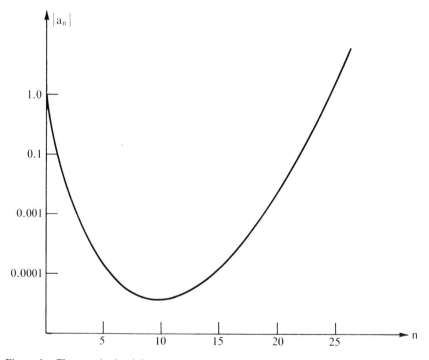

Figure 1. The magnitude of the terms $|a_n| = (0.1)^n n!$ in the hypergeometric series for $E(0.1)$.

As a final example of mathematical perversity, we refer to Part II, Section 3.9, where we exhibit a nontrivial function which is analytic in an annulus, and which cannot be properly approximated by polynomials in z in the annulus. Another similar function is our demonstration example of Section 1.1. Here we showed the success of low-order Padé approximants in practice, and rapid convergence at high order may be proved using the Stieltjes series methods of Chapter 5. For the most part in this book, we are concerned primarily with how, one way or another, various natural obstacles can be overcome.

1.4 The Baker Definition, the *C*-Table, and Block Structure

To motivate the discussion of uniqueness and the modern definition, of which the analysis is due to Baker, we must consider what can go wrong with the basic approach discussed so far. The classic example which demonstrated the shortcoming of a simple-minded approach is the construction of [1/1] for $1+z^2$.

We require

$$\frac{p_0+p_1z}{q_0+q_1z}=1+z^2+O(z^3). \tag{4.1}$$

Hence

$$p_0+p_1z=q_0+q_1z+q_0z^2+O(z^3).$$

Therefore

$$p_0=q_0, \qquad p_1=q_1, \quad \text{and} \quad q_0=0. \tag{4.2}$$

Consequently, the approximant is

$$\frac{0+q_1z}{0+q_1z}=1. \tag{4.3}$$

However, $1\neq1+z^2+O(z^3)$, and we rapidly and correctly conclude that (4.1) has no solution. What we did achieve is a "solution" of

$$p_0+p_1z=(q_0+q_1z)(1+z^2)+O(z^3), \tag{4.4}$$

but that is not what (4.1) requires. This solution, given by (4.2), is also given by the determinantal method:

$$P^{[1/1]}(z) = \begin{vmatrix} c_1 & c_2 \\ c_0 z & c_0 + c_1 z \end{vmatrix} = \begin{vmatrix} 0 & 1 \\ z & 1 \end{vmatrix} = -z$$

and

$$Q^{[1/1]}(z) = \begin{vmatrix} c_1 & c_2 \\ z & 1 \end{vmatrix} = \begin{vmatrix} 0 & 1 \\ z & 1 \end{vmatrix} = -z.$$

For a long time, the accepted solution was to take an analogue of (4.4) as the agreed definition of Padé approximants. Specifically, the classical definition, also called the Frobenius and Padé Frobenius definition, is that if $p_L(z)$, $q_M(z)$ are polynomials of orders L, M respectively, and if

$$q_M(z)f(z) - p_L(z) = O(z^{L+M+1}), \tag{4.5}$$

then $p_L(z)/q_M(z)$ is a Padé approximant of $f(z)$. Equation (4.5) is the general form of (4.4), and it is remarkable that polynomials $q_M(z)$, $p_L(z)$ of orders M, L can always be found to satisfy (4.5). However, our specific example emphasizes that if $Q_M(0) = 0$, then in this case

$$f(z) \neq \frac{p_L(z)}{q_M(z)} + O(z^{L+M+1}).$$

Padé defined a deficiency index, ω_{LM}, which is the least integer for which

$$f(z) = \frac{p_L(z)}{q_M(z)} + O(z^{L+M-\omega_{LM}+1}),$$

and ω_{LM} is a measure of the shortcoming of the approximation. Quite simply, the rational function $p_L(z)/q_M(z)$ does not approximate $f(z)$ through order $L+M$ in certain circumstances, and then we prefer to say that the Padé approximant does not exist.

In the general theory of rational interpolation, it is well known that there are certain unattainable values. If values $f(z_i)$ at certain points z_i are specified, the specification may be inconsistent, and in such circumstances, the rational interpolants are declared not to exist. Our approach is entirely in line with this attitude.

Because the accuracy-through-order requirement is fundamental, a definition which preserves it is essential, and we use the modern definition which was fully analysed by Baker [1973b].

DEFINITION (Baker). If polynomials $A^{[L/M]}(z)$, $B^{[L/M]}(z)$, of degrees L, M respectively, can be found such that

$$\frac{A^{[L/M]}(z)}{B^{[L/M]}(z)} = f(z) + O(z^{L+M+1}) \tag{4.6}$$

with

$$B^{[L/M]}(0) = 1, \tag{4.7}$$

then we define

$$[L/M] = \frac{A^{[L/M]}(z)}{B^{[L/M]}(z)}.$$

The notation emphasizes that numerator and denominator depend on both L and M. An entirely equivalent specification of the definition is to replace (4.6) by

$$A^{[L/M]}(z) - f(z)B^{[L/M]}(z) = O(z^{L+M+1})$$

provided that (4.7) is retained. The notation of (4.6) and (4.7) is exclusively reserved for this purpose throughout the work, and without further explanation.

If, with Equation (1.8), $Q^{[L/M]}(0) \neq 0$, then the rescaling

$$A^{[L/M]}(z) = \frac{P^{[L/M]}(z)}{Q^{[L/M]}(0)}$$

and

$$B^{[L/M]}(z) = \frac{Q^{[L/M]}(z)}{Q^{[L/M]}(0)}$$

implies that the two definitions correspond up to an unimportant numerical factor. Consequently, and charitably speaking, the distinction between the definitions is sometimes taken for granted. However, if $Q^{[L/M]}(0) = 0$, precise terminology is mandatory. Because the vanishing of $Q^{[L/M]}(0)$ is so important, a special symbol is exclusively reserved for this quantity:

DEFINITION.

$$C(L/M) = Q^{[L/M]}(0) = \begin{vmatrix} c_{L-M+1} & c_{L-M+2} & \cdots & c_L \\ c_{L-M+2} & c_{L-M+3} & \cdots & c_{L+1} \\ \vdots & \vdots & & \vdots \\ c_L & c_{L+1} & \cdots & c_{L+M-1} \end{vmatrix}. \tag{4.8}$$

Table 1. The C-Table.

$C(0/0)$	$C(1/0)$	$C(2/0)$	$\cdots$
$C(0/1)$	$C(1/1)$	$C(2/1)$	$\cdots$
$C(0/2)$	$C(1/2)$	$C(2/2)$	$\cdots$
$\vdots$	$\vdots$	$\vdots$	$\ddots$

Summary. If $C(L/M)\neq 0$, the classical Padé–Frobenius and the Baker definitions are entirely equivalent. If $C(L/M)=0$, it is possible that the $[L/M]$ Padé approximant does not exist. However, polynomials satisfying (4.5) exist, defining a rational fraction which historically was called the Padé approximant.

It is convenient to display $\{C(L/M), L, M=0,1,2,\ldots,\}$ in a table (Table 1), called the C-table. This is an array of values of determinants, and should be distinguished from the Padé table.

Example. Let

$$f(z)=\frac{1+2z+z^2+z^3}{1+z+z^3}.$$

This is given by the power series

$$f(z)=1+z-z^4+z^5-z^6+2z^7-3z^8+4z^9-\cdots, \tag{4.9}$$

and the C-table of Table 2 results. The most conspicuous features of the table are the square blocks of zeros. In fact, the zero at $C(4/4)$ is the start of an infinite square block. Before proceeding with the proof of these statements, let us investigate how Table 1 was actually constructed. A substantial amount of elementary algebra is needed to construct Table 2 from the basic

Table 2. The C-Table for $(1+2z+z^2+z^3)/(1+z+z^2)$.

L\M	0	1	2	3	4	5	6	7	8	9	$\cdots$
0	1	1	1	1	1	1	1	1	1	1	$\cdots$
1	1	1	0	0	-1	1	-1	2	-3	4	$\cdots$
2	-1	-1	0	0	-1	0	1	-1	-1		$\cdots$
3	-1	-1	1	-1	1	-1	1	-1			$\cdots$
4	1	2	0	1	0	0	0	0	0	0	$\cdots$
5	1	4	2	1	0	0	0	0	0	0	$\cdots$
$\vdots$	$\vdots$	$\vdots$	$\vdots$	$\vdots$	$\vdots$	$\vdots$	$\vdots$	$\vdots$	$\vdots$	$\vdots$	$\ddots$

definition (4.8) and the series (4.9); this algebra was not the method used for constructing most of the entries.

The first row is, by definition $C(L/0)=1$.

The second row is $C(L/1)=c_L$.

The first column is $C(0/M)=(-1)^{M(M-1)/2}c_0^M$.

Most of the remaining entries were calculated from the identity

$$C(L/M+1)=\frac{C(L+1/M)C(L-1/M)-C(L/M)^2}{C(L/M-1)}, \qquad (4.10)$$

which is valid if $C(L/M-1)\neq0$. Equation (4.10) is called a $\left(\begin{smallmatrix}*\\ *\ *\\ *\end{smallmatrix}\right)$ star identity, showing how it relates the entries in the C-table. In order fully to understand the consequences of (4.10), it is worthwhile reconstructing Table 1 from the initializing values. We proceed with Sylvester's theorem, of which (4.10) is a corollary.

THEOREM 1.4.1. *Let A be a matrix, and let A_{rp} denote the matrix with row r and column p deleted. Also let $A_{rs;pq}$ denote the matrix A with rows r and s and columns p and q deleted. Provided $r<s$ and $p<q$,*

$$\det A \det A_{rs;pq}=\det A_{rp}\det A_{sq}-\det A_{rq}\det A_{sp}.$$

Proof. Suppose that A is an $(n+2)\times(n+2)$ matrix, and we consider deletion of its last two rows and columns. Take $r=p=n+1$ and $s=q=n$. Write the matrices in block form, e.g. $A_{n-1,n;n-1,n}=M$,

$$A=\begin{pmatrix} M & h & g \\ f & e & d \\ c & b & a \end{pmatrix},$$

$$A_{n,n}=\begin{pmatrix} M & h \\ f & e \end{pmatrix}, \qquad A_{n,n-1}=\begin{pmatrix} A & g \\ f & d \end{pmatrix}.$$

Next we consider a $(2n+2)\times(2n+2)$ block matrix, with determinant given by

$$\begin{vmatrix} M & h & g & 0 \\ f & e & d & 0 \\ c & b & a & c \\ 0 & 0 & 0 & M \end{vmatrix}=\begin{vmatrix} M & h & g & 0 \\ f & e & d & 0 \\ c & b & a & c \\ M & h & g & M \end{vmatrix}=\begin{vmatrix} M & h & g & 0 \\ f & e & d & 0 \\ 0 & b & a & c \\ 0 & h & g & M \end{vmatrix}$$

$$
= \begin{vmatrix} M & h & g & 0 \\ f & e & d & 0 \\ 0 & 0 & a & c \\ 0 & 0 & g & M \end{vmatrix} + \begin{vmatrix} M & 0 & g & 0 \\ f & 0 & d & 0 \\ 0 & b & a & c \\ 0 & h & g & M \end{vmatrix}
$$

$$
= \begin{vmatrix} M & h \\ f & e \end{vmatrix} \cdot \begin{vmatrix} a & c \\ g & M \end{vmatrix} - \begin{vmatrix} M & g \\ f & d \end{vmatrix} \cdot \begin{vmatrix} b & c \\ h & M \end{vmatrix},
$$

which is interpreted as

$\det A \det A_{n+1,n+2;n+1,n+2}$

$$
= \det A_{n+2,n+2} \det A_{n+1,n+1} - \det A_{n+1,n+2} \det A_{n+2,n+1}.
$$

By interchanging rows r, $n-1$ and s, $n-2$ and columns p, $n-1$ and q, $n-2$, the theorem is proved.

COROLLARY.

$$
C(L/M+1)C(L/M-1) = C(L+1/M)C(L-1/M) - C(L/M)^2. \tag{4.11}
$$

Proof. Let $\det A = C(L/M+1)$ as given by (4.8). With $r=p=1$ and $s=q=M+1$, the result and consequently (4.10), if $C(L/M-1) \neq 0$, are proved.

This relation (4.11) is the key to the block structure of the C-table.

THEOREM 1.4.2. *Zero entries occur in the C-table in square blocks which are entirely surrounded by nonzero entries (except at infinity).*

Proof. We identify the top left-hand corner of the block by requiring that

$$
C(l/m) = 0, \qquad C(l/m-1) \neq 0, \quad \text{and} \quad C(l-1/m) \neq 0. \tag{4.12a}
$$

It is obvious how to redefine l or m if either of the latter two conditions does not hold. From the $l-1/m$ star identity,

$$
C(l-1/m-1)C(l-1/m+1) = C(l-2/m)C(l/m) - C(l-1/m)^2,
$$

we deduce that

$$
C(l-1/m-1)C(l-1/m+1) = -C(l-1/m)^2.
$$

Hence

$$
C(l-1/m-1) \neq 0 \quad \text{and} \quad C(l-1/m+1) \neq 0.
$$

Similarly, the $l/m-1$ star identity yields $C(l+1/m-1) \neq 0$, and the relevant portion of the C-table is shown in Table 3. Suppose that

$$C(l/m+1)\neq0. \tag{4.12b}$$

Then the $l/m+1$ star identity establishes that $C(l+1/m+1)\neq0$, and the $l+1/m$ star identity establishes that $C(l+1/m)\neq0$. Hence the theorem is proved for a unit square block. The only alternative to (4.12b) is that

Table 3. Consequences of (4.12)

$C(l-1/m+1)$ $\neq0$	$C(l/m-1)$ $\neq0$	$C(l+1/m-1)$ $\neq0$
$C(l-1/m)$ $\neq0$	$C(l/m)$ $=0$	
$C(l-1/m+1)$ $\neq0$		

Table 4. The Left Edge of a Block

$\neq0$	$\neq0$	$\neq0$
$\neq0$	0	
$\neq0$	0	
$\neq0$	0	
$\neq0$	0	
$\neq0$	$\neq0$	$\neq0$

$C(l/m+1)=0$. Suppose that $C(l/m+k)=0$ for $k=0,1,\ldots,\kappa-1$ and $C(l/m+\kappa)\neq0$. Using the $l-1/m+k$ star identity, we establish iteratively that $C(l-1/m+k+1)\neq0$ for $k=0,1,\ldots,\kappa-1$. Since $C(l/m+\kappa)\neq0$ and $C(l-1/m+\kappa)\neq0$, we deduce that $C(l+1/m+\kappa)\neq0$. Thus we have a column of zeros bordered on top, bottom and left by nonzero elements, as shown in Table 4. Similarly, we establish that the block is rectangular and, if finite, it is entirely bordered by nonzero elements as shown in Table 5. To

Table 5. A Hypothetical Section of the C-Table

$\neq0$	$\neq0$	$\neq0$	$\neq0$	$\neq0$	$\neq0$
$\neq0$	0	0	0	0	$\neq0$
$\neq0$	0	0	0	0	$\neq0$
$\neq0$	0	0	0	0	$\neq0$
$\neq0$	$\neq0$	$\neq0$	$\neq0$	$\neq0$	$\neq0$

establish that the blocks are square, we consider a block with r rows and s columns. First we prove that $r \geq s$, and to do this we choose a simple example which makes the general case obvious. Suppose we know that $C(2/2) = C(3/2) = C(4/2) = 0$. Then

$$\begin{vmatrix} c_1 & c_2 \\ c_2 & c_3 \end{vmatrix} = \begin{vmatrix} c_2 & c_3 \\ c_3 & c_4 \end{vmatrix} = \begin{vmatrix} c_3 & c_4 \\ c_4 & c_5 \end{vmatrix} = 0.$$

These statements are interpreted as implying that a linear combination of $\begin{pmatrix} c_1 \\ c_2 \end{pmatrix}$ and $\begin{pmatrix} c_2 \\ c_3 \end{pmatrix}$ vanishes, and also of $\begin{pmatrix} c_2 \\ c_3 \end{pmatrix}$ and $\begin{pmatrix} c_3 \\ c_4 \end{pmatrix}$, and also of $\begin{pmatrix} c_3 \\ c_4 \end{pmatrix}$ and $\begin{pmatrix} c_4 \\ c_5 \end{pmatrix}$. Using the implied multipliers, we deduce that

$$\begin{vmatrix} c_0 & c_1 & c_2 \\ c_1 & c_2 & c_3 \\ c_2 & c_3 & c_4 \end{vmatrix} = \begin{vmatrix} x & c_1 & x \\ 0 & c_2 & 0 \\ 0 & c_3 & 0 \end{vmatrix} = 0, \tag{4.13}$$

where the x denotes an unknown and totally immaterial entry. Likewise,

$$\begin{vmatrix} c_1 & c_2 & c_3 \\ c_2 & c_3 & c_4 \\ c_3 & c_4 & c_5 \end{vmatrix} = \begin{vmatrix} c_2 & c_3 & c_4 \\ c_3 & c_4 & c_5 \\ c_4 & c_5 & c_6 \end{vmatrix} = 0. \tag{4.14}$$

Finally, by interpreting (4.13) and (4.14) as asserting the existence of linearly dependent column vectors, we find that

$$C(2/4) = \begin{vmatrix} 0 & c_0 & c_1 & c_2 \\ c_0 & c_1 & c_2 & c_3 \\ c_1 & c_2 & c_3 & c_4 \\ c_2 & c_3 & c_4 & c_5 \end{vmatrix} = \begin{vmatrix} x & c_0 & c_1 & x \\ 0 & c_1 & c_2 & 0 \\ 0 & c_2 & c_3 & 0 \\ 0 & c_3 & c_4 & 0 \end{vmatrix} = 0.$$

Likewise, $C(2/4) = C(3/4) = C(4/4) = 0$. We claim that it is now obvious that if we have a block of the C-table with s columns, $L = l, l+1, \ldots, l+s-1$, then there are at least s rows. In other words, a block with r rows and s columns has $r \geq s$.

To prove the converse we refer forward to Hadamard's theorem of Section 1.6. We may always reset the problem for a function with $c_0 \neq 0$. We now assume that $c_0 \neq 0$, so that we may consider the C-table for the reciprocal function $g(z) = 1/f(z)$. Let $C'(M/L)$ be the entry for the (M/L) Hankel determinant of $g(z)$. From (6.9) we know that $C(L/M) = 0$ implies that $C'(M/L) = 0$ and vice versa. For the reciprocal function, the block with r' rows and s' columns satisfies our previous rule, $r' \geq s'$. But $r' = s$ and $s' = r$, and consequently the converse is proved. Hence $r = s$, and we have proved that all blocks of the C-table are square, and are entirely bordered by nonzero entries.

This completes the preliminary to Padé's theorem. This theorem has been modernized by Baker, but the content is essentially unchanged. The style of proof is quite different from that of the previous century.

THEOREM 1.4.3. *The Padé table consists of uniquely determined entries given by*

$$[L/M] = \frac{A^{[L/M]}(z)}{B^{[L/M]}(z)} = \frac{P^{[L/M]}(z)}{Q^{[L/M]}(z)}$$

following the definitions (1.8), (1.9), (4.6), and (4.7), provided $C(L/M) \neq 0$. Otherwise, suppose that $C(L/M) = 0$, in each entry of an $r \times r$ block of the C-table. Corresponding to this, blocks of the Padé table are $(r+1) \times (r+1)$ blocks, for which $C(\lambda/\mu) \neq 0$, $C(\lambda+i/\mu) \neq 0$, and $C(\lambda/\mu+i) \neq 0$ for $i = 1, 2, \ldots, r$ and $C(\lambda+i/\mu+j) = 0$ for all $i, j = 1, 2, \ldots, r$. An $[L/M]$ Padé approximant in the block obeys

either $[L/M] = [\lambda/\mu]$,   if $L + M \leq \lambda + \mu + r$,

or $[L/M]$ does not exist,   if $L + M > \lambda + \mu + r$.

Proof. The proof is divided into three parts.

Part 1.   If $C(L/M) \neq 0$, the Padé equations are linearly independent; with $b_0 = 1$ the coefficients are uniquely determined. In particular, this observation excludes the possibility that $A^{[L/M]}(z)$ and $B^{[L/M]}$ have a common factor.

Part 2.   The Padé approximants on the top edge and left-hand side of the block of the Padé table satisfy

$$[\lambda+i/\mu] = [\lambda/\mu+i] = [\lambda/\mu] \qquad \text{for} \quad i = 1, 2, \ldots, r$$

Proof. This result follows by noting that these approximants are uniquely determined, using part 1 and the block structure of the C-table. For $i = 1$, the result follows by noting that

the coefficient of $z^{\lambda+1}$ in $P^{[\lambda+1/\mu]}(z)$ is $\pm C(\lambda+1/\mu+1) = 0$, and
the coefficient of $z^{\mu+1}$ in $P^{[\lambda/\mu+1]}(z)$ is $\pm C(\lambda+1/\mu+1) = 0$.

In fact, by induction, it follows that because the coefficient of $z^{\lambda+i}$ in $P^{[\lambda+i/\mu]}(z)$ is $\pm C(\lambda+i/\mu+1) = 0$, that

$$[\lambda+i/\mu] = [\lambda+i-1/\mu] \qquad \text{for} \quad i = 1, 2, \ldots, r.$$

Similarly,

$$[\lambda/\mu+i]=[\lambda/\mu+i-1] \qquad \text{for} \quad i=1,2,\ldots,r,$$

and part 2 is proved.

Part 3. $[L/M]$ Padé approximants in the relevant block of the Padé table of the theorem satisfying

$$L+M\leqslant\lambda+\mu+r$$

exist and equal $[\lambda/\mu]$. Otherwise

$$L+M>\lambda+\mu+r$$

and the corresponding Padé approximants in the block do not exist.

Proof. Because the Padé approximants on the top edge, $[\lambda+i/\mu]$, $i=0,1,\ldots,r$, are identical, we know that the following $\mu+r$ equations for the denominator coefficients are satisfied:

$$\begin{pmatrix} c_{\lambda-\mu+1} & \cdots & c_{\lambda+1} \\ \vdots & & \vdots \\ c_{\lambda} & \cdots & c_{\lambda+\mu} \\ \vdots & & \vdots \\ c_{\lambda+r} & \cdots & c_{\lambda+\mu+r} \end{pmatrix} \begin{pmatrix} q_\mu \\ \vdots \\ q_0 \end{pmatrix} = 0, \qquad (4.15)$$

and because $[\lambda+r+1/\mu]$ is not in this block,

$$c_{\lambda+r+1}q_\mu+c_{\lambda+r+2}q_{\mu-1}\cdots c_{\lambda+r+\mu+1}q_0\neq0. \qquad (4.16)$$

Suppose that $[L/M]$ is a Padé approximant in the block, but not on the top edge or left-hand side. Thus $\lambda+1\leqslant L\leqslant\lambda+r$ and $\mu+1\leqslant M\leqslant\mu+r$ and $B^{[L/M]}(0)=1$. The explicit solution shows that this situation is impossible: if an explicit solution exists, $Q^{[L/M]}(0)=0$ and $[L/M]=[L-1/M-1]$. Using part 2, we see that if a Padé approximant in the block exists, then it reduces to $[\lambda/\mu]$, and has denominator coefficients obeying (4.15). Hence we see that the accuracy-through-order conditions are satisfied for $L+M\leqslant\lambda+\mu+r$, and so the Padé approximants exist and are reducible. Also, from (4.16), if $L+M>\lambda+\mu+r$, the accuracy at order $z^{\lambda+\mu+r+1}$ is not satisfied, and so the Padé approximant does not exist. The theorem is now completely proved.

An elementary view of a set of linear equations is that either the equations are consistent and have a solution, or they are inconsistent and have no solution. This view is mirrored by the previous theorem, in which linear equations either do or do not determine Padé approximants; of

Table 6. Part of the c-Table Showing a 3×3 Block and part of the Padé Table showing the corresponding 4×4 block.

(λ/μ) $\neq 0$	$\neq 0$	$\neq 0$	$\neq 0$		$[\lambda/\mu]$	Red.	Red.	Red.
$\neq 0$	0	0	0	$\leftrightarrow$	Red.	Con.	Con.	Inc.
$\neq 0$	0	0	0		Red.	Con.	Inc.	Inc.
$\neq 0$	0	0	0		Red.	Inc.	Inc.	Inc.

Red. denotes a regular Padé approximant with nonsingular equations, but which happens to reduce to order $[\lambda/\mu]$.

Con. denotes a Padé approximant with a singular Hankel determinant, but which is determined by a consistent set of equations, and also reduces to $[\lambda/\mu]$.

Inc. denotes a nonexistent Padé approximant. The equations for it are inconsistent.

course, the theorem also considers the question of uniqueness. Table 6 summarizes graphically the link between a block in the C-table and the Padé table.

Up until now, we have tended to take the question of infinite blocks of the C-table and the Padé table for granted. This casualness is for the good reason that an infinite block in the Padé table turns out to be uniquely associated with a rational fraction. It is certainly a consequence of Theorem 1.4.2 that if either side of any block in the C-table is of infinite length, then so is the other side of the block.

THEOREM 1.4.4. *Suppose that a function $f(z)$ is analytic at the origin and is uniquely determined by its Maclaurin series: $f(z) = \sum_{i=0}^{\infty} c_i z^i$. Then the existence of an infinite block in the Padé table of $\sum_{i=0}^{\infty} c_i z^i$ is a necessary and sufficient condition for $f(z)$ to be rational; let $f(z)$ be of type $[\lambda/\mu]$. Then the corresponding block in the C-table is defined by $C(\lambda/\mu+1) \neq 0$, $C(\lambda+1/\mu) \neq 0$, $C(\lambda+i/\mu+j) = 0$, $i,j = 1,2,\ldots,\infty$. This condition in turn is necessary and sufficient for the representation*

$$c_k = \sum_{i=1}^{N} \sum_{\tau=1}^{\mu_i} r_{i\tau} \binom{-\mu_i}{k} (-z_i)^{-k} \quad \text{for all} \quad k \geq \lambda - \mu. \quad (4.17)$$

Proof. Suppose that there is an infinite block in the Padé table, containing approximants which, according to Theorem 1.4.3, all reduce to $[\lambda'/\mu']$. Then, working term by term, we find that

$$B^{[\lambda'/\mu']}(z) \sum_{i=0}^{\infty} c_i z^i - A^{[\lambda'/\mu']}(z) = 0.$$

Hence, by the hypothesis of the theorem,

$$f(z) = \frac{A^{[\lambda'/\mu']}(z)}{B^{[\lambda'/\mu']}(z)}$$

and $\lambda = \lambda'$, $\mu = \mu'$. Let

$$b(z) = B^{[\lambda/\mu]}(z) = \prod_{i=1}^{N} \left(1 - \frac{z}{z_i}\right)^{\mu_i}, \qquad \text{where} \quad \sum_{i=1}^{N} \mu_i = \mu.$$

Let

$$\pi_1(z) = \sum_{k=0}^{\lambda-\mu-1} c_k z^k \quad \text{if } \lambda > \mu \quad \text{and} \quad \pi_1(z) = 0 \quad \text{otherwise.}$$

Then a polynomial $\pi_2(z)$, of degree μ at most, exists such that $f(z)$ decomposes into partial fractions:

$$f(z) = \pi_1(z) + \frac{\pi_2(z)}{b(z)} = \pi_1(z) + \sum_{i=1}^{N} \sum_{\tau=1}^{\mu_i} r_{i\tau}\left(1 - \frac{z}{z_i}\right)^{-\tau}.$$

Hence c_k is given by (4.17) for $k \geq \lambda - \mu$.

The converse, in which we suppose that $f(z)$ is rational, is a consequence of the uniqueness property of Padé approximants, given by Theorem 1.4.4, and the rest is obvious.

Having established that certain Padé approximants do not exist in certain circumstances, the following theorem establishes that, in every row, column, or diagonal, there are an infinite number of extant approximants.

THEOREM 1.4.5. *Let $\sum_{i=0}^{\infty} c_i z^i$ be a formal power series. In its Padé table, an infinite number of Padé approximants exist*

    (i) *in any row $[L/M]$, M fixed, $L \to \infty$,*
    (ii) *in any column $[L/M]$, L fixed, $M \to \infty$, and*
    (iii) *in any paradiagonal $[M+J/M]$, J fixed, $M \to \infty$.*

Proof. Consider the rows. In any row, there are either a finite number of blocks and consequently an infinite number of well-defined approximants, or else an infinite number of blocks. In any block, there is at least one extant Padé approximant in any row. Hence an infinite number of Padé approximants exist in any row of the Padé table. The argument is the same for columns and paradiagonals.

Note that the theorem makes no mention of convergence of these approximants, but it does mean that the discussion of convergence of rows, columns, and diagonals is more than a rhetorical exercise.

Exercise 1. Find a function for which the [3/2] Padé approximant does not exist.

Exercise 2. Without using a computer, calculate the C-table of $(1+z-z^3)/(1+z+z^2)$.

Exercise 3. If $C(L/M)\neq 0$, show that $P^{[L/M]}(z)$ and $Q^{[L/M]}(z)$ have no common factor.

Exercise 4. By using continuity of $Q^{[L/M]}(z)$ in the parameter c_{L-M+1} and the theory of the continuous denominator [Roth, 1965], renormalize $Q^{[L/M]}(z)$ when $C(L/M)=0$.

Exercise 5. Show that blocks in the Padé table may be contiguous but cannot overlap.

1.5 Duality and Invariance

In this section, we state some algebraic properties of Padé approximants which are attractive features of a class of approximating functions. The theorems are easy to prove, and the interpretation of the theorems is important. All the theorems of the section concern algebraic properties of power series, and no convergence property is in any way directly implied.

THEOREM 1.5.1 (Duality). *Let* $g(z)=\{f(z)\}^{-1}$ *and* $f(0)\neq 0$. *Then* $[L/M]_g(z)=\{[M/L]_f\}^{-1}$ *provided either Padé approximant exists.*

Proof. Suppose that $[M/L]_f$ exists, and define

$$\frac{p_M(z)}{q_L(z)}=[M/L]_f(z).$$

Because $f(0)\neq 0$, $[M/L]_f(0)\neq 0$, and hence $p_M(z), q_L(z)$ are polynomials of degrees at most M, L respectively, and $p_M(0)\neq 0$. Hence

$$g(z)-\frac{q_L(z)}{p_M(z)}=\frac{p_M(z)-f(z)q_L(z)}{f(z)p_M(z)}$$

$$=O(z^{L+M+1}),$$

and so $[L/M]_g = q_L(z)/p_M(z)$ as required. The proof, given that $[L/M]_g$ exists, is similar.

The duality may be summarized by saying that the Padé approximant of the reciprocal function is the reciprocal of the Padé approximant of the function; it may be glibly restated by saying that Padé approximation and reciprocation commute.

If a class of functions and the reciprocal functions have a common property, e.g. they are meromorphic, then the duality property shows a valuable symmetry feature of the Padé approximants as approximating functions.

THEOREM 1.5.2 (Homographic invariance under argument transformations). *Let* $f(z) = \sum_{i=0}^{\infty} c_i z^i$. *Define an origin preserving bilinear transformation of the argument*

$$w = \frac{az}{1+bz},$$

and thereby a new function $g(w) = f(z)$. *Then*

$$[M/M]_g(w) = [M/M]_f(z)$$

provided either Padé approximant exists.

Proof. Suppose that $[M/M]_g(w)$ exists. Then

$$[M/M]_g(w) = \frac{\displaystyle\sum_{i=0}^{M} a_i w^i}{\displaystyle\sum_{i=0}^{M} b_i w^i} = g(w) + O(w^{2M+1}).$$

Let

$$A_M(z) = (1+bz)^M \sum_{i=0}^{M} a_i \left(\frac{az}{1+bz}\right)^i$$

and

$$B_M(z) = (1+bz)^M \sum_{i=0}^{M} b_i \left(\frac{az}{1+bz}\right)^i.$$

Then

$$\frac{A_M(z)}{B_M(z)} = f(z) + O(z^{2M+1}),$$

where $A_M(z)$, $B_M(z)$ are polynomials in z of order M. Hence $A_M(z)/B_M(z)$ $=[M/M]_f(z)$.

The proof given that $[M/M]_f$ exists is similar, and the theorem is proved.

Notice that the proof is only valid for diagonal approximants, and so this homographic invariance only holds for the diagonal sequences. Theorem 1.5.2 is usually called the theorem of Baker, Gammel, and Wills [1961]; a closely related result was previously proved by Edrei [1939].

The homographic invariance theorem for argument transformations is the cornerstone of the optimistic approach to Padé approximants. This optimism is entirely validated by practical experience, but not as yet by proven theorems. The fundamental reason for this optimism is that the mapping $w=az/(1+bz)$ shown in Figure 1 allows any circle Γ in the z-plane enclosing the origin as an interior point to be mapped onto a given circle

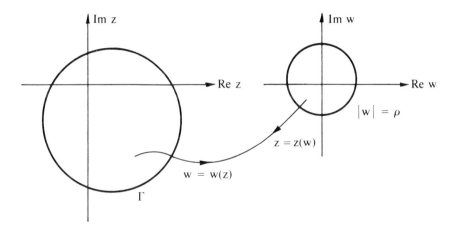

Figure 1. The mapping $w=az/(1+bz)$ and its inverse.

$|w|=\rho$ centered on the origin. If a sequence of diagonal approximants may be proved to converge to $g(w)$ within $|w|\leqslant\rho$, then convergence of the same sequence for $f(z)$ follows in the interior of Γ (see the quasitheorem of Section 6.7). We also understand the acceleration of convergence using the sequence of *diagonal* Padé approximants, mentioned in Chapter 3, in terms of a generalized Euler transformation [Thacher, 1974]. Other implications of this theorem are discussed in Section 6.7.

THEOREM 1.5.3 (Homographic invariance of value transformations). *Given a function $f(z)=\sum_{i=0}^{\infty} c_i z^i$, we define*

$$g(z)=\frac{a+bf(z)}{c+df(z)}.$$

If $c + df(0) \neq 0$, *then*

$$[M/M]_g(z) = \frac{a + b[M/M]_f(z)}{c + d[M/M]_f(z)}$$

provided $[M/M]_f(z)$ *exists.*

Proof. Because $c + df(0) \neq 0$, we may find polynomials $p_M(z)$ and $q_M(z)$ of degree M at most such that

$$\frac{p_M(z)}{q_M(z)} = \frac{a + b[M/M]_f(z)}{c + d[M/M]_f(z)}$$

with $q_M(0) \neq 0$. Then

$$\begin{aligned}
\frac{p_M(z)}{q_M(z)} - g(z) &= \frac{a + b[M/M]_f(z)}{c + d[M/M]_f(z)} - \frac{a + bf(z)}{c + df(z)} \\
&= \frac{(bc - ad)\{[M/M]_f(z) - f(z)\}}{\{c + d[M/M]_f(z)\}\{c + df(z)\}} \\
&= O(z^{2M+1}).
\end{aligned}$$

Therefore $p_M(z)/q_M(z)$ is the $[M/M]$ Padé approximant of $g(z)$.

Homographic invariance of the values, like invariance under argument transformations, is generally only valid for diagonal approximants. An interesting feature of this result is that the value ∞ of a Padé approximant is treated on a par with any other value; this is significant in the context of convergence on the Riemann sphere (see Section 6.4). The transformation $g = (a + bf)/(c + df)$ can be broken down into successive elementary transformations, each with a simple interpretation: the mapping

$$g = \frac{a + bf}{c + df}$$

is a composite of translations and inversion given by

$$g_1 = \alpha + f, \qquad g_2 = \beta/g_1, \quad \text{and} \quad g = \gamma + g_2.$$

Details of the interpretation are to be found in EPA, p. 113.

THEOREM 1.5.4 (Truncation theorem). *Let* $f(z) = \sum_{i=0}^{\infty} c_i z^i$, *and let*

$$g(z) = \sum_{i=0}^{\infty} g_i z^i = \left\{ f(z) - \sum_{i=0}^{k-1} c_i z^i \right\} z^{-k}.$$

Then $[L-k/M]_g(z) = \{[L/M]_f - \sum_{i=0}^{k-1} c_i z^i\} z^{-k}$ *for* $k \geq 1$, $L - k \geq M - 1$, *provided either Padé approximant exists.*

Proof. Suppose that $[L/M]_f = A^{[L/M]}(z)/B^{[L/M]}(z)$ exists; we leave the proof for the alternative case as an exercise. Define

$$p_{L-k}(z) - \left\{[L/M]_f \sum_{i=0}^{k-1} c_i z^i\right\} z^{-k} B^{[L/M]}(z).$$

By construction, $p_{L-k}(z)$ is a polynomial of order $L-k$ under the conditions of the theorem. Hence

$$\frac{p_{L-k}(z)}{B^{[L/M]}(z)} - g(z) = \{[L/M]_f - f(z)\} z^{-k} = O(z^{L+M-k+1}),$$

and therefore

$$\frac{p_{L-k}(z)}{B^{[L/M]}(z)} = [L-k/M]_g(z).$$

This theorem is used repeatedly in Chapter 5, where we prove a series of results for $[M-1/M]$ Padé approximants and generalize them to $[M+J/M]$ Padé approximants for $J \geq -1$, using a method equivalent to the truncation theorem.

THEOREM 1.5.5 (Unitarity) [Gammel and McDonald, 1966]. *Let* $f(z) = \sum_{i=0}^{\infty} c_i z^i$ *be unitary, by which we mean that*

$$f(z)[f(z)]^* = 1.$$

If $[M/M] = [M/M]_f(z)$ *is a diagonal Padé approximant of* $f(z)$, *then*

$$[M/M][M/M]^* = 1.$$

Proof. Let

$$[M/M] = \frac{A^{[M/M]}(z)}{B^{[M/M]}(z)}.$$

Then

$$[M/M][M/M]^* = \{f(z) + O(z^{2M+1})\}\{[f(z)]^* + O(z^{2M+1})\}$$
$$= 1 + O(z^{2M+1}).$$

Hence

$$A^{[M/M]}(z)\big\{A^{[M/M]}(z)\big\}* - \big\{B^{[M/M]}(z)\big\}*B^{[M/M]}(z) = O(z^{2M+1}).$$

The left-hand side of this expression is a polynomial of order $2M$ of most; therefore

$$A^{[M/M]}(z)\big\{A^{[M/M]}(z)\big\}* = \big\{B^{[M/M]}(z)\big\}*B^{[M/M]}(z),$$

and hence

$$[M/M][M/M]* = 1.$$

This theorem is summarized by saying that diagonal Padé approximants preserve unitarity. It is important in S-matrix theory (see Part II, Chapter 4) because the S-matrix is unitary. A timely note of caution is that complex conjugation may destroy analyticity, and it is prudent to define $g(z) = [f(z)]* = 1/f(z)$ before discussing the analytic structure of $[f(z)]*$.

The invariance theorems of this section have justified the value of Padé approximants as practical approximating functions. In fact, one obvious esthetic test of the merit of any generalization of Padé approximants is whether the generalizations have such useful invariance properties.

Exercise 1 Let $f(z)$ satisfy the conditions of Theorem 1.5.5. Define $f(z) = 1 + it(z)$, so that $t(z)$ obeys a unitarity condition

$$t(z) - \{t(z)\}* = 2it(z)\{t(z)\}*.$$

Prove that formation of $[L/M]$ Padé approximants of $t(z)$ with $L \leqslant M$ preserves this unitarity property [Masson, 1967a, b].

Exercise 2 Let

$$g(z) = \frac{bf(z)}{c + df(z)}.$$

Assuming that $L \leqslant M$, $[L/M]_f$ exists, and $c + df(0) \neq 0$, prove that*

$$[L/M]_g = \frac{b[L/M]_f}{c + d[L/M]_f}.$$

---

*D. Bessis, private communication.

1.6 Bigradients and Hadamard's Formula

Hitherto we have considered the problem of finding rational functions which satisfy

$$f(z) = \sum_{i=0}^{\infty} c_i z^i = \frac{P^{[L/M]}(z)}{Q^{[L/M]}(z)} + O(z^{L+M+1}) \qquad (6.1)$$

where c_i are given coefficients and $c_0, c_1, \ldots, c_{L+M}$ are actually needed. A natural generalization of this problem is that of finding polynomials $P^{[L/M]}(z)$, $Q^{[L/M]}(z)$ of orders L, M respectively which satisfy

$$f(z) = \frac{g(z)}{d(z)} = \frac{\displaystyle\sum_{i=0}^{\infty} g_i z^i}{\displaystyle\sum_{i=0}^{\infty} d_i z^i} = \frac{P^{[L/M]}(z)}{Q^{[L/M]}(z)} + O(z^{L+M+1}). \qquad (6.2)$$

Here g_i, d_i are the given coefficients, of which we need to know $g_0, g_1, \ldots, g_{L+M}$ and $d_0, d_1, \ldots, d_{L+M}$. If $d(z)$ is a polynomial of order m and $m < L + M$, we complete the definition by taking $d_{m+1} = d_{m+2} = \cdots = d_{L+M} = 0$. We assume that $d_0 \neq 0$, and leave the question of what modifications are needed in the presence of degeneracy as an exercise. With these conventions, and in the absence of degeneracy,

$$P^{[L/M]}(z) =$$

$$-\begin{vmatrix}
d_0 & 0 & \cdot & \cdot & \cdot & \cdot & 0 & 0 & \cdot & \cdot & \cdot & 0 & g_0 \\
d_1 & d_0 & \cdot & \cdot & \cdot & \cdot & 0 & 0 & \cdot & \cdot & \cdot & g_0 & g_1 \\
\cdot & \cdot & \cdot & & & & \cdot & & & & & \cdot & \cdot \\
\cdot & & & & & & \cdot & & & & & \cdot & \cdot \\
\cdot & & & & & \cdot & g_0 & \cdot & \cdot & \cdot & g_{M-1} & g_M \\
d_L & d_{L-1} & \cdot & \cdot & \cdot & \cdot & d_0 & g_1 & \cdot & \cdot & \cdot & g_M & g_{M+1} \\
d_{L+1} & d_L & \cdot & \cdot & \cdot & \cdot & d_1 & \cdot & & & & \cdot & \cdot \\
\cdot & \cdot & \cdot & & & & \cdot & \cdot & & & & \cdot & \cdot \\
\cdot & \cdot & & & & \cdot & \cdot & \cdot & & & & \cdot & \cdot \\
d_{L+M} & d_{L+M-1} & \cdot & \cdot & \cdot & \cdot & d_M & g_L & \cdot & \cdot & \cdot & g_{L+M-1} & g_{L+M} \\
1 & z & \cdot & \cdot & \cdot & \cdot & z^L & 0 & \cdot & \cdot & \cdot & 0 & 0
\end{vmatrix},$$

$$\qquad (6.3)$$

$$Q^{[L/M]}(z)=$$

$$\begin{vmatrix}
d_0 & 0 & \cdot & \cdot & \cdot & \cdot & 0 & 0 & \cdot & \cdot & \cdot & 0 & g_0 \\
d_1 & d_0 & & & & & 0 & 0 & & & & g_0 & g_1 \\
\cdot & \cdot & & \cdot & & & \cdot & \cdot & & & & \cdot & \cdot \\
\cdot & \cdot & & & \cdot & & \cdot & \cdot & & & & \cdot & \cdot \\
\cdot & \cdot & & & & \cdot & \cdot & \cdot & & \cdot & & \cdot & \cdot \\
\cdot & & & & & \cdot & \cdot & g_0 & \cdot & \cdot & \cdot & g_{M-1} & g_M \\
d_L & d_{L-1} & \cdot & \cdot & \cdot & \cdot & d_0 & g_1 & \cdot & \cdot & \cdot & g_M & g_{M+1} \\
d_{L+1} & d_L & \cdot & \cdot & \cdot & \cdot & d_1 & \cdot & & & & \cdot & \cdot \\
\cdot & \cdot & & & & & \cdot & \cdot & & & & \cdot & \cdot \\
\cdot & \cdot & & & & & \cdot & \cdot & & & & \cdot & \cdot \\
\cdot & \cdot & & \cdot & & & \cdot & \cdot & & & & \cdot & \cdot \\
d_{L+M} & d_{L+M-1} & \cdot & \cdot & \cdot & \cdot & d_M & g_L & \cdot & \cdot & \cdot & g_{L+M-1} & g_{L+M} \\
0 & 0 & \cdot & \cdot & \cdot & \cdot & 0 & z^M & \cdot & \cdot & \cdot & z & 1
\end{vmatrix}.$$

$$(6.4)$$

We may verify that the proposed solution of (6.3) and (6.4) satisfies (6.2) by forming

$$g(z)Q^{[L/M]}(z)-d(z)P^{[L/M]}(z)\equiv E(z)$$

$$=\begin{vmatrix}
d_0 & 0 & \cdot & \cdot & \cdot & \cdot & 0 & 0 & \cdot & \cdot & \cdot & 0 & g_0 \\
d_1 & d_0 & & & & & 0 & 0 & & & & g_0 & g_1 \\
\cdot & \cdot & & \cdot & & & \cdot & \cdot & & & & \cdot & \cdot \\
\cdot & \cdot & & & \cdot & & \cdot & \cdot & & \cdot & & \cdot & \cdot \\
\cdot & \cdot & & & & \cdot & g_0 & \cdot & \cdot & \cdot & g_{M-1} & g_M \\
d_L & d_{L-1} & \cdot & \cdot & \cdot & \cdot & d_0 & \cdot & & & & \cdot & \cdot \\
\cdot & \cdot & & & & & \cdot & \cdot & & & & \cdot & \cdot \\
\cdot & \cdot & & \cdot & & & \cdot & \cdot & & & & \cdot & \cdot \\
d_{L+M} & d_{L+M-1} & \cdot & \cdot & \cdot & \cdot & d_M & g_L & \cdot & \cdot & \cdot & g_{L+M-1} & g_{L+M} \\
d(z) & zd(z) & \cdot & \cdot & \cdot & \cdot & z^L d(z) & z^M g(z) & \cdot & \cdot & \cdot & zg(z) & g(z)
\end{vmatrix}.$$

$$(6.5)$$

By subtraction of the first row of (6.5) from its last row, followed by subtraction of z times the second row from the new last row, etc., we find that

$$E(z) = \sum_{k=L+M+1}^{\infty} z^k$$

$$\times \begin{vmatrix} d_0 & 0 & \cdot & \cdot & \cdot & 0 & 0 & \cdot & \cdot & \cdot & g_0 \\ d_1 & \cdot & & & & 0 & 0 & & & & \cdot \\ \cdot & & \cdot & & & \cdot & \cdot & & \cdot & & \cdot \\ \cdot & & & \cdot & & \cdot & g_0 & \cdot & \cdot & \cdot & g_M \\ d_L & \cdot & \cdot & \cdot & \cdot & d_0 & & \cdot & & & \cdot \\ \cdot & & & & & \cdot & & \cdot & & & \cdot \\ \cdot & & & & & \cdot & & \cdot & & & \cdot \\ d_{L+M} & \cdot & \cdot & \cdot & \cdot & d_M & g_L & \cdot & \cdot & \cdot & g_{L+M} \\ d_k & \cdot & \cdot & \cdot & \cdot & d_{k-L} & g_{k-M} & \cdot & \cdot & \cdot & g_k \end{vmatrix},$$

which proves (6.2).

The determinants in (6.3) and (6.4) are called polynomial bigradients. An ordinary bigradient $\Delta_{L,M}(d_i, g_i)$ is an $(L+M) \times (L+M)$ determinant formed from the coefficients d_i arranged in its first L columns with a negative gradient and from the coefficients g_i arranged in the next M columns with a positive gradient. Specifically, define

$$\Delta_{L,M}(d_i, g_i) =$$

$$\begin{vmatrix} d_0 & 0 & \cdot & \cdot & \cdot & \cdot & 0 & 0 & \cdot & \cdot & \cdot & 0 & g_0 \\ d_1 & d_0 & \cdot & \cdot & \cdot & \cdot & 0 & 0 & \cdot & \cdot & \cdot & g_0 & g_1 \\ \cdot & \cdot & & \cdot & & & \cdot & \cdot & & & & \cdot & \cdot \\ \cdot & \cdot & & & \cdot & & \cdot & \cdot & & & & \cdot & \cdot \\ \cdot & \cdot & & & & \cdot & \cdot & g_0 & \cdot & \cdot & \cdot & g_{M-2} & g_{M-1} \\ d_{L-1} & d_{L-2} & \cdot & \cdot & \cdot & \cdot & d_0 & g_1 & \cdot & \cdot & \cdot & g_{M-1} & g_M \\ d_L & d_{L-1} & \cdot & \cdot & \cdot & \cdot & d_1 & \cdot & & & & \cdot & \cdot \\ \cdot & \cdot & & & & & \cdot & \cdot & & & & \cdot & \cdot \\ \cdot & \cdot & & & & & \cdot & \cdot & & & & \cdot & \cdot \\ d_{L+M-1} & d_{L+M-L} & \cdot & \cdot & \cdot & \cdot & d_M & g_L & \cdot & \cdot & \cdot & g_{L+M-2} & g_{L+M-1} \end{vmatrix}$$

$$(6.6)$$

As a preliminary to proving Hadamard's theorem, we define

$$e(z) = \frac{1}{d(z)} = \sum_{i=0}^{\infty} e_i z^i.$$

From (6.2), it follows that

$$e(z)g(z)=f(z).$$

Consequently,

$$\sum_{j=0}^{i} e_j d_{i-j}=\delta_{i0}, \qquad i=0,1,2,\dots,$$

and

$$\sum_{j=0}^{i} e_j g_{i-j}=c_i, \qquad i=0,1,2,\dots.$$

The equations justify the following identities:

$$\begin{pmatrix} e_0 & 0 & 0 \\ e_1 & e_0 & 0 \\ e_2 & e_1 & e_0 \end{pmatrix} \begin{pmatrix} d_0 & 0 & 0 \\ d_1 & d_0 & 0 \\ d_2 & d_1 & d_0 \end{pmatrix} = \begin{pmatrix} 1 & 0 & 0 \\ 0 & 1 & 0 \\ 0 & 0 & 1 \end{pmatrix} \qquad (6.7)$$

and

$$\begin{pmatrix} e_0 & 0 & 0 \\ e_1 & e_0 & 0 \\ e_2 & e_1 & e_0 \end{pmatrix} \begin{pmatrix} 0 & 0 & g_0 \\ 0 & g_0 & g_1 \\ g_0 & g_1 & g_2 \end{pmatrix} = \begin{pmatrix} 0 & 0 & c_0 \\ 0 & c_0 & c_1 \\ c_0 & c_1 & c_2 \end{pmatrix}. \qquad (6.8)$$

Now let us consider the $(L+M)\times(L+M)$ matrix

$$E = \begin{pmatrix} e_0 & & & \\ e_1 & e_0 & & \text{\Large 0} \\ \vdots & & \ddots & \\ e_{L+M-1} & e_{L+M-2} & \cdots & e_0 \end{pmatrix},$$

for which $\det E = e_0^{L+M}$. Using the matrix E as a left multiplier for the bigradient matrix, and regarding (6.7) and (6.8) as truncations of large matrix operations, we find that

$$e_0^{L+M}\Delta_{L,M}(d_i, g_i)=$$

$$\begin{vmatrix} 1 & 0 & \cdot & \cdot & \cdot & 0 & 0 & \cdot & \cdot & \cdot & 0 & c_0 \\ 0 & 1 & \cdot & \cdot & \cdot & 0 & 0 & \cdot & \cdot & \cdot & c_0 & c_1 \\ \cdot & \cdot & \cdot & & & \cdot & \cdot & & & & \cdot & \cdot \\ \cdot & \cdot & & \cdot & & \cdot & \cdot & & & & \cdot & \cdot \\ \cdot & \cdot & & & \cdot & c_0 & \cdot & \cdot & \cdot & & c_{M-2} & c_{M-1} \\ 0 & 0 & \cdot & \cdot & \cdot & 1 & c_1 & \cdot & \cdot & \cdot & c_{M-1} & c_M \\ 0 & 0 & \cdot & \cdot & \cdot & 0 & \cdot & & & & \cdot & \cdot \\ \cdot & \cdot & & & & \cdot & & & & & \cdot & \cdot \\ \cdot & \cdot & & & & \cdot & & & & & \cdot & \cdot \\ \cdot & \cdot & & & & \cdot & & & & & \cdot & \cdot \\ 0 & 0 & \cdot & \cdot & \cdot & 0 & c_L & \cdot & \cdot & \cdot & c_{L+M-2} & c_{L+M-1} \end{vmatrix} \cdot$$

Hence $\Delta_{L,M}(d_i, g_i)=d_0^{(L+M)}C(L/M)$, with the Hankel determinant defined by (4.8). This result enables us to prove Hadamard's theorem.

THEOREM 1.6.1. *Let* $f(z)=\sum_{i=0}^{\infty}c_iz^i$, *and let* $C(L/M)$ *be its Hankel determinant defined in the usual way by* (4.8). *Let* $\{f(z)\}^{-1}=\sum_{i=0}^{\infty}c_i'z^i$, *and let* $C'(l/m)$ *be its Hankel determinant. Then*

$$C'(M/L)=(-1)^{(L+M)(L+M-1)/2}C(L/M)c_0^{-(L+M)}. \tag{6.9}$$

Proof. For the bigradient (6.6) relating to (6.1) and (6.2) we have established that

$$\Delta_{L,M}(d_i, g_i)=d_0^{L+M}C(L/M).$$

Had we started with the totally compatible *Ansatz*

$$\{f(z)\}^{-1}=\frac{d(z)}{g(z)}=\sum_{i=0}^{\infty}c_i'z^i \tag{6.10}$$

we would have constructed the bigradient

$$\Delta_{M,L}'(g_i, d_i)=g_0^{L+M}C'(M/L).$$

The two bigradients are, in fact, the same except for the order of the columns, and careful inspection shows that $\frac{1}{2}(L+M)(L+M-1)$ column interchanges are needed to identify the bigradients. Hence

$$C'(M/L)=(-1)^{(L+M)(L+M-1)/2}\left(\frac{d_0}{g_0}\right)^{L+M}C(L/M).$$

Since $c_0=g_0/d_0$, (6.9) is proved. By virtue of the algebraic nature of the result, (6.9) holds good whenever it is well defined, which is whenever c_0, g_0 and d_0 are nonzero.

Trudi's theorem is probably the best-known result which uses bigradients explicitly. It also identifies the nature of degeneracies encountered with bigradients.

THEOREM 1.6.2 [Trudi, 1862]. *Let $g(z)$ be a polynomial of degree l, and let $d(z)$ be a polynomial of degree m. Define the bigradient $\Delta_{l,m}(d_i, g_i)$ as in (6.6). If*

$$\Delta_{l,m}(d_i, g_i)=\Delta_{l-1,m-1}(d_i, g_i)=\cdots=\Delta_{l-j+1,m-j+1}(d_i, g_i)=0$$

and

$$\Delta_{l-j,m-j}(d_i, g_i)\neq0,$$

then $d(z)$ and $g(z)$ have a common polynomial divisor of order j.

Proof. If $\Delta_{l,m}(d_i, g_i)=0$, then $C(l/m)=0$ and the Padé equations for $f(z)=g(z)/d(z)$ are degenerate. However, $C(l-j/m-j)\neq0$, and so a nondegenerate $[l-j/m-j]$ Padé approximant of $f(z)$ exists. By the theory of Section 4, this Padé approximant is also the $[l/m]$ Padé approximant of $f(z)$, and so the theorem follows.

For further properties of bigradients, we refer to Householder [1970, 1971], Householder and Stewart [1969], and Padé [1900].

Exercise If $g(z)$, $d(z)$ are both polynomials, under what circumstances is the problem of determining $P^{[L/M]}(z)$ and $Q^{[L/M]}(z)$ in (6.2) well posed?

CHAPTER 2

Direct Application

2.1 Direct Calculation of Padé Approximants

In this section we consider the calculation of the denominators of Padé approximants directly from the explicit determinantal formula (1.1.8) for $Q^{[L/M]}(z)$, for $L \geq M-1$. We are able to do this principally for the class of functions which may be represented as

$$f(z) = {}_2F_1(\alpha, 1; \gamma; z).$$ \hfill (1.1a)

Results for functions with the representations

$$f(z) = {}_1F_1(1; \gamma; z)$$ \hfill (1.1b)

and

$$f(z) = {}_2F_0(\alpha, 1; z)$$ \hfill (1.1c)

follow as corollaries. In Section 1.2, we showed that explicit calculation of the numerator and denominator for each Padé approximant of the exponential function is possible using the direct method, but existence of a simple explicit form for the numerator polynomial is special to this case. Once the denominator of an $[L/M]$ Padé approximant is constructed, the formula (1.1.9) leads directly to the numerator coefficients; equivalently, we may consider $A^{[L/M]}(z)$ to be defined by truncation of $B^{[L/M]}f(z)$ beyond terms of order z^L.

Of course, we do not suggest that numerical calculations are to be made using determinantal formulas, and in Section 2.4 we consider these problems. Nor do we suggest that the method of this section is always the best

ENCYCLOPEDIA OF MATHEMATICS and Its Applications, Gian-Carlo Rota (ed.). Vol. 13: George A. Baker, Jr., and Peter R. Graves-Morris, Padé Approximants: Basic Theory, Part I   ISBN 0-201-13512-4

when the coefficients c_i of $f(x)$ are given algebraically. Use of the Q.D. algorithm (see Sections 3.6, 4.4) may well be simpler, but it leads to a continued-fraction representation of the Padé approximant. Explicit formulas for the numerator and denominator polynomials would have to be derived from the recursion formulas (4.2.7).

The coefficients of the Maclaurin expansion of (1a) are

$$c_i = \frac{(\alpha)_i}{(\gamma)_i} = \frac{(\alpha+i-1)!}{(\alpha-1)!}\frac{(\gamma-1)!}{(\gamma+i-1)!} = \frac{\Gamma(\alpha+i)}{\Gamma(\alpha)}\frac{\Gamma(\gamma)}{\Gamma(\gamma+i)}. \qquad (1.2)$$

Substituting in (1.4.8) for $L \geq M-1$ we find that

$$C(L/M) = \begin{vmatrix} \dfrac{(\alpha)_{L-M+1}}{(\gamma)_{L-M+1}} & \cdots & \dfrac{(\alpha)_L}{(\gamma)_L} \\ \vdots & & \vdots \\ \dfrac{(\alpha)_L}{(\gamma)_L} & \cdots & \dfrac{(\alpha)_{L+M-1}}{(\gamma)_{L+M-1}} \end{vmatrix}. \qquad (1.3)$$

Reduction of (1.3) is simplified by defining p to be the product of the leading elements of the rows

$$p = \prod_{i=1}^{M} \frac{(\alpha)_{L-M+i}}{(\gamma)_{L-M+i}},$$

so that

$$C(L/M) = p \begin{vmatrix} 1 & \dfrac{\alpha+L-M+1}{\gamma+L-M+1} & \cdots & \dfrac{(\alpha+L-M+1)_{M-1}}{(\gamma+L-M+1)_{M-1}} \\ 1 & \dfrac{\alpha+L-M+2}{\gamma+L-M+2} & \cdots & \dfrac{(\alpha+L-M+2)_{M-1}}{(\gamma+L-M+2)_{M-1}} \\ \vdots & \vdots & & \vdots \\ 1 & \dfrac{\alpha+L}{\gamma+L} & \cdots & \dfrac{(\alpha+L)_{M-1}}{(\gamma+L)_{M-1}} \end{vmatrix}. \qquad (1.4)$$

Sequential row subtraction and expansion about the $(1,1)$ element leads to

$$C(L/M) = \frac{p(\alpha-\gamma)^{M-1}(M-1)!}{\displaystyle\prod_{i=1}^{M-1}(\gamma+L-M+i)_2}\begin{vmatrix} 1 & \dfrac{\alpha+L-M+2}{\gamma+L-M+3} & \cdots & \dfrac{(\alpha+L-M+2)_{M-2}}{(\gamma+L-M+3)_{M-2}} \\[2ex] 1 & \dfrac{\alpha+L-M+3}{\gamma+L-M+4} & \cdots & \dfrac{(\alpha+L-M+3)_{M-2}}{(\gamma+L-M+4)_{M-2}} \\[2ex] \vdots & \vdots & & \vdots \\[2ex] 1 & \dfrac{\alpha+L}{\gamma+L+1} & \cdots & \dfrac{(\alpha+L)_{M-2}}{(\gamma+L+1)_{M-2}} \end{vmatrix}.$$

$$(1.5)$$

By using this technique recursively,

$$C(L/M) = \frac{p\displaystyle\prod_{i=1}^{M-1}\left\{(\alpha-\gamma-i+1)^{M-i}(M-i)!\right\}}{\displaystyle\prod_{k=1}^{M-1}\prod_{i=k}^{M-1}(\gamma+L-i+2k-2)_2}.$$

$$(1.6)$$

To calculate the coefficient of z^j in $Q^{[L/M]}(z)$, we need the determinant following, in which the dotted lines enclose a deleted column:

$$(-)^j q_j^{[L/M]} = \begin{vmatrix} \dfrac{(\alpha)_{L-M+1}}{(\gamma)_{L-M+1}} & \cdots & \vdots & \dfrac{(\alpha)_{L-j+1}}{(\gamma)_{L-j+1}} & \vdots & \cdots & \dfrac{(\alpha)_{L+1}}{(\gamma)_{L+1}} \\[2ex] \vdots & & \vdots & \vdots & \vdots & & \vdots \\[2ex] \dfrac{(\alpha)_L}{(\gamma)_L} & \cdots & \vdots & \dfrac{(\alpha)_{L-M-j}}{(\gamma)_{L-M-j}} & \vdots & \cdots & \dfrac{(\alpha)_{L-M}}{(\gamma)_{L-M}} \end{vmatrix}$$

$$= p\begin{vmatrix} 1 & \dfrac{\alpha+L-M+1}{\gamma+L-M+1} & \cdots & \vdots & \dfrac{(\alpha+L-M+1)_{M-j}}{(\gamma+L-M+1)_{M-j}} & \vdots & \cdots & \dfrac{(\alpha+L-M+1)_M}{(\gamma+L-M+1)_M} \\[2ex] 1 & \dfrac{\alpha+L-M+2}{\gamma+L-M+2} & \cdots & \vdots & \dfrac{(\alpha+L-M+2)_{M-j}}{(\gamma+L-M+2)_{M-j}} & \vdots & \cdots & \dfrac{(\alpha+L-M+2)_M}{(\gamma+L-M+2)_M} \\[2ex] \vdots & \vdots & & \vdots & \vdots & \vdots & & \vdots \\[2ex] 1 & \dfrac{\alpha+L}{\gamma+L} & \cdots & \vdots & \dfrac{(\alpha+L)_{M-j}}{(\gamma+L)_{M-j}} & \vdots & \cdots & \dfrac{(\alpha+L)_M}{(\gamma+L)_M} \end{vmatrix}.$$

Making $M-j$ simplifications of the type leading from (1.4) to (1.5), i.e., expansion by top left-hand element after row reduction, we are led to define the product of common factors

$$p_j = p \frac{\displaystyle\prod_{i=1}^{M-j} \{(\alpha-\gamma-i+1)^{M-i}(M-i+1)!\}}{(M-j)! \displaystyle\prod_{k=1}^{M-j} \prod_{i=k}^{M-1} (\gamma+L-i+2k-2)}, \tag{1.7}$$

and then

$$(-)^j q_j^{[L/M]} = p_j \begin{vmatrix} \dfrac{\alpha+L-j+1}{\gamma+L+M+1-2j} & \cdots & \dfrac{(\alpha-L-j+1)_j}{(\gamma+L+M+1-2j)_j} \\ \vdots & & \vdots \\ \dfrac{\alpha+L}{\gamma+L+M-j} & \cdots & \dfrac{(\alpha+L)_j}{(\gamma+L-M-j)_j} \end{vmatrix}$$

Using the same technique of reduction, we obtain

$$(-)^j q_j^{[L/M]} = p_j \frac{\displaystyle\prod_{i=1}^{j} \{(\alpha+L-i+1)(j-i)!\} \prod_{i=M-j+1}^{M} (\alpha-\gamma-i+1)^{M-i}}{\displaystyle\prod_{i=1}^{j} (\gamma+L+M-j-i+1) \prod_{k=1}^{j-1} \prod_{i=k}^{j-1} (\gamma+L+M+2k-i-j-1)} \tag{1.8}$$

From (1.6), (1.7), and (1.8) there is substantial cancellation, and

$$(-)^j q_j^{[L/M]} = C(L/M) \frac{M!}{(M-j)!} \frac{(\alpha+L)\cdots(\alpha+L-j+1)}{(\gamma+L+M-1)\cdots(\gamma+L+M-j)}.$$

Hence, for $L \geqslant M-1$,

$$\begin{aligned} Q^{[L/M]}(z) = C(L/M) &\left\{ 1 + \frac{(-\alpha-L)(-M)z}{(1-\gamma-L-M)1!} + \right. \\ &\left. \frac{(-\alpha-L)(-\alpha-L+1)(-M)(-M+1)z^2}{(1-\gamma-L-M)(2-\gamma-L-M)2!} + \cdots \right\} \\ = C(L/M) &\,_2F_1(-M, -\alpha-L, 1-\gamma-L-M; z), \tag{1.9} \end{aligned}$$

completing the derivation of the Padé denominator. We may derive a corollary from this result, by noting that

$$_1F_1(\beta,\gamma:z)= \lim_{\alpha\to\infty} {_2F_1}\left(\alpha,\beta,\gamma;\frac{z}{\alpha}\right)$$

Consequently the Padé denominators of

$$f(z)={_1F_1}(1,\gamma;z)$$

are given by

$$B^{[L/M]}(z)= \lim_{\alpha\to\infty} {_2F_1}\left(-M,-\alpha-L,1-\gamma-L-M;\frac{z}{\alpha}\right)$$
$$= {_1F_1}(-M;1-\gamma-L-M,-z), \qquad L\geqslant M-1. \qquad (1.10)$$

Furthermore, by choosing $\gamma=1$, we obtain the special case of the exponential function of Section 1.2.

Likewise, the Padé denominators of the asymptotic expansion of $_2F_0(\alpha,1;z)$ are given by replacing z with γz and letting $z\to\infty$. We find

$$B^{[L/M]}(z)={_2F_0}(-M,-\alpha-L;z), \qquad L\geqslant M-1.$$

There is another class of formal series, derived from coefficients

$$c_i= \prod_{j=0}^{i-1} \frac{A-q^{j+l}}{B-q^{j+m}}, \qquad i=0,1,2,\dots, \qquad (1.11)$$

which generate the power series

$$f(z)=1+\frac{A-q^l}{B-q^m}z+\frac{(A-q^l)(A-q^{l+1})}{(B-q^m)(B-q^{m+1})}z^2+\cdots, \qquad (1.12)$$

for which explicit expressions for the $[L/M]$ Padé approximants (with $L\geqslant M-1$) can be given. As in (1.4), we construct the determinant $C(L/M)$ by defining

$$C(L/M)=D(M,L-M+1+l,L-M+1+m)\prod_{j=0}^{L-M}\left\{\frac{A-q^{j+l}}{B-q^{j+m}}\right\}^M,$$

where

$$D(M,\mu,\nu)=\begin{vmatrix} 1 & \cdots & \prod_{r=0}^{M-1}\dfrac{A-q^{\mu+r}}{B-q^{\nu+r}} \\ \vdots & & \vdots \\ \prod_{r=0}^{M-1}\dfrac{A-q^{\mu+r}}{B-q^{\nu+r}} & \cdots & \prod_{r=0}^{2M-2}\dfrac{A-q^{\mu+r}}{B-q^{\nu+r}} \end{vmatrix}.$$

These determinants may be evaluated recursively from the recurrence relation

$$D(M, \mu, \nu) = \frac{(q-1)(q^2-1)\cdots(q^{M-1}-1)(Aq^\nu - Cq^\mu)^{M-1}(A-q^\mu)^{M-1}}{(C-q^\nu)(C-q^{\nu+1})\cdots(C-q^{\nu+M-1})(C-q^{\nu+1})^{M-1}}$$

$$\times D(M-1, \mu+1, \nu+2)q^{(M-1)(M-2)/2}/(C-q^\nu)^{M-1}$$

Hence, using the same methods as for the hypergeometric function (1.2), explicit expressions for the Padé approximants may be obtained.

If the coefficients were given by

$$c_i' = \prod_{j=0}^{i-1} (A - q^{j+l}) \tag{1.13}$$

instead of by (1.11), substitute $w = z/B$ and let $B \to \infty$.

If the coefficients were given by

$$c_i'' = \prod_{j=0}^{i-1} (B - q^{j+m})^{-1}, \tag{1.14}$$

substitute $w = Az$ and let $A \to \infty$. Thus results from series generated by (1.13) or (1.14) are special cases of results derived from (1.12). We pursue the direct calculation no further, because the approach based on the Q.D. algorithm (4.4.17) is algebraically simpler. We refer to Wynn [1967] for explicit formulas using the Q.D. algorithm.

2.2   Decipherment of Singularities from Padé Approximants

According to the theory of analytic continuation of functions of a complex variable, all the properties of a function, analytic at a point, are contained in its power-series expansion at that point. While in principle the function may be continued by reexpansion about other points, the central practical problem is to decipher the properties of the function from the given series coefficients. Padé approximants can be used very effectively in determining quantitative results about functions when the analytic properties are qualitatively known, and they can also be used to deduce considerable information about the singularity structure of a function from its Taylor series coefficients. Before embarking on the proper use of Padé

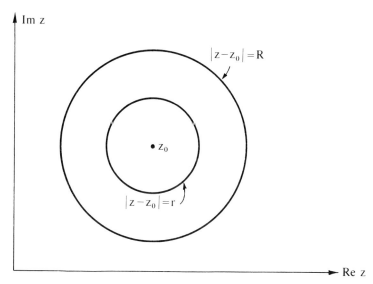

Figure 1.   An annulus of analyticity.

approximants, we review briefly the nature of the permissible singularities of an analytic function.

First, let $f(z)$ be analytic and single-valued in $|z-z_0|<R$ except at $z=z_0$. Then, for all r such that $0<r<R$, $f(z)$ is analytic and single-valued in the annulus $r<|z-z_0|<R$. Laurent's theorem states that in these circumstances the expansion

$$f(z)= \sum_{n=0}^{\infty} a_n(z-z_0)^n+ \sum_{n=1}^{\infty} b_n(z-z_0)^{-n} \tag{2.1}$$

is convergent within the annulus. Laurent's theorem follows from Cauchy's theorem, and gives a specific representation for a_n and b_n,

$$a_n=\frac{1}{2\pi i}\int_{|z|=R}\frac{f(z)\,dz}{(z-z_0)^{n+1}}, \qquad b_n=\frac{1}{2\pi i}\int_{|z|=r}f(z)(z-z_0)^{n-1}\,dz \tag{2.2}$$

Equation (2.1) permits unique categorization of singularities at $z=z_0$.

If $b_n=0$, $n=1,2,\dots$, then $z=z_0$ is a regular point and $f(z)$ is analytic at $z=z_0$.

If $b_n=0$, $n=2,3,\dots$, and $b_1\neq0$, then $z=z_0$ is a simple pole.

If $b_n=0$, $n=m+1,m+2,\dots$, and $b_m\neq0$, then z_0 is an mth-order pole, and for $m>1$, $f(z)$ is said to have a multipole.

If, for all m and some $n>m$, $b_n\neq0$, then z_0 is an essential singularity of $f(z)$.

Provided z_0 is not an essential singularity, (2.1) suggests that $f(z)$ may be approximated by Padé approximants. However, even if z_0 is an essential singularity, provided it is not approached too closely, the essential singularity resembles a finite sum of multipoles. This is because b_n given by (2.2) decrease rapidly with n; in fact $|b_n| < r^n$ for any $r > 0$ and n sufficiently large.

Example.

$$f(z) = \exp\left(\frac{1}{1+z}\right) = \sum_{n=0}^{\infty} \frac{1}{n!}\left(\frac{1}{1+z}\right)^n. \tag{2.3}$$

$f(z)$ has an essential singularity at $z = -1$, but the coefficients $b_n = 1/n!$ decrease very rapidly. This qualitative fact can be exploited rigorously as in the proof of Theorem 6.5.4.

Secondly, $f(z)$ may have branch points. If $f(z)$ is analytically continued by expansion and reexpansion and remains a single-valued function of z, there are no branch points. However, if there is any point z_0 for which analytic continuation by a clockwise and counterclockwise path of arbitrarily small radius yields different values at the same point, then $f(z)$ is not single-valued and z_0 is a branch point. A single-valued function can be obtained for $f(z)$ in various ways. One may form the Mittag–Leffler star by drawing a straight line from each branch point to infinity along a ray from the origin. It sometimes happens that a single-valued function can be formed by connecting two or more branch points by branch cuts. For

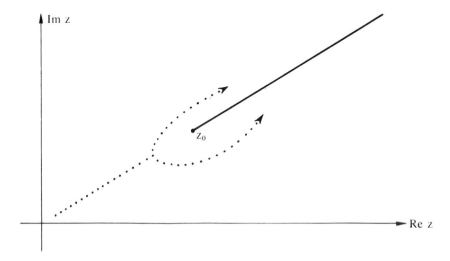

Figure 2.   A branch point at z_0, an associated branch cut, and two paths of continuation.

example $f(z)=\sqrt{(z-a)(z-b)}$ may be uniquely defined with a branch cut from a to b. This line is a discontinuity of the given function.

Finally, there may exist natural boundaries. For example

$$f(z)=1+z+z^2+z^4+z^8+\cdots=\sum_{i=0}^{\infty}z^{(2^i)} \tag{2.4}$$

has a natural boundary on $|z|=1$. To prove this assertion, let us note that $f(z)$ is well defined by a convergent Maclaurin expansion in $|z|<1$. Yet if

$$z=z_{pq}=\exp(2\pi ip/2^q), \qquad \text{where p and q are integers,}$$

then

$$f(z)=\sum_{n=0}^{q-1}z^{(2^i)}+\sum_{n=q}^{\infty}1, \qquad \text{which diverges.}$$

Since the points $\{z_{pq}\}$ are dense on the unit circle, continuation to $|z|>1$ is no longer straightforward. Sometimes such analytic continuation is possible, and sometimes Padé approximants converge to this continued function: references to Padé approximants and quasianalytic functions are given in the selected bibliography.

In order to interpret the results of using Padé approximants on a new function, we consider how Padé approximants represent the singularities in the previous categories. If $f(z)$ has a simple pole, then a simple zero in the denominator of the Padé approximant near the pole is expected. If $f(z)$ has a multiple pole, a cluster of zeros of the Padé denominator is expected; these zeros should tend to coalesce at the multipole with increasing order of approximation. For an essential singularity, we recall *Weierstrass's theorem*: Let $f(z)$ have an essential singularity at $z=z_0$. Given $\rho>0$, $\varepsilon>0$, and any value v, there exists a point in the circle $|z-z_0|<\rho$ at which $|f(z)-v|<\varepsilon$ [Titchmarsh, 1939, p. 51]. In other words, this theorem shows that $f(z)$ tends to any desired limit v as $z\rightarrow z_0$ through a suitable sequence of z-values. Thus we expect a clustering of poles and zeros of Padé approximants at an essential singularity. In a low order of approximation, of course, a multiple pole and an essential singularity will appear the same.

To simulate a branch cut, we expect to find a path delineated by roughly alternating poles and zeros of the Padé approximants. To appreciate this, consider

$$f(z)\equiv\int_{P}\frac{\rho(u)\,du}{z-u}=\frac{-1}{2\pi i}\int_{\Gamma}\frac{f(u)}{z-u}. \tag{2.5}$$

where P is a path from a to b in the complex plane, Γ is a contour enclosing

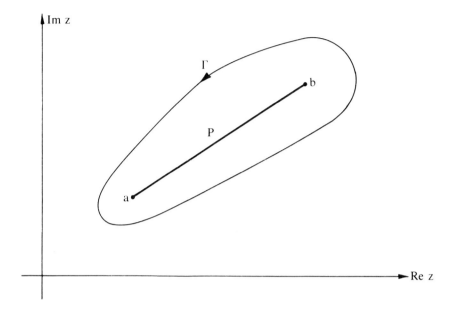

Figure 3. The path P and the contour Γ in the complex plane.

P, and the second part of (2.5) is a consequence of Cauchy's theorem. $f(z)$ is uniformly differentiable and so is analytic except on the path P. $f(z)$ may be approximated by a Riemann sum

$$f(z) \approx \sum_{i=1}^{N} \frac{\rho(u_i)\,\delta u_i}{z - u_i}, \qquad z \notin P,$$

which consists of a sequence of poles with residues $\rho(u_i)\,\delta u_i$. This behavior is what one expects the Padé approximants to reproduce.

Example.

$$f(z) = \frac{1}{\sqrt{1+z}} = \frac{1}{\pi} \int_{-\infty}^{-1} \frac{dx}{(z-x)\sqrt{-1-x}}$$

$f(z)$ is defined with a cut on $-\infty < z \leqslant -1$, and the poles of the Padé approximant are located on that cut, as shown in Figure 4.

Since there are many ways of defining the cuts to define, in turn, a single-valued function, and the Mittag–Leffler star is but one, it is interesting to speculate that the limiting distribution of the poles and zeros of a suitable sequence of Padé approximants delineate a natural cut structure (or principal Riemann surface) from the Maclaurin series. We presently expect

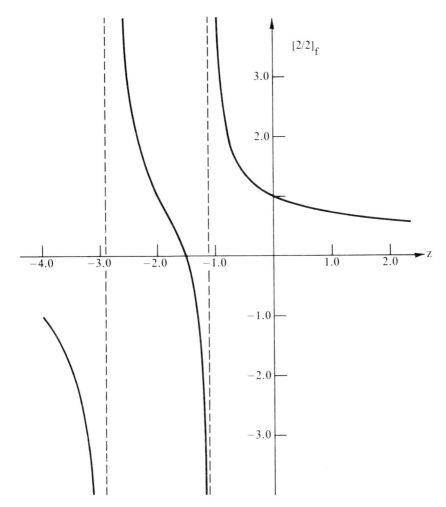

Figure 4. The [2/2] Padé approximant of $f(z)=(1-z)^{-1/2}$ showing poles at $z_{\pm}=-2\pm 2/\sqrt{5}$ lying on the branch cut of $f(z)$, $-\infty<z\leqslant-1$.

that the cuts so defined are such as to minimize their capacity in the inverse z-plane, a point which we discuss briefly in Section 6.7.

The foregoing account states what we expect from Padé approximants to analytic functions with various singularities. In ideal circumstances precisely these results are obtained; in practice, whether one expects them or not, *defects* occur for all but the simplest functions. A *defect* is the name given to an extraneous pole and a nearby zero. We consider, nonrigorously, a function $f(z)$ which is smooth near $z=\alpha$ and let

$$g(z)=f(z)+\frac{\varepsilon}{z-\alpha}, \qquad |\varepsilon|\ll 1.$$

If ε is very small and $|z-\alpha|$ is not small, $g(z) \approx f(z)$. Only in a neighborhood of $z=\alpha$ do $g(z)$ and $f(z)$ differ appreciably. We expect $g(\alpha)=\infty$, but also we have

$$g(z) \approx f(\alpha) + \frac{\varepsilon}{z-\alpha} = 0$$

if $z-\alpha = -\varepsilon/f(\alpha)$, which is very small. Hence there is a zero close to the pole. There are real difficulties in the numerical detection of defects; we refer to Abd-Elall et al. [1970] for details. In short, the addition of an "insignificant" pole to $f(z)$ produces a defect, and this problem has to be expected with Padé approximation. We regard the nearby zero of numerator and denominator as canceling approximately, which puts the defects in their proper perspective. Defects are easily recognized by their transient nature. They tend to appear and disappear as one looks at one approximant and then the next. They contrast strongly with the more stable patterns seen in conjunction with the true singularities of the function. We will return to this topic in the next section; in Section 6.5, we prove that the residue of the pole of a defect is small for high-order Padé approximation of meromorphic functions.

The following example shows why it is normally wise to ignore results derived from Padé approximants with defects close to the origin. The function $1+z^2$ has no [1/1] Padé approximant (see Section 1.4). Consider the function [Zinn-Justin, 1970]

$$f(z) = 1 + \varepsilon z + z^2,$$

for which the [1/1] Padé approximant is

$$[1/1] = \frac{1+(\varepsilon - 1/\varepsilon)z}{1-z/\varepsilon}.$$

The pole of the [1/1] approximant occurs at $z=\varepsilon$, and its residue is $-\varepsilon^3$. If $|\varepsilon|$ is small, we see that the pole is close to the origin and its residue is small. We understand a Padé approximant with a defect near the origin as a nearly degenerate approximant, and we are suspicious of drawing any implications from the values of such an approximant.

Given a function defined by a formal power series as the starting point, the first step towards deciphering the information contained in the series is formation of the Padé approximants lying in at least a broad band about the central diagonal of the Padé table. Unless there are reasons to the contrary, as much of the Padé table as possible should be constructed. Normally, most computing effort goes into construction of the coefficients rather than formation of the Padé approximants, and so construction of all available Padé approximants is relatively inexpensive. The next step is the examination of the distribution of poles and zeros. Are these persistent, or do they form defects? It should then be possible to decide which poles and zeros closest to the origin represent true singularities of the function. It is sometimes possible at this stage to detect that the function has asymptotic behavior z^{J_0} for some J_0 and in some half plane by virtue of the stability of $[M+J_0/M]$ approximants, and poles receding to infinity for $L-M<J_0$.

Once the general nature of the structure of the function has been determined, either by the preceding analysis or by other qualitative information (see Part II, Section 1.3), one is in a position to make a more refined analysis. The presence of poles is normally sufficiently clear. Their influence on the function depends on their residues, their multiplicity, and their proximity to the origin. The type of structure which has been further analysed profitably is principally the branch point and branch cut. An important strategy is the manipulation of the series to a form which can be exactly represented by Padé approximants, except for small corrections. We consider the particular case where

$$f(z)=A(z)(1-\mu z)^{-\gamma}+B(z) \tag{2.6}$$

is expected to represent the function. $f(z)$ has a cut from $z=\mu^{-1}$ to ∞, and $A(z)$, $B(z)$ are to be analytic at $z=\mu^{-1}$ and are expected to have little structure. The following methods have been used to good effect. [Baker, 1961; Baker et al., 1967; Hunter and Baker, 1973; Baker, 1977].

(i) Form Padé approximants to

$$F_1(z)=\frac{d}{dz}\ln f(z)$$

$$\simeq \frac{\gamma}{\mu^{-1}-z} \qquad \text{near} \quad z=\mu^{-1}. \tag{2.7}$$

The use of the appropriate Padé approximant of $F_1(z)$ determines the pole position $z=\mu^{-1}$, and $-\gamma$ is its residue. The approximants we have defined for $f(z)$ are not necessarily rational. They are commonly called D-log Padé approximants (see Section 5.3). The estimates of μ and γ are called "unbiased" because no assumed values of μ, γ, etc. are used: μ, γ are determined directly from the series coefficients.

(ii) Form Padé approximants to

$$F_2(z)=(\mu^{-1}-z)\frac{d}{dz}\ln f(z)\approx\gamma \tag{2.8}$$

for an assumed value of μ, and obtain a biased estimate of γ by evaluating the Padé approximant at $z=\mu^{-1}$.

(iii) Form Padé approximants to

$$F_3(z) = [f(z)]^{1/\gamma} \approx \frac{\left[A(\mu^{-1})\right]^{1/\gamma}}{1 - \mu z} \tag{2.9}$$

by assuming a value for γ, and obtain biased estimates of $A(\mu^{-1})$ and μ^{-1} from the roots and residues of the Padé approximants.

(iv) Form Padé approximants to

$$F_4(z) = (1 - \mu z)^{\gamma} f(z) \approx A(\mu^{-1}) \tag{2.10}$$

by assuming values for μ, γ, and obtain a biased estimate of $A(\mu^{-1})$ by evaluating the Padé approximant at $z = \mu^{-1}$.

(v) Form Padé approximants to

$$F_5(z) = \frac{\dfrac{d}{dz}\left(\ln \dfrac{d}{dz} f(z)\right)}{\dfrac{d}{dz} \ln f(z)} \approx 1 + \frac{1}{\gamma}, \tag{2.11}$$

and evaluate the Padé approximants at the assumed value of μ^{-1}. This process yields a biased estimate of γ, but as a practical matter it is frequently relatively insensitive to the choice of μ^{-1} [Baker et al., 1967].

(vi) In cases where there are two or more series with the same branch point, and it is the branch point closest to the origin, one may obtain an unbiased estimate of the difference of the exponents by the method of "critical-point renormalization" as follows: we have

$$\begin{aligned}
f(x) &= A(z)(1 - \mu z)^{-\gamma} + B(z) = \sum_{j=0}^{\infty} c_j z^j, \\
g(z) &= C(z)(1 - \mu z)^{-\varepsilon} + D(z) = \sum_{j=0}^{\infty} c_j' z^j,
\end{aligned} \tag{2.12}$$

and $A(z)$, $B(z)$, $C(z)$, and $D(z)$ are analytic at $z = \mu^{-1}$. We define

$$h(z) = \sum_{j=0}^{\infty} \frac{c_j}{c_j'} z^j \propto (1 - z)^{-\gamma + \varepsilon - 1} \qquad \text{for} \quad z \approx 1. \tag{2.13}$$

Form Padé approximants to

$$F_6(z) = (1 - z) \frac{d}{dz} \ln h(z) \approx \gamma - \varepsilon + 1 \tag{2.14}$$

to obtain an unbiased value of $\varepsilon - \gamma$ by evaluating the Padé approximants at

$z=1$. In (2.12) it is possible* to choose $g(z)=[f(z)]^p$, where p is an integer, $p \geqslant 2$. Then $\varepsilon = p\gamma$ and $F_6(z) \sim 1 + \gamma(1-p)$. A similar scheme was proposed by Sheludyak and Rabinovich [1979].

(vii) The Baker–Hunter method [1973] sometimes allows one to detect subdominant confluent singularities. This procedure is a subtle one, and sometimes the method is interfered with by strong, nonconfluent singularities. Suppose, instead of (2.6), that

$$f(z)= \sum_{n=0}^{\infty} c_n z^n \approx \sum_{j=1}^{m} A_j(1-\mu z)^{-\gamma_j} \qquad (2.15)$$

near $z=\mu^{-1}$, and that $\gamma_1 > \gamma_2 > \cdots > \gamma_m$. Then $f(z)$ has m confluent singularities at $z=\mu^{-1}$. Make the change of variable

$$z=\mu^{-1}(1-e^{-\xi}), \qquad (2.16)$$

so that

$$g(\xi)=f(z(\xi))= \sum_{n=0}^{\infty} g_n \xi^n \approx \sum_{j=1}^{m} A_j e^{\gamma_j \xi}. \qquad (2.17)$$

A transform of $g(\xi)$ is defined by

$$G(s)= \int_0^{\infty} e^{-\xi} g(s\xi)\, d\xi. \qquad (2.18)$$

From (2.17), we find

$$G(s)= \sum_{n=0}^{\infty} n!\, g_n s^n \approx \sum_{j=1}^{m} \frac{A_j}{1-\gamma_j s}, \qquad (2.19)$$

and hence we may compute biased estimates of γ_j and A_j to determine at least some of the stronger confluent, subdominant singularities.

By way of a caution, we add that when confluent singularities exist, they can, and often do, bias the results of the other methods (i)–(vi) listed above as well as appreciably slow down the rate of convergence. Furthermore, it has been frequently observed that other significant singularities interfere not only with the confluent-singularity analysis, but dramatically reduce the rate of convergence. Where feasible, steps to suppress their competing influence are rewarding.

---

*D. Bessis, private communication.

2.3   Apparent Errors

With any approximation scheme, one must be able to estimate the approximation error. Using Padé approximants, there are three principal sources of error: (i) the given coefficients c_j are known only to limited accuracy, (ii) accuracy is lost in forming the coefficients of the Padé approximant and in forming its value, and (iii) the Padé approximant is not the function itself, leading to the fundamental approximation error. There is little to be done about (i) except to note that accuracy in the given coefficients is essential. The Padé approximant is necessarily misguided by errors in the given series, no matter what their source. The variation of the coefficients of the given series within their accuracy limits provides a very useful and instant error estimate of the accuracy of Padé approximation. For (ii), loss of accuracy in the formation of Padé approximants is, in practice, usually inexcusable. Double precision on a modern computer gives 20 or 30 decimal places, and this precision should be used if necessary. Of course, the approximation problem is often ill conditioned, and the accuracy of the input coefficients is usually better than the accuracy of the output coefficients even with exact arithmetic. A rough and ready working guide is that one extra decimal place of working accuracy should be kept for every decimal place of accuracy required in the answer (cf. Section 2.4). As a purely empirical anthropological observation, most inexperienced Padé approximators use sufficiently high order approximants—often too high—and insufficient working accuracy to justify them.

We now turn to the problem of estimating the mathematical error of the approximation. In practice, this estimate must be based on likely hypotheses, and estimating these apparent errors is currently an art as well as a science.

In the previous section, we encountered defects in the approximants. They are nearby pole and zero combinations which are significant when they occur within the region of interest. The region is frequently most conveniently taken to be the largest circle containing the origin and whatever singularities are of interest, but excluding, with a margin to spare, all other nonpolar singularities. A defect would then be any pole in that region with a residue less than some preassigned value, say 0.003 times the expected residue. We are inclined to exclude these defective Padé approximants from the set of Padé approximants expected to be useful approximations. By so doing, we expect to obtain a set which is uniformly bounded on a bounded region which excludes singularities of the given function. Provided that there are sufficiently many Padé approximants in this set, the theorems of Section 6.4 assure convergence. In the assessment of apparent errors, it should be noted that, as a practical matter, the occurrence of defects seems often to cause successive Padé approximants ($\Delta L = 1$ or $\Delta M = 1$ or both) to be approximately equal, and one can be misled about the rate of convergence if the defects are not detected either directly or indirectly. The occurrence of defects can be thought of as a near miss at the existence of a block in the Padé table, where the determinant does not vanish but is anomalously small. In the case where there is a block, certain consecutive Padé approximants are, of course, exactly equal (Section 1.4). The existence of defects shows that it is important to analyze as much as possible of the Padé table to decide which poles and zeros are significant, and which are defects indicating that the Padé approximant in question is unreliable. Furthermore, it is plain that a blind calculation of Padé approximants at any one value (e.g. by using the ε-algorithm) ignores much of the information provided by the Padé approximants themselves about their convergence in the z-plane.

Let us now turn our attention to functions which have a dominant singularity of the form, see (2.6),

$$A(1-\mu z)^{-\gamma},\tag{3.1}$$

and let us suppose that we have estimated this singularity, probably by using the methods of Section 2.2, to be

$$A'(1-\mu'z)^{-\gamma'}.\tag{3.2}$$

The three parameters A, μ, and γ are in principle determined by three equations, which originate from accuracy-through-order equations. Let these equations be accurate to order z^J, z^{J+1}, and z^{J+2}, and then we know that

$$A\binom{-\gamma}{j}\mu^j=A'\binom{-\gamma'}{j}\mu'^j(1+\eta_j),\qquad j=J,J+1,J+2,\tag{3.3}$$

where η_j are small percentage errors. Obviously, if $\eta_j=0$, the parameters are identical, and so the η_j are to be regarded as the source of the error in the approximation scheme. We consider percentage errors, because μ determines the scale of the z-plane; if μ is very different from unity, the magnitude of the terms in (3.3) can vary rapidly with j. If the approximation is a good one, we may use first-order expansions

$$A'=A+\delta A,\qquad \gamma'=\gamma+\delta\gamma,\qquad \mu'=\mu+\delta\mu.\tag{3.4}$$

A first-order analysis and logarithmic differentiation of (3.3) gives

$$\frac{\delta A}{A}+j\frac{\delta\mu}{\mu}+\delta\gamma\left(\sum_{k=0}^{j-1}\frac{1}{\gamma+k}\right)=\eta_j,\qquad j=J,J+1,J+2.\tag{3.5}$$

These three linear equations can easily be solved to give

$$-\frac{\delta\mu}{\mu}=(2\gamma+2J+1)\eta_{J+1}-(\gamma+J+1)\eta_{J+2}-(\gamma+J)\eta_J, \qquad (3.6a)$$

$$\delta\gamma=(\gamma+J)\left(\eta_{J+1}-\eta_J-\frac{\delta\mu}{\mu}\right), \qquad (3.6b)$$

$$\frac{\delta A}{A}=\eta_J-J\frac{\delta\mu}{\mu}-\left(\sum_{k=0}^{J-1}\frac{1}{\gamma+k}\right)\delta\gamma. \qquad (3.6c)$$

From (3.6a), we see that $\delta\mu/\mu$ is of order J compared to η_j, provided there is no unusual cancellation. From (3.6b) it follows that $\delta\gamma$ is of order $J\delta\mu/\mu$ or $J^2\eta_j$. Finally (3.6c) shows that $\delta A/A$ is of order $J^2\ln J\eta_j$. We conclude that the errors in this determination of A, μ, and γ are in the ratio

$$\frac{\delta\mu}{\mu}:\delta\gamma:\frac{\delta A}{A}=1:J:J\ln J, \qquad (3.7)$$

where J is the order of the active terms in the series.

The determination of the apparent absolute size of the errors is more difficult. We will be explicit about our procedure when it is used in conjunction with a method such as (i) or (iv) of the previous section. In those cases the parameter μ is just a zero of the reciprocal of the function, and so $\delta\mu/\mu$ is proportional to the error of estimation of (e.g.) $1/F_1(x)$. Thus we can relate the error $\delta\mu/\mu$ to the errors in a table of values, which are in practice relatively easy to compute.

First let us look at the structure of the difference between two adjacent Padé approximants. By the ($*\,*$) identity (3.4.5),

$$[L/M]-[L-1/M]=\frac{C(L/M+1)}{C(L-1/M)}\frac{z^{L+M}}{B^{[L/M]}(z)B^{[L-1/M]}(z)}, \qquad (3.8)$$

where the determinants are defined by (1.4.8). Since $[L/M]$ is exact through order z^{L+M}, the right-hand side of (3.8) gives the error in the coefficient of z^{L+M} in $[L-1/M]$. The next step is to relate this error to the magnitude of the η's. The hypothesis involved here is that there are no unusual cancellations, so that a reasonable estimate for the magnitude of η is obtained.

Since $B^{[L/M]}(0)=1$, one can estimate the coefficient of z^{L+M}, and at the same time to some extent take into account higher-order terms, by forming a table of the left-hand side of (3.8) over the range $0<z\leqslant\mu^{-1}$ and fitting it to the monomial z^{L+M} at the point in that range which gives the largest coefficient.

Now, in the treatment of method (i) of Section 2.2 (F_1), a different reduction of (3.5) is convenient, namely

$$\delta\gamma = (\gamma+J+1)(\gamma+J)(2\eta_{J+1}-\eta_J-\eta_{J+2}),$$

$$\frac{\delta\mu}{\mu} = -\delta\gamma(\gamma+J)^{-1}+\eta_{J+1}-\eta_J,$$

$$\frac{\delta A}{A} = -\frac{J\delta\mu}{\mu}-\left(\sum_{k=0}^{J-1}\frac{1}{\gamma+k}\right)\delta\gamma+\eta_J.$$

Here $\delta\gamma=0$ as $\gamma=\gamma'=1$. Thus $\delta\mu/\mu$ is of order η, and $\delta A/A$, which by (2.7) plays the role of $\delta\gamma$, is of order $J\delta\mu/\mu$, in harmony with (3.7). Thus by this analysis we have, by (3.8), that η is given by

$$A\mu^{L+M}\eta \approx \frac{C(L/M+1)}{C(L-1/M)},$$

where we have used

$$\binom{-1}{n}=1.$$

Exercise Why is the variation in value of a diagonal Padé approximant with respect to small variations in c_0 usually a poor estimate of the accuracy of the formation of the approximant?

2.4 Numerical Calculation of Padé Approximants

Quite probably, the first problem to be tackled by anyone interested in Padé approximants involves the calculation of values of a set of diagonal approximants using the coefficients of the Maclaurin series as data. Explicit calculation is only feasible by hand for the lowest-order approximants, and so a good numerical algorithm becomes an early requirement. In this section we discuss the methods available and the qualities which are required of a "good" method.

We start by considering the direct method of solution (see Section 2.1), and defer the detailed discussion of the various distinct objectives and the other methods available until this discussion can be put in context. Numerical solution of the linear equations

$$
\begin{vmatrix}
c_{L-M+1} & c_{L-M+2} & \cdots & c_L \\
c_{L-M+2} & c_{L-M+3} & \cdots & c_{L+1} \\
\vdots & \vdots & & \vdots \\
c_L & c_{L+1} & \cdots & c_{L+M-1}
\end{vmatrix}
\begin{vmatrix}
b_M \\
b_{M-1} \\
\vdots \\
b_1
\end{vmatrix}
=
\begin{vmatrix}
-c_{L+1} \\
-c_{L+2} \\
\vdots \\
-c_{L+M}
\end{vmatrix}
\qquad (4.1)
$$

is the core of the problem. We assume that the data coefficients $\{c_i,$ $i=0,1,2,\ldots,L+M\}$ are real or complex numbers known to a certain accuracy. If $i<0$ in (1), we take $c_i=0$. The simplest approach to (1) which can be recommended is to solve the linear system using Gaussian elimination with full pivoting; a FORTRAN subprogram which calculates $b_1, b_2,\ldots, b_M$ in this way and completes the calculation of an $[L/M]$ Padé approximant at a (real) point is given in the appendix. The function PADE($\cdot\cdot$) is written for use with real, single-precision coefficients $c_0, c_1,\ldots, c_{L+M}$, and may readily be adapted to suit other cases.

An obvious question to put is whether Gaussian elimination with full pivoting is the best method of solving (4.1). If we decide to use the direct method of calculating Padé approximants, most of the work is entailed in the actual solution of (4.1) for the coefficients $b_1, b_2,\ldots, b_M$. For our purposes, we need a method which will recognize whether the system of equations is degenerate, which occurs if $C(L/M)=0$. Unless the data coefficients $c_0, c_1,\ldots, c_{L+M}$ are integers (or rational fractions with an exact binary representation) for which symbolic methods are suitable [Geddes, 1979], it is likely that rounding errors will prevent our computer program from discovering that $C(L/M)=0$, even if it is true. If the data coefficients correspond to a degenerate case in which $C(L/M)=0$, any results of the calculation which we obtain are generated by rounding error and are likely to be misleading. We stress that this sort of case does arise: for example, if $f(z)$ is a geometric series, all its $[L/2]$ Padé approximants are degenerate. Our numerical algorithm must be able to recognize similar cases in which a lower order of Padé approximation is quite possibly what is required. In short, one of our requirements is an accurate method of solving the system (4.1); a method which is more accurate than one might at first sight expect is needed to discriminate against exactly degenerate equations. The accuracy requirement alone is achieved using the Gaussian elimination method with double-precision arithmetic, as in the program in the appendix of Part I, or even higher precision if necessary, but this head-on attack is not quite the best approach.

We wish to ascertain not only whether the data coefficients correspond to an exactly degenerate problem, but also whether the given data constitute an ill-posed problem or correspond within error to a degenerate Padé approximation problem. We wish to know if the system can become degenerate when the coefficients are varied within their given accuracy limits. The method consisting of equilibration, partial pivoting for generation of an LU decomposition and subsequent iterative refinement of the solution is especially useful in near-degenerate cases. We refer to Wilkinson [1963, 1965] for details of this numerical method of solving general linear equations and its error analysis. Basically, this sophisticated method consists of finding the LU decomposition of the coefficient matrix in (4.1), which is used to generate a sequence of approximate solution vectors. The iterative refinement converges geometrically for stable well-posed problems, and also for the ill-posed problems which occur so frequently in Padé approximation. If the iteration does not converge properly, we conclude that the solution is generated by rounding error in the data coefficients, which almost certainly corresponds to a degenerate approximant. The numerical convergence characteristics of the iterative refinement permit better discrimination between the ill-conditioned but nondegenerate coefficient matrices and singular coefficient matrices corresponding to degenerate approximants. This method is implemented in the (Mark 6) N.A.G. library algorithm.

Whatever method of solution of (4.1) is chosen, the complete algorithm must contain an error exit for use when the requested approximant is degenerate according to agreed criteria. If the algorithm includes pointwise evaluation of an approximant, the algorithm should contain an error exit in case the approximant is being evaluated at a pole.

We have emphasized that a good numerical algorithm should contain a fairly sophisticated degeneracy test. If the Padé approximant corresponding to given data does not exist, we expect good algorithms to recognize this situation and to detect the degeneracy. We call such algorithms *reliable*. We might also expect that small variations of the data coefficients within their accuracy limits should not significantly affect the computed value of the Padé approximants. However, we expect to face ill-posed problems in Padé approximation, and so, more realistically, we expect the vanishingly small variation of the data coefficients to correspond to vanishingly small variation in the computed numerator and denominator coefficients. This reflects the idea of mathematical continuity, and methods having this property are called *stable*. With ill-posed problems, we recognize that there will be magnification of rounding errors in the solution. Beyond these considerations, we expect our algorithm to be efficient, but efficiency is not as important as reliability and stability. Experienced Padé approximators are wary of degenerate and near-degenerate cases; they prefer to use more sophisticated approximation methods (see Part II, Chapter 1) rather than to resort to high-order approximants. Furthermore, the actual calculation of the Padé approximants is usually the least time consuming part of a complete calculation, in which it is the computation of the coefficients to high accuracy which is expensive in computer time. We note that solution of (4.1) by elimination requires $O(\frac{1}{3}M^3)$ operations, unless iterative refinement is necessary. We consider methods of lower order, which are more efficient when M is large, later in the section and discuss the penalties incurred by increased efficiency.

Our concern for accuracy of the numerical algorithm stems primarily from experience. In the previous section, we mentioned the empirical rule of keeping M extra decimal places of precision for computation of an $[L/M]$ approximant. Of course, one cannot say *a priori* that M, $M-1$, $2M$, or any other number of extra decimal places are generally required. We simply sound a clear warning about the minimum number of guarding figures likely to be required, and suggest that any calculations of $[L/M]$ approximants performed with less than M guarding figures be looked at critically. From a numerical point of view, the Padé approximant derives its capacity to extrapolate certain power series beyond their circle of convergence from using the information contained in the tails of the decimal expansion of the data: naturally, accurate data are of paramount importance. We understand some of these ideas in terms of the following example. We consider a Stieltjes series (see Chapter 5; this and other material of the section may be more suitable for a second reading of this work) in which the coefficients are given by

$$c_j = \int_0^a u^j \, d\phi(u), \qquad j = 0, 1, 2, \ldots \tag{4.2}$$

where $a > 0$ and $\phi(u)$ is bounded and nondecreasing. (Our sign convention here differs from (5.1.2) in that $f(-z)$ is a Stieltjes function.) A special case of (4.2) is given by

$$c_j = \int_0^1 u^j \, du = \frac{1}{j+1}, \qquad j = 0, 1, 2, \ldots \tag{4.3}$$

For Stieltjes series, various sequences of Padé approximants are known to converge systematically, and Stieltjes functions form the major class of functions for which we have a complete convergence theory. We see that the equations (4.1) with coefficients given by (4.3) are

$$
\begin{pmatrix}
\dfrac{1}{L-M+2} & \dfrac{1}{L-M+3} & \cdots & \dfrac{1}{L+1} \\[2ex]
\dfrac{1}{L-M+3} & \dfrac{1}{L-M+4} & \cdots & \dfrac{1}{L+2} \\[2ex]
\vdots & \vdots & & \vdots \\[2ex]
\dfrac{1}{L+1} & \dfrac{1}{L+2} & \cdots & \dfrac{1}{L+M}
\end{pmatrix}
\begin{pmatrix}
b_M \\[2ex]
b_{M-1} \\[2ex]
\vdots \\[2ex]
b_1
\end{pmatrix}
=
\begin{pmatrix}
\dfrac{-1}{L+2} \\[2ex]
\dfrac{-1}{L+3} \\[2ex]
\vdots \\[2ex]
\dfrac{-1}{L+M+1}
\end{pmatrix}
\tag{4.4}
$$

for $L \geqslant M - 1$. The coefficient matrix of (4.4) is a Hilbert segment which is notorious for its ill-conditioning. In fact, we have encountered ill-conditioning in an ideal case which is in no way degenerate. In this context, it is interesting to consider the condition number of the Hankel matrix

$$
H_M =
\begin{pmatrix}
c_{L-M+1} & c_{L-M+2} & \cdots & c_L \\
c_{L-M+2} & c_{L-M+3} & \cdots & c_{L+1} \\
\vdots & \vdots & & \vdots \\
c_L & c_{L+1} & \cdots & c_{L+M-1}
\end{pmatrix},
\tag{4.5}
$$

in which the elements of H_M are defined by (4.2). The condition number of H_M is defined by

$$\kappa(H_M) = \|H_M^{-1}\|_2 \|H_M\|_2.$$

A matrix with a large condition number corresponds to an ill-posed problem, and very crudely, one may expect that $\log_{10} \kappa(H_M)$ is the number of significant figures lost in the solution of (1) with rounded data. In the general case defined by (4.2) and (4.5), Taylor [1978] has shown that

$$\liminf_{M \to \infty} \kappa(H_M)^{1/M} \geqslant 4,$$

provided $d\phi(a-) > 0$. For the particular case of the Hilbert segment defined by (4.3) and (4.5),

$$\liminf_{M \to \infty} \kappa(H_M)^{1/M} \geqslant 16.$$

These results strongly support our rule of thumb that *approximately* M decimal places of accuracy are lost in the calculation of an $[L/M]$ approximant. In reality, this means that approximately M extra decimal places of precision are required of the data coefficient $c_{L-M+1}, c_{L-M+2}, \ldots, c_{L+M}$ than is expected of the solution coefficients $b_0, b_1, \ldots, b_M$. It is unusual for any further significant loss of accuracy to occur in the calculation of the numerator coefficients using (2.1b).

   Hitherto in this section, we have tacitly assumed that our problem is the calculation of the coefficients of a particular Padé approximant with a view to calculating the approximant at a prespecified value of z. In practice, it is much more likely that our task consists of the tabulation of a particular $[L/M]$ approximant at a number of values of z, or else that it consists of pointwise evaluation of a whole sequence of Padé approximants at a prespecified value of z. Which is the best algorithm to choose for Padé approximation depends on the problem at hand. It is normal to distinguish between the coefficient problem and the value problem. If tabulation of values of a particular approximant is required, it is probably best to calculate the coefficients $a_0, a_1, \ldots, a_L, b_0, b_1, \ldots, b_M$ first, and then to evaluate the approximants. If pointwise evaluation of a whole sequence is required, it may be better to consider methods such as using the ε-algorithm or the Q.D. algorithm in which a whole sequence is calculated recursively.

   The direct method of calculating Padé approximants is stable and reliable, but may not be the most efficient such method. There are several "$O(\alpha M^2)$" methods, where α is typically 4 or 6, which have computational efficiency gained at the expense of reliability. This means that they are suitable for use in contents where the existence and nondegeneracy of the approximants is not in question. Possibly, any of them may be developed into a reliable method at some future time. Our list of $O(\alpha M^2)$ methods, each with its own individual merits, for the coefficient problem is

(i) Kronecker's algorithm for the numerator and denominator coefficients.

(ii) The Q.D. algorithm for the corresponding continued-fraction representation.

(iii) Baker's algorithm for the numerator and denominator coefficients.

(iv) Watson's algorithm for the numerator and denominator coefficients.

(v) Toeplitz and Hankel matrix methods for the denominator coefficients.

Kronecker's algorithm [1881]. This is discussed in Part II, Section 1.1. The application to Padé approximation is the case in which all the interpolation points are confluent:

$$z_0 = z_1 = z_2 = \cdots = z_{L+M} = 0.$$

The coefficients of the Newton interpolating polynomial in this case are simply the Maclaurin series coefficients $c_0, c_1, \ldots, c_{L+M}$. Kronecker's algorithm is essentially based on exploitation of the $(_* * ^*)$ identity (3.5.12) to develop an antidiagonal sequence in the Padé table. Consequently, the algorithm is well defined only if all elements of the antidiagonal sequence exist and are nondegenerate. However, in every degenerate case, Kronecker's algorithm can be extended using the Euclidean algorithm [Claessens (thesis), 1976; McEliece and Shearer, 1978; Cordellier, 1979b; Graves-Morris, 1979].

The Q.D. algorithm. This algorithm develops a continued-fraction representation for a sequence of Padé approximants, and is fully discussed in Section 4.4. We require that all the elements of the corresponding fraction be finite and nonzero, which is equivalent to requiring that the convergents of the continued fraction be nondegenerate.

Baker's algorithm [1973a]. We construct the sequence of numerators $\{\eta_i(z), i=0,1,2,\ldots\}$ and denominators $\{\theta_i(z), i=0,1,2,\ldots\}$ of Padé approximants in a staircase sequence given by

$$\frac{\eta_{2j}(z)}{\theta_{2j}(z)} = [L+M-j/j] \quad \text{and} \quad \frac{\eta_{2j+1}(z)}{\theta_{2j+1}(z)} = [L+M-j-1/j].$$

The coefficients of $\eta_i(z), \theta_i(z)$ are related recursively using formulas based on the $(_*^*)$ and $(^{**})$ identities (3.5.11). Essentially, this algorithm is a fast method for calculating an antidiagonal staircase sequence, and we refer to EPA, p. 77, and Pindor [1976] for details.

Watson's algorithm [1973].   We construct the sequence of numerators $\{\eta_i(z),\ i=0,1,2,\dots\}$ and denominators $\{\theta_i(z),\ i=0,1,2,\dots\}$ of Padé approximants in a staircase sequence given by

$$\frac{\eta_{2j}(z)}{\theta_{2j}(z)} = [L+j/j] \quad \text{and} \quad \frac{\eta_{2j+1}(z)}{\theta_{2j+1}(z)} = [L+j+1/j]$$

The coefficients of $\eta_i(z), \theta_i(z)$ are calculated recursively using formulae based on the $\binom{*}{**}$ and $\binom{**}{*}$ identities (3.5.11) and the accuracy-through-order condition. This is a fast method for calculating staircase sequences parallel to the diagonal. We refer to EPA p. 79, Watson [1973], and Padé [1894]. Gragg's algorithm [1974] for the denominator polynomials is somewhat similar, but does not exploit the accuracy-through-order condition.

Toeplitz and Hankel matrix methods.   The equations (4.1) have a Hankel coefficient matrix. The structure of an $N\times N$ Hankel matrix H is such that its (i, j) element is given by

$$H_{ij} = h_{i+j}, \qquad i, j = 1, 2, \dots, N.$$

By reordering the equations (4.1), the direct method is also seen to be based on a solution of

$$
\begin{pmatrix}
c_L & c_{L+1} & \cdots & c_{L+M-1} \\
c_{L+1} & c_L & \cdots & c_{L+M-2} \\
\vdots & \vdots & & \vdots \\
c_{L-M+1} & c_{L-M+2} & \cdots & c_L
\end{pmatrix}
\begin{pmatrix}
b_M \\
b_{M-1} \\
\vdots \\
b_1
\end{pmatrix}
=
\begin{pmatrix}
-c_{L+M} \\
-c_{L+M-1} \\
\vdots \\
-c_{L+1}
\end{pmatrix}.
$$

$$(4.6)$$

The coefficient matrix of (4.6) is a Toeplitz matrix T. The structure of an $N\times N$ Toeplitz matrix T is such that its (i, j) element is given by

$$T_{ij} = t_{i-j}, \qquad i, j = 1, 2, 3, \dots, N.$$

Special methods exist which exploit the structure of Toeplitz and Hankel matrices to enable them to be solved in nondegenerate cases by an $O(\alpha N^2)$ method. Such methods do not use pivoting, and the calculation is an iterative scheme based on an embedding principle. The Padé numerator coefficients are calculated by back substitution in (2.1b). We refer to Bareiss [1969], Trench [1964, 1965], Rissanen [1973], Zohar [1974], Bultheel [1979], Graves-Morris [1979], Brent et al. [1980], and Bultheel and Wuytack [1980] for details.

We refer to Brezinski [1976] and Bultheel [1979, 1980a] for details of reliable calculation of a paradiagonal sequence of Padé approximants. We refer to Luke [1980] for a note on the accuracy of the value problem for small values of $|z|$, and to Pindor [1979a] for some interesting conjectures on the value problem. An up-to-date bibliography is given by Wuytack [1979].

Exercises 1. Find the [3/3] Padé approximant of Euler's series (5.5.9) using each of the methods of this section.

Exercise 2. If multiplication and division are regarded as the significant arithmetic operations, show that Baker's algorithm is an $O(4M^2)$ method, and that Kronecker's algorithm is an $O(4M^2)$ method for $M \ll L$.

CHAPTER 3

Padé Approximants and Numerical Methods

3.1 Aitken's Δ^2 Method as $[L/1]$ Padé Approximants

One of the best-known and simplest techniques of accelerating the convergence of a sequence is Aitken's Δ^2 method. Given a sequence of real or complex numbers,

$$\mathbb{S} = \{ S_n, n=0,1,2,\dots \}, \tag{1.1}$$

such that $S_n \to S$ as $n \to \infty$, the problem is to find a new sequence which converges faster to S.

Define

$$\Delta S_n = S_{n+1} - S_n,$$
$$\Delta^2 S_n = \Delta(\Delta S_n) = S_{n+2} - 2S_{n+1} + S_n,$$

which are the usual forward differences, and the new sequence

$$\mathbb{T} = \{ T_n, n=0,1,2,\dots \},$$

where

$$T_n = S_n - \frac{(\Delta S_n)^2}{\Delta^2 S_n}. \tag{1.2}$$

It is clear from (1.2) why Aitken's method is called the Δ^2 method. There are many reasons for expecting in general that $\mathbb{T}$ converges to a limit, that this

ENCYCLOPEDIA OF MATHEMATICS and Its Applications, Gian-Carlo Rota (ed.). Vol. 13: George A. Baker, Jr., and Peter R. Graves-Morris, Padé Approximants: Basic Theory, Part I   ISBN 0-201-13512-4

limit is S, and that convergence has been accelerated. But we must add an early word of caution: Aitken's method does not work for any arbitrary convergent sequence $\tilde{S}$. Like all algorithms of numerical analysis, Aitken's method has its own domain of validity, and in certain circumstances it should not be used. An important example is where all the S_n are identical, so that the T_n are undefined. A more insidious example is the one in which the S_n are equal up to rounding errors, so that the T_n are meaningless noise. This is a notorious situation to beware of. But, in general, the method is safe if it is empirically convergent, and it has wide applicability.

Basically, Aitken's method [1926] is designed to treat sequences with geometric convergence. Suppose that

$$S_n = S - a\alpha^n \tag{1.3}$$

with $a \neq 0$ and $|\alpha| < 1$. Then

$$\Delta S_n = a\alpha^n(1-\alpha),$$
$$\Delta^2 S_n = -a\alpha^n(1-\alpha)^2,$$

and from (2),

$$T_n = (S - a\alpha^n) + (a\alpha^n) = S. \tag{1.4}$$

We see that in this case, Aitken's method yields the exact answer at every stage. More generally, for a sequence $\tilde{S}$ which is dominated by one geometrically convergent component, we expect that Aitken's method accelerates convergence by "taking out the geometrically convergent part".

As a practical example (see Part II, Section 3.1) we consider the numerical evaluation of

$$S = \int_0^1 x^{-1/2}(1-x)^{-1/2} \exp x \, dx. \tag{1.5}$$

The integrand of (5) is infinite at the end points, but the integral is well defined. We define S_n as the value of the integral obtained by using 2^n equally spaced integration points. These Riemann sums, obtained by doubling the number of integration points at each successive evaluation, converge to S and are obtained with great ease. It turns out that Aitken's algorithm is a very effective technique of estimating S.

The connection between Padé approximants and Aitken's Δ^2 method is made, as in Section 1.3, by using the series derived from the sequence. Define

$$c_{n+1} = \Delta S_n = S_{n+1} - S_n, \qquad n = 0, 1, 2, \ldots,$$
$$c_0 = S_0. \tag{1.6}$$

It follows that S_n are the partial sums of the series, and of course the series converges to S. We form the power series

$$f(z)= \sum_{i=0}^{\infty} c_i z^i. \tag{1.7}$$

Remember that formal power series have a radius of convergence which may be zero, finite, or infinite. We wish to evaluate $f(1)=S$. The method of finding $f(1)$ using the second row of the Padé table is to evaluate

$$[L/1]_f(1), \qquad L=0,1,2,\dots, \tag{1.8}$$

and to determine the limit as $L \to \infty$.
    From (1.1.8)–(1.1.10),

$$
\begin{aligned}
[L/1]_f(1) &=
\begin{vmatrix} c_L & c_{L+1} \\ \displaystyle\sum_{i=0}^{L-1} c_i & \displaystyle\sum_{i=0}^{L} c_i \end{vmatrix}
\div
\begin{vmatrix} c_L & c_{L+1} \\ 1 & 1 \end{vmatrix} \\[2mm]
&= \frac{(S_L-S_{L-1})S_L-(S_{L+1}-S_L)S_{L-1}}{(S_L-S_{L-1})-(S_{L+1}-S_L)} \\[2mm]
&= \frac{S_{L-1}(S_{L+1}-2S_L+S_{L-1})-(S_L-S_{L-1})^2}{S_{L+1}-2S_L+S_{L-1}} \\[2mm]
&= S_{L-1}-\frac{(\Delta S_{L-1})^2}{\Delta^2 S_{L-1}},
\end{aligned}
$$

which agrees with (1.2) and shows Aitken's method to be the equivalent of using $[L/1]$ Padé approximants. An even more rapid proof of this result is given by taking $M=1$ in (1.3.8).
    A few sequences of numerical analysis are of the special type

$$S_{n+1}=f(S_n) \tag{1.9}$$

The function f is called a one-point iteration function in this context [Traub, 1964]. For example, the geometric sequence

$$S_n=S-a\alpha^n$$

in (1.3) is generated by

$$
\begin{aligned}
S_{n+1} &=S-a\alpha^{n+1} \\
&=S+\alpha(S_n-S) \\
&=f(S_n)
\end{aligned}
$$

with the identification $f(z)=S+\alpha(z-S)$. We see that in this case we have a convergent sequence when $\alpha=f'(S)$ satisfies $|\alpha|<1$. Further, the geometric sequence corresponds to $f(z)$ being linear. If $\alpha=1$, then $S_n=S-a$, and Aitken's method is inapplicable in this situation. We get further confirmation of the power of Aitken's method and so also of the $[L/1]$ Padé method from the following theorem.

THEOREM 3.1.1 [Henrici, 1964]. *Let* $S_{n+1}=f(S_n)$ *define a convergent real sequence with limit* S, *let* $f(x)$ *be twice differentiable at* S, *and let* $f'(S)\neq1$. *Then, with the definition* (1.2),

$$T_n-S=O((S_n-S)^2).$$

Proof.

$$S_{n+1}-S_n=f(S_n)-S_n$$
$$=f(S)+(S_n-S)f'(S)+(S-S_n)^2f''(\xi_n)-S_n \quad (1.10)$$

for some ξ_n lying between S and S_n. Continuity of $f(x)$ and convergence of the sequence imply that $f(S)=S$. Hence (1.10) may be written as

$$\Delta S_n=A(S_n-S)+O((S-S_n)^2), \quad (1.11)$$

where

$$A=f'(S)-1\neq0.$$

From (1.10),

$$S_{n+1}-S=(S_n-S)f'(S)+(S-S_n)^2f''(\xi_n).$$

Similarly, from (1.10) and (1.11),

$$S_{n+2}-S_{n+1}=(S_{n+1}-S)(f'(S)-1)+(S-S_{n+1})^2f''(\xi_{n+1})$$
$$=A(S_{n+1}-S)+O((S-S_n)^2). \quad (1.12)$$

Therefore from (1.11) and (1.12)

$$\Delta^2S_n=A\,\Delta S_n+O((S-S_n)^2).$$

Hence

$$
\begin{aligned}
T_n &= S_n - \frac{(\Delta S_n)\{A(S_n-S)+O((S-S_n)^2)\}}{A\,\Delta S_n(1+O(S-S_n)^2)} \\
&= S_n - (S_n - S) + O((S-S_n)^2) \\
&= S + O((S-S_n)^2),
\end{aligned}
$$

so proving the theorem.

The previous theorem makes quite precise the statement that Aitken's method and the $[L/1]$ Padé method accelerate convergence of a sequence dominated by a geometrically convergent component, of the type given by (1.9). In the next section we turn our attention to generalizing these basic ideas. For further details of the general theory, we refer to Brezinski [1977].

If both sequences $\mathfrak{S}$ and $\mathfrak{T}$ given by (1.1) and (1.2) converge, then they converge to the same limit [Lubkin, 1952]. For an account of recent progress with convergence theory, we refer to Cordellier [1979a] and Germain-Bonne [1979].

Exercise 1. Show that Newton's method of finding a zero of a function $\phi(z)$ takes the form (1.9), with the one-point iteration function

$$
f(z) = z - \frac{\phi(z)}{\phi'(z)}.
$$

Exercise 2. Is it a good idea to accelerate convergence of the sequence generated by Newton's method of the previous exercise if
  (i) z_0 is a simple root of $\phi(z)$?
  (ii) z_0 is a multiple root of $\phi(z)$?

Exercise 3. Consider the series

$$
\sum_{i=0}^{\infty} c_i = \frac{1}{2} + \frac{1}{3} - \frac{5}{6} + \frac{1}{4} + \frac{1}{5} - \frac{9}{20} + \cdots
$$

where

$$
c_{3m-3} = \frac{1}{2m}, \qquad c_{3m-2} = \frac{1}{2m+1}, \quad \text{and} \quad c_{3m-1} = -\frac{4m+1}{4m^2+2m}
$$

for $m=1,2,3,\ldots$ [Marx, 1963]. Define $S_n = \Sigma_{i=0}^n c_i$, and T_n by (1.2). Prove that

  (i) $S_{3m-1}=0$ for $m=1,2,3,\ldots$,
  (ii) $S_n \to 0$ as $n \to \infty$,
  (iii) $T_{3m} \to 0$, $T_{3m-1} \to 1$, and $T_{3m-2} \to 0$ as $m \to \infty$.

Notice the implication that $\{S_n\}$ converges, yet $\{T_n\}$ diverges by oscillation.

3.2 Acceleration and Overacceleration of Convergence

It is natural to ask how the accelerated sequence derived from Aitken's Δ^2 method may be improved upon. The natural answer is to iterate Aitken's method. This answer is not entirely satisfactory, because of the lack of justification based on principle, as the following remarks will make clear.

Aitken's scheme works well if the original sequence converges geometrically; the accelerated sequence takes full account of the dominant terms in the original sequence, and one should wonder what is the reason for accelerating again. Let us suppose that the original sequence is a geometric sequence rounded to given accuracy. Then the accelerated sequence, according to (1.4), is the limit but contains the small rounding errors. Further acceleration by Aitken's method (and the Padé method for that matter) requires differencing, and consequently, the results depend entirely on rounding errors in the original sequence. Thus, some sort of theoretical basis or an empirical numerical criterion is an essential prerequisite before iterating acceleration schemes. It is all too tempting to try to extract too much information by accelerating a few terms of a sequence too fast.

Consider the partial sums

$$S_n = \sum_{r=0}^{n} (1+\varepsilon_r)(0.5)^r, \qquad n=1,2,3,\ldots,$$

where the numbers $\{\varepsilon_r\}$ represent floating-point rounding errors with $|\varepsilon_i| \leqslant \varepsilon$. We consider estimation of the quantity S_∞ using the Padé-approximant method by forming diagonal approximants to

$$f(z) = \sum_{r=0}^{\infty} (1+\varepsilon_r)z^r$$

and evaluating the approximants at $z=0.5$. (Note the trivial variation on the method described in Sections 3.1, and 1.1, where the approximants are evaluated at $z=1$.) Working to first order in ε in (1.1.8) and (1.1.9), we find the error bounds

$$[1/1] = 2 \pm 7\varepsilon,$$

showing that this approximant is sensitive, but not unduly sensitive, to rounding errors in the data coefficients. However, when we come to consider the [2/2] Padé approximant, we find that

$$Q^{[2/2]}(z) = z^2(\varepsilon_2 + \varepsilon_4 - 2\varepsilon_3) - z(\varepsilon_1 + \varepsilon_4 - \varepsilon_2 - \varepsilon_3) + (\varepsilon_1 + \varepsilon_3 - 2\varepsilon_2)$$

to first order. We see that the value of $Q^{[2/2]}(z)$ is completely controlled by rounding error in this case; the zeros of $Q^{[2/2]}(z)$, which are the poles of the approximant, are distributed all over the complex plane. (We do not suggest that the distribution is random, and we would expect more zeros of $Q^{[2/2]}(z)$ near $|z|=1$ than near $z=0$, for example.) Since the value of the [2/2] Padé approximant depends primarily on rounding error, whereas the [1/1] approximant is accurate within errors, use of the [2/2] approximant is an example of overacceleration of convergence. In this case, the moral is that we should use the lower-order approximant.

The Padé method has the following interpretation (among others): the given sequence has a certain number of geometric components which dominate. Let us suppose that

$$S_n = a \sum_{r=0}^{n} \alpha^r + b \sum_{r=0}^{n} \beta^r + \text{much smaller terms} \qquad (2.1)$$

with $a \neq 0$, $b \neq 0$, $|\alpha| < 1$, $|\beta| < 1$, and $n = 0, 1, 2, \ldots$. This expression may be rewritten as

$$S_n = \frac{a}{1-\alpha}(1 - \alpha^{n+1}) + \frac{b}{1-\beta}(1 - \beta^{n+1}) + \text{much smaller terms.} \quad (2.2)$$

Note that S_n is derived from the series

$$c_0 \equiv S_0 = a + b + \text{a much smaller term,}$$
$$c_n \equiv S_n - S_{n-1} = a\alpha^n + b\beta^n + \text{much smaller terms,} \qquad n = 1, 2, 3, \ldots$$

$$(2.3)$$

The third row of the Padé table takes account of the explicit leading terms in (2.1), (2.2), and (2.3), whereas direct calculation shows that the once iterated Aitken method does not. We assumed in (2.1) that $a \neq 0$ and $b \neq 0$, so that there are genuinely two geometric components which dominate the remainder, and this assumption is crucial. We are a bit vague about the size of the remainder terms, so as not to prejudice the development, and to admit possibilities such as $c_n = \alpha^n + (-\alpha)^n$ for which the odd terms vanish. We assume that $|\alpha| < 1$ and $|\beta| < 1$ which is conventional but not entirely necessary. The Padé method does make sense of well-posed problems with divergent sequences, such as (2.1) with $|\alpha| > 1$ or $|\beta| > 1$. If $\alpha = 1$ or $\beta = 1$, corresponding to a divergent sequence with one component derived from an arithmetic progression, the Padé method gives $S = \infty$ with an obvious interpretation.

As in Section 1.3, to justify the Padé method, we form the function

$$f(z) \equiv \sum_{r=0}^{\infty} c_r z^r = \frac{a}{1 - \alpha z} + \frac{b}{1 - \beta z} + \text{correction terms}, \qquad (2.4)$$

which we wish to evaluate at $z = 1$. Formation of $[L/2]$ approximants is suggested by the explicit dominant terms of (2.4), and we apply de Montessus's theorem. Borrowing from Section 6.2, we quote the theorem in context here.

Let $R > |\alpha|^{-1}$ and $R > |\beta|^{-1}$. We assume that the remainder terms in (2.3) are small, and explicitly we require that $c_n = o(R^{-n})$. Hence, $f(z)$ is meromorphic with precisely two poles within $|z| < R$. Since $|\alpha| > 1$, then $R > 1$. Now de Montessus's theorem asserts that $[L/2]$ approximants converge to $f(z)$ at $z = 1$, which is not a pole of $f(z)$, by assumption. Hence, the $[L/2]$ approximants converge for sequences such as (2.1) or series such as (2.3), with the stated hypothesis about the residuals.

3.3   The ε-Algorithm and the η-Algorithm

In this section we describe the ε-algorithm for sequence transformations and show that one of its columns is the sequence of Aitken's Δ^2 method. Then we describe the η-algorithm, which is the corresponding algorithm for series transformations.

The ε-algorithm originates with Shanks [1955] and Wynn [1956]. It involves the two-dimensional array called the ε-table (Table 1). The subscript k of $\varepsilon_k^{(j)}$ denotes the column, and the superscript j measures the progression down the column. The table is constructed iteratively from its first two elements. Define $\varepsilon_{-1}^{(j)}$ to be zero and $\varepsilon_0^{(0)}$ to be the given sequence, for $j = 0, 1, 2, \ldots$. Then all the other elements may be calculated from the

Table 1.   The ε-Table.

$$
\begin{array}{ccccccc}
\varepsilon_{-1}^{(0)} & & & & & & \\
& \varepsilon_0^{(0)} & & & & & \\
\varepsilon_{-1}^{(1)} & & \varepsilon_1^{(0)} & & & & \\
& \varepsilon_0^{(1)} & & \varepsilon_2^{(0)} & & & \\
\varepsilon_{-1}^{(2)} & & \varepsilon_1^{(1)} & & \vdots & & \\
& \varepsilon_0^{(2)} & & \vdots & & & \\
\varepsilon_{-1}^{(3)} & & \vdots & & & & \\
\vdots & & & & & &
\end{array}
$$

ε-algorithm, which is

$$\varepsilon_{k+1}^{(j)} = \varepsilon_{k-1}^{(j+1)} + \left[\varepsilon_k^{(j+1)} - \varepsilon_k^{(j)}\right]^{-1}. \tag{3.1}$$

To see more clearly how this rule should be applied, we note that it connects the elements in the rhombus pattern of Figure 1, which shows how the

$$\varepsilon_k^{(j)}$$

$$\varepsilon_{k-1}^{(j+1)} \qquad\qquad \varepsilon_{k+1}^{(j)}$$

$$\varepsilon_k^{(j+1)}$$

Figure 1. A rhombus pattern.

right-hand member $\varepsilon_{k+1}^{(j)}$ is derived from the other three members. It is now plain that the ε-algorithm allows the whole ε-table to be calculated. It is further plain that if $\varepsilon_k^{(j)} = \varepsilon_k^{(j+1)}$, i.e. two successive members of the same column are equal, the element $\varepsilon_{k+1}^{(j)}$ does not exist. We assume, unless explicitly stated otherwise, that all elements exist. Otherwise the table is said to be degenerate. We will show that the sequence of the fourth column, namely $\{\varepsilon_2^{(j)}, j=0,1,2,\dots\}$, is the same as that obtained from Aitken's Δ^2 rule.

From (3.1),

$$\varepsilon_1^{(j)} = \left[\varepsilon_0^{(j+1)} - \varepsilon_0^{(j)}\right]^{-1}$$
$$= \left[S_{j+1} - S_j\right]^{-1} = \left[\Delta S_j\right]^{-1}.$$

Again from (3.1),

$$\varepsilon_2^{(j)} = \varepsilon_0^{(j+1)} + \left[\varepsilon_1^{(j+1)} - \varepsilon_1^{(j)}\right]^{-1}$$

$$= S_{j+1} + \frac{1}{\left[\Delta S_{j+1}\right]^{-1} - \left[\Delta S_j\right]^{-1}}$$

$$= S_j + \Delta S_j + \frac{\Delta S_j \Delta S_{j+1}}{\Delta S_j - \Delta S_{j+1}}$$

$$= S_j - \frac{\left(\Delta S_j\right)^2}{\Delta^2 S_j}, \qquad \text{where} \quad \Delta^2 S_j = \Delta S_{j+1} - \Delta S_j.$$

This formula is precisely Aitken's Δ^2 method (1.2) applied to the sequence $\{S_j, j=0,1,2,\dots\}$, and is also the result of using the second row of the Padé table as a Padé method for sequence acceleration.

After we have established Wynn's identity in Section 3.4, we then show in Section 3.5 that the ε-table and the Padé table are identified by the formula

$$\varepsilon_{2k}^{(j)} = [k+j/k]_f(1).$$

What we have just achieved is the proof of this result for $k=1$ and $j=0,1,2,\ldots$.

Table 2. Even Columns of an ε-Table for π.

$n=0$	4.0000000				
1	2.6666667	3.1666667			
2	3.4666667	3.1333333	3.1423423		
3	2.8952381	3.1452381	3.1413919	3.1416149	
4	3.3396825	3.1396825	3.1416627	3.1415873	3.1415933
5	2.9760462	3.1427129	3.1415634	3.1415943	3.1415925
6	3.2837385	3.1408813	3.1416065	3.1415921	
7	3.0170718	3.1420718	3.1415854		
8	3.2523659	3.1412548			
9	3.0418396				

Example 1. We consider Gregory's notoriously slowly convergent series for π,

$$\pi = 4 - \tfrac{4}{3} + \tfrac{4}{5} - \tfrac{4}{7} + \cdots .$$

In Table 2, we exhibit the even columns of the ε-table for this series. The first column seems scarcely convergent, whereas the correctness of the final extrapolations is instantly recognizable.

Table 3. Even Columns of an ε-Table for $\ln 3$.

$n=0$	.000000				
1	2.000000	1.000000			
2	.000000	1.142857	1.090909		
3	2.666667	1.066667	1.101449	1.098039	
4	−1.333333	1.128205	1.097046	1.098805	1.098570
5	5.066667	1.066667	1.099725	1.098521	1.098626
6	−5.600000	1.136842	1.097674	1.098667	
7	12.685714	1.049351	1.099507		
8	−19.314286	1.165714			
9	37.574603				

Example 2. We consider a familiar divergent series, namely the one given by the Maclaurin series of $\ln(1+z)$ with $z=2$:

$$\ln 3 = 2 - 2 + \frac{8}{3} - 4 + \cdots .$$

This has a remarkable ε-table, with even columns given by Table 3. For comparison, $\ln 3 = 1.098612\ldots$. We see, by example, that the ε-algorithm may be used to sum divergent series.

As an amusing paradox, we briefly mention Hardy's puzzle, which has an entirely straightforward solution. Let

$$f(z) = 1 + \frac{1}{2}\left(\frac{2z}{1+z^2}\right)^2 + \frac{1}{2}\cdot\frac{3}{4}\cdot\left(\frac{2z}{1+z^2}\right)^4 + \cdots. \qquad (3.2)$$

Consider the domains $\mathcal{D}$ and $\mathcal{E}$ defined by

$$z \in \mathcal{D} \qquad \text{provided} \quad |z-i| < \sqrt{2} \text{ and } |z+i| < \sqrt{2},$$

$$z \in \mathcal{E} \qquad \text{provided} \quad |z-i| > \sqrt{2} \text{ and } |z+i| > \sqrt{2}.$$

Define

$$g(z) = \frac{1+z^2}{1-z^2}.$$

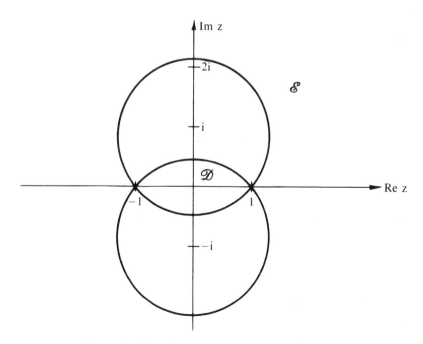

Figure 2. The z-plane, showing the domains $\mathcal{D}$ and $\mathcal{E}$.

Notice that (3.2) is a binomial series which converges if $|2z|<|1+z^2|$. The boundary of convergence is given by

$$(zz^* + iz^* - iz - 1)(zz^* + iz - iz^* - 1) = 0$$

showing that (3.2) converges for $z \in \mathcal{D}$ and $z \in \mathcal{E}$. Therefore

$$f(z) = \frac{1+z^2}{1-z^2} = g(z) \qquad \text{if} \quad z \in \mathcal{D}$$

and

$$f(z) = \frac{z^2+1}{z^2-1} = -g(z) \qquad \text{if} \quad z \in \mathcal{E}.$$

The puzzle is to decide what result is given by the ε-algorithm for (3.2) with $z = 2i$. Does it converge to $g(2i) = -\frac{3}{5}$ or $-g(2i) = +\frac{3}{5}$?

At this stage, since the expansion parameter is invariant under $z \to 1/z$, the correct answer should seem clear, and it can be empirically verified by calculating $\varepsilon_2^{(0)}$, which is quite close. Proof of the veracity of this solution requires the technique of the theorem of Baker, Gammel, and Wills in Section 6.7.

As a final remark on the ε-algorithm, we note the possibility that the sequence $\mathcal{S} = \{\varepsilon_k^{(0)}, k = 0, 1, 2, \ldots\}$ does not converge, and yet an iteration of the algorithm using the derived sequence $\mathcal{S}$ as a new initializing sequence $\{\varepsilon_0''^{(J)}, J = 0, 1, 2, \ldots\}$ may lead to a convergent sequence $\{\varepsilon_k''^{(0)}\}$, e.g. as in Table 1 of Part II, Section 1.3. But no theoretical justification for iterating the ε-algorithm is known yet.

The ε-algorithm may be regarded as a sequence-to-sequence transformation: it is an algorithm for transforming the elements of the given series in the second column to the elements on the principal diagonal. Specifically, the sequence $\{\varepsilon_0^{(J)}, j = 0, 1, 2, \ldots\}$ is transformed to a new sequence $\{\varepsilon_k^{(0)}, k = 0, 1, 2, \ldots\}$. Bauer's η-algorithm is the equivalent series-to-series transformation.

THE η-ALGORITHM [Bauer, 1959]. The series c_i, $i = 0, 1, 2, \ldots$, is given. The η-algorithm is initialized by assigning these values to the second column of the η-table:

$$\eta_0^{(i)} = c_i, \qquad i = 0, 1, 2, \ldots . \tag{3.3}$$

The elements of the first column of the η-table are defined by the artificial values

$$\eta_{-1}^{(i)} = \infty, \qquad i = 0, 1, 2, \ldots . \tag{3.4}$$

$$\eta_{-1}^{(0)}$$
$$\eta_0^{(0)}$$
$$\eta_{-1}^{(1)} \qquad \eta_1^{(0)}$$
$$\eta_0^{(1)} \qquad \eta_2^{(0)}$$
$$\eta_{-1}^{(2)} \qquad \eta_1^{(1)} \qquad \eta_3^{(0)}$$
$$\eta_0^{(2)} \qquad \eta_2^{(1)}$$
$$\eta_{-1}^{(3)} \qquad \eta_1^{(2)}$$
$$\eta_0^{(3)}$$
$$\eta_{-1}^{(4)} \qquad \vdots$$
$$\vdots$$

Figure 3. The η-table.

The recurrence scheme is defined by

$$\frac{1}{\eta_{2k+1}^{(i)}} = \frac{1}{\eta_{2k-1}^{(i+1)}} + \frac{1}{\eta_{2k}^{(i+1)}} - \frac{1}{\eta_{2k}^{(i)}} \tag{3.5}$$

and

$$\eta_{2k}^{(i)} = \eta_{2k-2}^{(i+1)} + \eta_{2k-1}^{(i+1)} - \eta_{2k-1}^{(i)}. \tag{3.6}$$

Equations (3.5), (3.6) are rhombus rules connecting the entries shown in Figure 4. These equations enable the rightmost entries of the rhombus to be

$$\eta_{2k}^{(i)}$$
$$\eta_{2k-1}^{(i+1)} \qquad \qquad \eta_{2k+1}^{(i)} \qquad\qquad\qquad \eta_{2k-2}^{(i+1)} \qquad\qquad \eta_{2k}^{(i)}$$
$$\eta_{2k}^{(i+1)} \qquad\qquad\qquad\qquad\qquad \eta_{2k-1}^{(i+1)}$$

Elements of (5)   Elements of (6)

Figure 4. Rhombus rules for the η-table.

calculated; hence the entire η-table of Figure 3 can be constructed from its first two columns given by (3.3), (3.4). With these definitions (3.3)–(3.6), the η-algorithm defines a transformation of the series $c_i = \eta_0^{(i)}$, $i = 0, 1, 2, \ldots$ to a new series $c_k' = \eta_k^{(0)}$, $k = 0, 1, 2, \ldots$. One purpose of this algorithm is the construction from a convergent series $\sum_{i=0}^{\infty} c_i$ of a new series $\sum_{k=0}^{\infty} c_k'$ which converges faster to the same limit. As an empirical example of this we consider Gregory's series again:

$$\frac{\pi}{4} = 1 - \frac{1}{3} + \frac{1}{5} - \frac{1}{7} + \frac{1}{9} - \cdots. \tag{3.7}$$

Table 4. The η-Table for Gregory's Series.

	1					
∞		$-\dfrac{1}{4}$				
	$-\dfrac{1}{3}$		$\dfrac{1}{24}$			
∞		$\dfrac{1}{8}$		$-\dfrac{1}{136}$		
	$\dfrac{1}{5}$		$-\dfrac{1}{120}$		$\dfrac{4}{3145}$	
∞		$-\dfrac{1}{12}$		$\dfrac{1}{444}$		$\vdots$	$\cdots$
	$-\dfrac{1}{7}$		$\dfrac{1}{336}$		$\vdots$	
∞		$\dfrac{1}{16}$		$\vdots$		
$\vdots$	$\dfrac{1}{9}$		$\vdots$			
	$\vdots$					

Example 3. The second column of the η-table is constructed from (3.3) using the terms of the series (3.7). The first column is artificial, and the other columns are constructed from the rhombus rules (3.5) and (3.6). From Table 4 we find that the series (3.7) has been transformed to

$$\frac{\pi}{4} = 1 - \frac{1}{4} + \frac{1}{24} - \frac{1}{136} + \frac{4}{3145} - \cdots; \tag{3.8}$$

we have assumed that the transformed series converges to the same limit. From the figures quoted, the series (3.8) appears to converge faster than (3.7), as expected. To justify these manipulations, we will prove first that the ε-algorithm and the η-algorithm are equivalent in the sense that

$$\varepsilon_{2k}^{(0)} = \sum_{r=0}^{2k} \eta_r^{(0)}. \tag{3.9}$$

In the context of Example 3, this means that the odd partial sums of (3.8) yield the diagonal estimates of π given by the principal diagonal of Table 2, as is easily checked.

THEOREM 3.3.1. *The identities*

$$\eta_{2k}^{(i)} = \varepsilon_{2k}^{(i)} - \varepsilon_{2k}^{(i-1)} = \left[\varepsilon_{2k+1}^{(i-1)} - \varepsilon_{2k-1}^{(i)} \right]^{-1}, \tag{3.10}$$

$$\eta_{2k+1}^{(i)} = \varepsilon_{2k+2}^{(i-1)} - \varepsilon_{2k}^{(i)} = \left[\varepsilon_{2k+1}^{(i)} - \varepsilon_{2k+1}^{(i-1)} \right]^{-1} \tag{3.11}$$

hold so long as the quantities involved are well defined.

Proof. As indicated in (3.10) and (3.11), we use the identity (3.1), which constitutes the ε-algorithm, whenever necessary. Since either (3.5) and (3.6) or (3.10) and (3.11) uniquely define the η-table, our method of proof consists of using (3.10) and (3.11) to establish (3.5) and (3.6). It is convenient to extend the domain of definition by assigning $\eta_0^{(-1)} = \varepsilon_0^{(-1)} = c_{-1} = 0$ in the second column. We show that the elements of the η-table defined by (3.10) and (3.11) are identical to those defined by (3.3)–(3.6). For the second column of the η-table, the definition (3.3) yields the same values as (3.10) with $k=0$, because

$$\varepsilon_0^{(i)} = \sum_{j=0}^{i} c_i.$$

For the third column of the table, (3.11) becomes

$$\left[\eta_1^{(i)}\right]^{-1} = \varepsilon_1^{(i)} - \varepsilon_1^{(i-1)}$$
$$= \left[\varepsilon_0^{(i+1)} + \varepsilon_0^{(i)}\right]^{-1} - \left[\varepsilon_0^{(i)} - \varepsilon_0^{(i-1)}\right]^{-1}$$
$$= \left[\eta_0^{(i+1)}\right]^{-1} - \left[\eta_0^{(i)}\right]^{-1},$$

which yields the same values as are defined by (3.5) when $k=0$. To justify (3.5) and (3.6) as defining equations for the fourth and subsequent columns, we consider an identity among elements in a rectangular array in the ε-table. Let A, B, C, D be the elements shown in Figure 5. The identity $(D-B)-$

$$A = \varepsilon_{2k-1}^{(i)} \qquad\qquad B = \varepsilon_{2k-1}^{(i-1)}$$
$$\varepsilon_{2k}^{(i)}$$
$$C = \varepsilon_{2k-1}^{(i+1)} \qquad\qquad D = \varepsilon_{2k+1}^{(i)}$$

Figure 5.   Five elements of the ε-table.

$(C-A)=(D-C)-(B-A)$ is interpreted by (3.10) and (3.11) as

$$\left[\eta_{2k+1}^{(i)}\right]^{-1} - \left[\eta_{2k-1}^{(i+1)}\right]^{-1} = \left[\eta_{2k}^{(i+1)}\right]^{-1} - \left[\eta_{2k}^{(i)}\right]^{-1},$$

proving (3.5). Let W, X, Y, Z be the elements shown in Figure 6. The identity $(Z-X)-(Y-W)=(Z-Y)-(X-W)$ is interpreted by (3.10) and

$$W = \varepsilon_{2k-2}^{(i)} \qquad\qquad\qquad X = \varepsilon_{2k}^{(i-1)}$$
$$\varepsilon_{2k-1}^{(i)}$$
$$Y = \varepsilon_{2k-2}^{(i+1)} \qquad\qquad\qquad Z = \varepsilon_{2k}^{(i)}$$

Figure 6.   Five elements of the ε-table.

(3.11) as

$$\eta_{2k}^{(i)} - \eta_{2k-2}^{(i)} = \eta_{2k-1}^{(i+1)} - \eta_{2k-1}^{(i)},$$

proving (3.6). Hence the theorem is proved.

It only remains to observe that an immediate consequence of (3.10) and (3.11) is that

$$\eta_{2k+2}^{(0)} = \varepsilon_{2k+2}^{(0)} - \varepsilon_{2k+2}^{(-1)},$$

$$\eta_{2k+1}^{(0)} = \varepsilon_{2k+2}^{(-1)} - \varepsilon_{2k}^{(0)}.$$

Hence $\eta_{2k+1}^{(0)} + \eta_{2k+2}^{(0)} = \varepsilon_{2k+2}^{(0)} - \varepsilon_{2k}^{(0)}$, and by summation $\varepsilon_{2k}^{(0)} = \Sigma_{r=0}^{2k} \eta_r^{(0)}$, proving (3.9). We have thereby established its equivalence to the ε-algorithm as a sequence to sequence transformation.

For further details about the ε-algorithm and η-algorithm, we refer to Bauer et al. [1963], Wynn [1960, 1961a], and Brezinski [1977]. For applications of the ε-algorithm to vector and matrix valued quantities, we refer to Wynn [1962b, 1964], McCleod [1971], Gekeler [1972], Mills [1975], and Brezinski [1977].

3.4 Wynn's Identity and the ε-Algorithm

Wynn's identity is an identity connecting neighboring Padé approximants in the Padé table:

$$([L/M+1]-[L/M])^{-1}+([L/M-1]-[L/M])^{-1}$$

$$= ([L-1/M]-[L/M])^{-1}+([L+1/M]-[L/M])^{-1}.$$

$$(4.1)$$

It is easy to remember this identity from the identification with compass points in the Padé table, shown in Figure 1.
With this mnemonic, the identity is written as

$$(S-C)^{-1}+(N-C)^{-1}=(W-C)^{-1}+(E-C)^{-1}.$$

It is valid when all the indicated Padé approximants exist and are nondegenerate. This section is mostly devoted to a self-contained proof of Wynn's

	$[L/M-1]$			N	
$[L-1M]$	$[L/M]$	$[L+1/M]$	$\rightarrow$	W C	E
	$[L/M+1]$			S	

Figure 1. Compass points in the Padé table.

identity; in Section 3.5 there is a more complete derivation of the various other identities. At the end of this section, we use Wynn's identity to prove that entries in even columns of the ε-table are, in fact, Padé approximants.

We will need two Frobenius identities in the course of this proof. Consider the determinant

$$
Q^{[L/M+1]}(z) = \begin{vmatrix}
c_{L-M} & c_{L-M+1} & \cdots & c_L & c_{L+1} \\
c_{L-M+1} & c_{L-M+2} & \cdots & c_{L+1} & c_{L+2} \\
\vdots & \vdots & & \vdots & \vdots \\
c_L & c_{L+1} & \cdots & c_{L+M} & c_{L+M+1} \\
z^{M+1} & z^M & \cdots & z & 1
\end{vmatrix}.
$$

We apply Sylvester's identity and consider the deletion of the first and last rows and the first and last columns. Each determinant defined by these deletions is of a standard type, and the identity is

$$
Q^{[L/M+1]}(z)C(L+1/M) = Q^{[L+1/M]}(z)C(L/M+1)
$$
$$
- zQ^{[L/M]}(z)C(L+1/M+1).
$$

We will use this identity in the form

$$
\frac{Q^{[L+1/M]}(z)}{C(L+1/M)} - \frac{Q^{[L/M+1]}(z)}{C(L/M+1)} = \frac{zQ^{[L/M]}(z)C(L+1/M+1)}{C(L+1/M)C(L/M+1)}. \tag{4.2}
$$

We also consider the following determinant for $Q^{[L-1/M]}(z)$, which is contrived to give the desired result:

$$
Q^{[L-1/M]}(z) = (-1)\begin{vmatrix}
c_{L-M} & c_{L-M+1} & \cdots & c_{L-1} & c_L & 0 \\
c_{L-M+1} & c_{L-M+2} & \cdots & c_L & c_{L+1} & 0 \\
\vdots & \vdots & & \vdots & \vdots & \vdots \\
c_{L-1} & c_L & \cdots & c_{L+M-2} & c_{L+M-1} & 0 \\
c_L & c_{L+1} & \cdots & c_{L+M-1} & c_{L+M} & 1 \\
z^M & z^{M-1} & \cdots & z & 1 & 0
\end{vmatrix}.
$$

Application of Sylvester's identity with deletion of the first and last rows and the first and last columns yields another identity among the above quantities:

$$
Q^{[L-1/M]}(z)C(L+1/M)
$$
$$
= Q^{[L/M-1]}(z)C(L/M+1) + Q^{[L/M]}(z)C(L/M),
$$

which we use in the form

$$\frac{Q^{[L-1/M]}(z)}{C(L/M+1)} - \frac{Q^{[L/M-1]}(z)}{C(L+1/M)} = \frac{Q^{[L/M]}(z)C(L/M)}{C(L/M+1)C(L+1/M)}. \quad (4.3)$$

We are now in a position to prove Wynn's identity. We consider first

$$[L+1/M]-[L/M] = \frac{P^{[L+1/M]}(z)Q^{[L/M]}(z) - P^{[L/M]}(z)Q^{[L+1/M]}(z)}{Q^{[L+1/M]}(z)Q^{[L/M]}(z)}.$$

The numerator has degree $L+M+1$, but the approximants agree to order z^{L+M} by their definition. Hence

$$[L+1/M]-[L/M] = \frac{z^{L+M+1}}{Q^{[L+1/M]}(z)Q^{[L/M]}(z)} \cdot \text{const.}$$

The origin of the constant is the coefficient of z^{L+1} in $P^{[L+1/M]}(z)$ and of z^M in $Q^{[L/M]}(z)$, because all other terms are of lower order. Hence, using (1.1.8) and (1.1.9),

$$[L+1/M]-[L/M] = \frac{C(L+1/M)C(L+1/M+1)z^{L+M+1}}{Q^{[L+1/M]}(z)Q^{[L/M]}(z)}, \quad (4.4)$$

and replacing L by $L-1$ gives

$$[L/M]-[L-1/M] = \frac{C(L/M)C(L/M+1)z^{L+M}}{Q^{[L/M]}(z)Q^{[L-1/M]}(z)}. \quad (4.5)$$

We consider next

$$[L/M+1]-[L/M] = \frac{P^{[L/M+1]}(z)Q^{[L/M]}(z) - Q^{[L/M+1]}(z)P^{[L/M]}(z)}{Q^{[L/M+1]}(z)Q^{[L/M]}(z)}.$$

The numerator has degree $L+M+1$, and the approximants agree to order z^{L+M} by their definition. Hence

$$[L/M+1]-[L/M] = \frac{\text{const} \cdot z^{L+M+1}}{Q^{[L/M+1]}(z)Q^{[L/M]}(z)}.$$

The origin of the constant is the coefficient of z^{M+1} in $Q^{[L/M+1]}(z)$ and z^L

in $P^{[L/M]}(z)$. Hence

$$[L/M+1]-[L/M]=\frac{C(L/M+1)C(L+1/M+1)z^{L+M+1}}{Q^{[L/M]}(z)Q^{[L/M+1]}(z)}, \quad (4.6)$$

and replacing M by $M-1$ gives

$$[L/M]-[L/M-1]=\frac{C(L/M)C(L+1/M)z^{L+M}}{Q^{[L/M-1]}(z)Q^{[L/M]}(z)}. \quad (4.7)$$

Equations (4.4)–(4.7) are all that is required for Wynn's identity. We find

$$\{[L+1/M]-[L/M]\}^{-1}-\{[L/M+1]-[L/M]\}^{-1}$$

$$=\frac{Q^{[L/M]}(z)}{C(L+1/M+1)z^{L+M+1}}\left\{\frac{Q^{[L+1/M]}(z)}{C(L+1/M)}-\frac{Q^{[L/M+1]}(z)}{C(L/M+1)}\right\}$$

$$=\frac{\{Q^{[L/M]}(z)\}^2}{z^{L+M}C(L+1/M)C(L/M+1)}, \quad (4.8)$$

using (4.4), (4.6), and then (4.2).

We also find that

$$\{[L-1/M]-[L/M]\}^{-1}-\{[L/M-1]-[L/M]\}^{-1}$$

$$=\frac{Q^{[L/M]}(z)}{C(L/M)z^{L+M}}\left\{-\frac{Q^{[L-1/M]}(z)}{C(L/M+1)}+\frac{Q^{[L/M-1]}(z)}{C(L+1/M)}\right\}$$

$$=\frac{-\{Q^{[L/M]}(z)\}^2}{z^{L+M}C(L+1/M)C(L/M+1)}, \quad (4.9)$$

using (4.5), (4.7), and (4.3). Since (4.8)$=-$(4.9), we deduce

$$\{[L+1/M]-[L/M]\}^{-1}+\{[L-1/M]-[L/M]\}^{-1}$$

$$=\{[L/M+1]-[L/M]\}^{-1}+\{[L/M-1]-[L/M]\}^{-1},$$

which is Wynn's identity.

Hitherto in this section, we have assumed the relevant entries of the Padé table to exist and be nondegenerate. A modification of (4.1) which takes explicit account of the presence of blocks is Cordellier's identity. This identity relates extant elements in the Padé table, four of which are at the vertices of a rectangle.

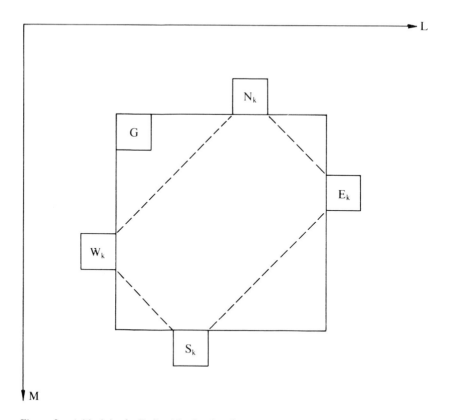

Figure 2. A block in the Padé table showing the entries connected by Cordellier's identity.

CORDELLIER'S IDENTITY. *Let* $G \equiv [L/M]$ *denote the nondegenerate entry of an* $n \times n$ *block in the Padé table. It is shown in the top left-hand corner of Figure 2. For any* k *in the range* $0 \leqslant k < n$, *we define*

$$N_k \equiv [L+k/M-1], \quad W_k \equiv [L-1/M+k], \quad S_k \equiv [L+n-k-1/M+n]$$
and $E_k \equiv [L+n/M+n-k-1]$.

The identity states that

$$(N_k - G)^{-1} + (S_k - G)^{-1} = (W_k - G)^{-1} + (E_k - G)^{-1}.$$

For a proof of this identity, which is based on the ideas of the Euclidean modification of Kronecker's algorithm, we refer to Cordellier [1979b].

We next show that the entries in the even columns of the ε-table are entries in the rows of the Padé table. Specifically we will prove that

$$\varepsilon_{2k}^{(j)} = [k+j/k]_f(1) \tag{4.10}$$

provided the indicated quantities exist. In Section 3.2, we defined the second column of the ε-table, namely $\{\varepsilon_0^{(j)}, j=0,1,2,\dots\}$, to be the sequence of $[j/0]$ approximants, which is the sequence of truncated Taylor series, both evaluated at $z=1$. We proved that the fourth column, namely $\{\varepsilon_2^{(j)}, j=0,1,2,\dots\}$ is the sequence of $[j+1/1]$ Padé approximants evaluated at $z=1$. We will prove (4.10) by induction, noting that we have already established the result for $k=0$ and $k=1$. We use the ε-algorithm repeatedly:

$$\varepsilon_{k+1}^{(j)} = \varepsilon_{k-1}^{(j+1)} + \left(\varepsilon_k^{(j+1)} - \varepsilon_k^{(j)}\right)^{-1}, \tag{4.11}$$

as indicated by the rhombus rule in Figure 3. We will prove the connection

		n						$\varepsilon_{2k}^{(j-1)}$		
	nw		ne				nw		ne	
w		c		e	$\equiv$	$\varepsilon_{2k-2}^{(j+1)}$		$\varepsilon_{2k}^{(j)}$		$\varepsilon_{2k+2}^{(j-1)}$
	sw		se				sw		se	
		s						$\varepsilon_{2k}^{(j+1)}$		

Figure 3. Part of the ε-table.

between the $[L/M]$, $[L/M\pm 1]$, and $[L\pm 1/M]$ Padé approximants, and we expect from (4.10) to involve $\varepsilon_{2k}^{(j)}$, $\varepsilon_{2k-2}^{(j+1)}$, $\varepsilon_{2k+2}^{(j-1)}$, $\varepsilon_{2k}^{(j+1)}$, and $\varepsilon_{2k}^{(j-1)}$ which are the corresponding epsilons. These are indicated in the figure. Note that because the columns of the ε-table will be shown to correspond to rows of the Padé table, the compass points do not correspond, and to emphasize this we use small letters in the ε-table.

Application of the ε-algorithm (4.11) gives the formulas

$$\begin{aligned}
\text{ne}-\text{nw} &= (c-n)^{-1},\\
\text{se}-\text{sw} &= (s-c)^{-1},\\
(\text{sw}-\text{nw})^{-1} &= c-w,\\
(\text{se}-\text{ne})^{-1} &= e-c.
\end{aligned}$$

Simple manipulation yields

$$(n-c)^{-1}+(s-c)^{-1}=(w-c)^{-1}+(e-c)^{-1},$$

which is Wynn's identity for Padé approximants with the identification (4.10). We have only to observe that the ε-algorithm is used to calculate columns of the ε-table working from left to right, and that Wynn's algorithm calculates Padé approximants working from the first and second rows down. Then we see by direct construction of individual elements by induction that the formula (4.10) is valid whenever the indicated quantities exist.

Notice that odd columns of the ε-table are not Padé approximants and also that even columns of the ε-table are Padé approximants on and above the diagonal, evaluated at the particular value $z = 1$. The connection between the ε-algorithm and the Padé table may be made directly using (1.1.8), (1.1.9), and (1.3.8) [Shanks, 1955; Wynn, 1961b], but the extension to Cordellier's identity is obscured.

Exercise For a normal Padé table, prove the inverse crossed rule

$$(N^{-1} - C^{-1})^{-1} + (S^{-1} - C^{-1})^{-1} = (W^{-1} - C^{-1})^{-1} + (E^{-1} - C^{-1})^{-1}$$

defined with the notation of Figure 1.*

3.5 Common Identities and Recursion Formulas

The identities we discuss in this section apply either to the Padé approximants themselves, or to the numerators and denominators; consequently there are two quite different types of relationships to be distinguished.

One of the most remarkable relationships which occurs in the theory is that the numerators and denominators of neighboring Padé approximants obey the same recurrence relations. This fact is the key to the connection with continued-fraction theory. The other relations we will prove have diverse applications elsewhere in this book.

We start with the basic definition

$$Q^{[L/M]}(z) = \begin{vmatrix} c_{L-M+1} & c_{L-M+2} & \cdots & c_L & c_{L+1} \\ c_{L-M+2} & c_{L-M+3} & \cdots & c_{L+1} & c_{L+2} \\ \vdots & \vdots & & \vdots & \vdots \\ c_L & c_{L+1} & \cdots & c_{L+M-1} & c_{L+M} \\ z^M & z^{M-1} & \cdots & z & 1 \end{vmatrix}. \quad (5.1)$$

This is an $(M+1) \times (M+1)$ determinant; its *general structure* is preserved after the deletion of the first or last columns and the first row. With either of these pairs of deletions, we end up with another $Q^{[l/m]}(z)$. This discussion is the precursor to applying Sylvester's identity with deletion of the first and last rows and columns. This action gives [Frobenius, 1881]

$$Q^{[L/M]}(z) C(L+1/M-1) = Q^{[L+1/M-1]}(z) C(L/M)$$
$$- z Q^{[L/M-1]} C(L+1/M). \quad (5.2)$$

*C. Brezinski, private communication.

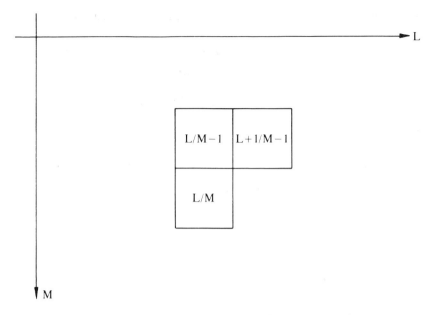

Figure 1. The locations in the Padé table of the denominators in (5.2).

Equation (5.2) is referred to as a ($^{**}_{}$) identity, because it connects de-nominators of Padé approximants with the configuration shown in Figure 1.

We obtain a similar result by allowing deletion of rows M, $M+1$ and the first and last columns of (5.1), which is

$$Q^{[L/M]}(z)C(L/M-1)=C(L/M)Q^{[L/M-1]}(z)$$

$$-C(L+1/M)zQ^{[L-1/M-1]}(z). \quad (5.3)$$

This is a ($^{**}_{*}$) identity.

By rewriting (5.2) as

$$C(L/M)zQ^{[L-1/M-1]}(z)-C(L-1/M)Q^{[L/M-1]}(z)$$

$$+C(L/M-1)Q^{[L-1/M]}(z)=0 \quad (5.4)$$

and writing (5.3) as

$$C(L+1/M)zQ^{[L-1/M-1]}(z)-C(L/M)Q^{[L/M-1]}(z)$$

$$+C(L/M-1)Q^{[L/M]}(z)=0, \quad (5.5)$$

we see that we may eliminate terms to obtain two new identities. These are

$$\{C(L/M)^2 - C(L+1/M)C(L-1/M)\}Q^{[L/M-1]}(z)$$
$$+ C(L+1/M)C(L/M-1)Q^{[L-1/M]}(z)$$
$$- C(L/M-1)C(L/M)Q^{[L/M]}(z) = 0$$

and

$$\{C(L/M)^2 - C(L+1/M)C(L-1/M)\}zQ^{[L-1/M-1]}(z)$$
$$+ C(L/M)C(L/M-1)Q^{[L-1/M]}(z)$$
$$- C(L-1/M)C(L/M-1)Q^{[L/M]}(z) = 0.$$

Using the simplest Sylvester identity, (1.4.11),

$$C(L/M+1)C(L/M-1) = C(L-1/M)C(L+1/M) - C(L/M)^2,$$

we find

$$-C(L/M+1)Q^{[L/M-1]}(z)$$
$$+ C(L+1/M)Q^{[L-1/M]}(z) - C(L/M)Q^{[L/M]}(z) = 0, \quad (5.6)$$

which is a $\binom{*}{**}$ identity, and

$$-C(L/M+1)zQ^{[L-1/M-1]}(z)$$
$$+ C(L/M)Q^{[L-1/M]} - C(L-1/M)Q^{[L/M]}(z) = 0, \quad (5.7)$$

which is a $\binom{*}{**}$ identity.

Equations (5.4)–(5.7) are the Frobenius identities for the Padé denominators. For the numerators,

$$P^{[L/M]}(z) = \begin{vmatrix} c_{L-M+1} & c_{L-M+2} & \cdots & c_L & c_{L+1} \\ c_{L-M+2} & c_{L-M+3} & \cdots & c_{L+1} & c_{L+2} \\ \vdots & \vdots & & \vdots & \vdots \\ c_L & c_{L+1} & & c_{L+M-1} & c_{L+M} \\ \sum_{j=M}^{L} c_{j-M}z^j & \sum_{j=M-1}^{L} c_{j-M+1}z^j & \cdots & \sum_{j=1}^{L} c_{j+1}z^j & \sum_{j=0}^{L} c_j z^j \end{vmatrix},$$

$$(5.8)$$

and Sylvester's identity with deletion of the first and last rows and columns

$$\begin{pmatrix} S^{[L-1/M-1]} & S^{[L/M-1]} \\ S^{[L-1/M]} & S^{[L/M]} \end{pmatrix} \leftrightarrow \begin{pmatrix} * & * \\ * & * \end{pmatrix}$$

Figure 2. Scheme for the Frobenius identities.

leads to

$$P^{[L/M]}(z)C(L+1/M-1)$$
$$= P^{[L+1/M-1]}(z)C(L/M) - zP^{[L/M-1]}C(L+1/M).$$
$$(5.9)$$

Equation (5.9) has a form precisely similar to (5.2). Thus it is normal to write the Frobenius identities using $S^{[L/M]}(z)$, where we choose

either $\qquad S^{[L/M]}(z) = Q^{[L/M]}(z) \qquad\qquad\qquad (5.10a)$

or $\qquad S^{[L/M]}(z) = P^{[L/M]}(z) \qquad\qquad\qquad (5.10b)$

or $\qquad S^{[L/M]}(z) = G(z)P^{[L/M]}(z) + H(z)Q^{[L/M]}(z) \qquad (5.10c)$

The generalization to (5.10c) is possible if $G(z)$ and $H(z)$ are functions of z only, and are independent of L and M; it is easily justified because the Frobenius identities are linear in the sense that (5.10c) is linear. Our conclusion is that with the definitions (5.10), we have identities among the elements $S^{[l/m]}$ of Figure 2 as follows:

FROBENIUS IDENTITIES.

$$C(L/M)zS^{[L-1/M-1]}(z) - C(L-1/M)S^{[L/M-1]}(z)$$
$$+ C(L/M-1)S^{[L-1/M]}(z) = 0 \qquad \begin{pmatrix} * & * \\ * & \end{pmatrix},$$

$$C(L+1/M)zS^{[L-1/M-1]}(z) - C(L/M)S^{[L/M-1]}(z)$$
$$+ C(L/M-1)S^{[L/M]}(z) = 0 \qquad \begin{pmatrix} * & * \\ & * \end{pmatrix},$$

$$C(L/M+1)S^{[L/M-1]}(z) - C(L+1/M)S^{[L-1/M]}(z)$$
$$+ C(L/M)S^{[L/M]}(z) = 0 \qquad \begin{pmatrix} & * \\ * & * \end{pmatrix},$$

$$C(L/M+1)zS^{[L-1/M-1]}(z) - C(L/M)S^{[L-1/M]}(z)$$
$$+ C(L-1/M)S^{[L/M]}(z) = 0 \qquad \begin{pmatrix} * & \\ * & * \end{pmatrix}. \qquad (5.11)$$

From these important results, we may obtain further identities. Using the symbol · to denote entries to be eliminated, we examine the configuration $\left(\begin{smallmatrix} \cdot & \cdot \\ * & * & * \end{smallmatrix}\right)$ using $\left(\begin{smallmatrix} \cdot \\ * \end{smallmatrix}\right)$ twice and $\left(\begin{smallmatrix} \cdot \\ \cdot & * \end{smallmatrix}\right)$. This leads to a single identity for the $\left(\begin{smallmatrix} \\ * & * & * \end{smallmatrix}\right)$ configuration which turns out to be

$$C(L+1/M)^2 S^{[L-1/M+1]}(z) + C(L/M+1)^2 S^{[L+1/M-1]}(z)$$
$$= \{C(L+1/M-1)C(L/M+2) - C(L-1/M+1)C(L+2/M)$$
$$+ zC(L+1/M)C(L/M+1)\} S^{[L/M]}(z), \tag{5.12}$$

which is useful for Kronecker's algorithm (see Section 2.4). Similarly, there is a $(*\ *\ _*)$ identity, which is

$$\frac{C(L/M)}{C(L+1/M+1)} S^{[L+1/M+1]}(z) + \frac{C(L+1/M+1)}{C(L/M)} z^2 S^{[L-1/M-1]}(z)$$
$$= \frac{\{C(L+1/M+2)C(L/M-1) - zC(L+2/M+1)C(L-1/M)\} S^{[L/M]}(z)}{C(L+1/M+1)C(L/M)}.$$

There are also $\left(\begin{smallmatrix} * \\ * \\ * \end{smallmatrix}\right)$ and $(*\ *\ *)$ identities, given in EPA.

This identity concludes our survey of identities for $Q^{[L/M]}(z)$ and $P^{[L/M]}(z)$. Next we turn to identities for the Padé approximants themselves. First, there are the fundamental two-term identities between neighboring approximants. Consider

$$f(z) - \frac{P^{[L/M]}(z)}{Q^{[L/M]}(z)} = O(z^{L+M+1}) \tag{5.13}$$

and

$$f(z) - \frac{P^{[L+1/M]}(z)}{Q^{[L+1/M]}(z)} = O(z^{L+M+2}). \tag{5.14}$$

If the Padé approximants are degenerate, then Equations (5.13), (5.14) are to be understood as being multiplied up, in which case they become correct. The results, being algebraic, are essentially unchanged. Subtracting (5.13) and (5.14),

$$\frac{P^{[L+1/M]}(z)}{Q^{[L+1/M]}(z)} - \frac{P^{[L/M]}(z)}{Q^{[L/M]}(z)} = O(z^{L+M+1}).$$

Therefore

$$Q^{[L/M]}(z)P^{[L+1/M]}(z) - Q^{[L+1/M]}(z)P^{[L/M]}(z) = O(z^{L+M+1}). \tag{5.15}$$

Since the left-hand side of (5.15) is a polynomial of order $L+M+1$, (5.15) becomes

$$Q^{[L/M]}(z)P^{[L+1/M]}(z)-Q^{[L+1/M]}(z)P^{[L/M]}(z)=Kz^{L+M+1}.$$

$$(5.16)$$

By inspection of the determinantal forms (5.1), (5.8), we find the leading coefficients to be as follows:

The coefficient of z^M in $Q^{[L/M]}(z)$ is $(-1)^M C(L+1/M)$.
The coefficient of z^L in $P^{[L/M]}(z)$ is $(-1)^M C(L/M+1)$.

Substituting in (5.16) for the leading coefficients, we obtain

$$C(L+1/M)C(L+1/M+1)=K,$$

and so (5.16) after division by $Q^{[L/M]}(z)Q^{[L+1/M]}(z)$ finally becomes

$$[L+1/M]-[L/M]=\frac{C(L+1/M)C(L+1/M+1)z^{L+M+1}}{Q^{[L/M]}(z)Q^{[L+1/M]}(z)},$$

$$(5.17)$$

which is a $(**)$ identity. By working in a precisely similar way, and using Sylvester's identity for (5.18), we find

$$[L+1/M+1]-[L/M]=\frac{C(L+1/M+1)^2 z^{L+M+1}}{Q^{[L+1/M+1]}(z)Q^{[L/M]}(z)} \quad \left(\begin{matrix}*\\ &*\end{matrix}\right), \quad (5.18)$$

$$[L/M+1]-[L/M]=\frac{C(L/M+1)C(L+1/M+1)z^{L+M+1}}{Q^{[L/M+1]}(z)Q^{[L/M]}(z)} \quad \left(\begin{matrix}*\\ &*\end{matrix}\right),$$

$$(5.19)$$

$$[L/M+1]-[L+1/M]=\frac{C(L+1/M+1)^2 z^{L+M+2}}{Q^{[L/M+1]}(z)Q^{[L+1/M]}(z)} \quad \left(\begin{matrix}&*\\ *\end{matrix}\right). \quad (5.20)$$

This identity concludes the derivation of the two-term identities (5.17)–(5.20). From these may be derived some invariants, called cross ratios, which are independent of z. We find

$$\frac{[L/M]-[L/M+1]}{[L/M]-[L+1/M]}\frac{[L+1/M]-[L+1/M+1]}{[L/M+1]-[L+1/M+1]}$$

$$=\frac{C(L/M+1)}{C(L+1/M)}\frac{C(L+2/M+1)}{C(L+1/M+2)}$$

which is a $\binom{*-*}{*-*}$ identity. It is not the only $\binom{**}{**}$ identity; there are others given by $\binom{*\times}{**}$ interconnections and by $\binom{*\times}{**}$ and by more complicated patterns. Again we refer to EPA for details.

3.6 The Q.D. Algorithm and the Root Problem

If we are given the formal expansion of $f(z)$,

$$f(z) = c_0 + c_1 z + c_2 z^2 + \cdots, \tag{6.1}$$

and we know that $f(z)$ is a meromorphic function which is analytic at the origin, it is natural to wonder if the Padé method is useful for locating the poles and zeros of $f(z)$. Indeed, a vast theory of root solving exists; we will give a simple method which works when the poles of $f(z)$ have distinct moduli, and so also do the zeros. In this case, we may order the poles as $u_1, u_2, \ldots$ with

$$|u_1| < |u_2| < \cdots < |u_M| < \cdots. \tag{6.2}$$

and the zeros as $v_1, v_2, \ldots$ with

$$|v_1| < |v_2| < \cdots < |v_L| < \cdots. \tag{6.3}$$

Then $f(z)$ has the representation

$$f(z) = \frac{\displaystyle\prod_{i=1}^{L} \left(1 - \frac{z}{v_i}\right)}{\displaystyle\prod_{j=1}^{M} \left(1 - \frac{z}{u_j}\right)} h(z), \tag{6.4}$$

where $h(z)$ is analytic and zero-free in $|z| < \min(|v_{L+1}|, |u_{M+1}|)$. The special cases where $f(z)$ has either no poles or no zeros cause no problems in principle. However, the condition that the poles and zeros respectively have distinct moduli is important both in principle and in practice. To find the poles of $f(z)$, which means the numerical values of u_j, we form $[L/M]$ Padé approximants to $f(z)$. Keeping M fixed, and with increasing order of approximation, we expect that

$$Q^{[L/M]}(z) \to C(L/M) \prod_{j=1}^{M} \left(1 - \frac{z}{u_j}\right) \qquad \text{as} \quad L \to \infty. \tag{6.5}$$

In fact, this result is a consequence of de Montessus's theorem.
  If we write

$$Q^{[L/M]}(z)=q_M^{[L/M]}z^M+q_{M-1}^{[L/M]}z^{M-1}+\cdots+q_0^{[L/M]}$$

and use the explicit determinantal formula (5.1) for $q_M^{[L/M]},q_0^{[L/M]}$, it
follows from the usual expression for the product of the roots that

$$\prod_{j=1}^{M}u_j=\lim_{L\to\infty}\frac{C(L-1/M)}{C(L/M)}.\tag{6.6}$$

Similarly, by considering $M-1$ roots,

$$\prod_{j=1}^{M-1}u_j=\lim_{L\to\infty}\frac{C(L/M-1)}{C(L+1/M-1)}.\tag{6.7}$$

We define

$$u(L/M)=\frac{C(L+1/M-1)C(L-1/M)}{C(L/M-1)C(L/M)}\tag{6.8}$$

For the particular case of $M=1$, the appropriate definition is

$$u(L/1)=c_{L-1}/c_L.\tag{6.9}$$

Then, from (6.6), (6.7), and (6.8),

$$u_M=\lim_{L\to\infty}u(L/M).\tag{6.10}$$

Recalling the duality theorem (Theorem 1.5.1), the reciprocal of $[L/M]$ is
the $[M/L]$ Padé approximant of $\{f(z)\}^{-1}$. Let $C'(M/L)$ denote the
Hankel determinant of the $[M/L]$ Padé approximant of $g(z)=\{f(z)\}^{-1}$.
With this notation, we find, as in (6.6), that

$$\prod_{j=1}^{L}v_j=\lim_{M\to\infty}\frac{C'(M-1/L)}{C'(M/L)},\tag{6.11}$$

and, corresponding to (6.8), we define

$$v(L/M)=\frac{C'(M-1/L)}{C'(M/L)}\frac{C'(M+1/L-1)}{C'(M/L-1)}\tag{6.12}$$

$$=\frac{C(L-1/M+1)C(L/M-1)}{C(L-1/M)C(L/M)},\tag{6.13}$$

where (6.13) follows from (6.12) by Hadamard's formula (1.6.9). For the particular case of $L=1$, the appropriate definition is

$$v(1/M)=g_{M-1}/g_M \qquad (6.14)$$

where $g(z)=\sum_{i=0}^{\infty}g_i z^i$. By the construction (6.12), the Lth zero of $g(z)$ is given by

$$\lim_{M\to\infty} v(L/M)=v_L. \qquad (6.15)$$

To determine the numerical values of u_m, v_l, we need some computational rules.

PRODUCT RULE.

$$u(L/M)v(L/M)=u(L/M+1)v(L+1/M). \qquad (6.16)$$

Proof. From the definitions (6.8), (6.13),

$$u(L/M)v(L/M)=\frac{C(L+1/M-1)C(L-1/M+1)}{C(L/M)^2}$$

$$=u(L/M+1)v(L+1/M).$$

ADDITION RULE.

$$u(L/M+1)+v(L+1/M)=u(L+1/M+1)+v(L+1/M+1) \qquad (6.17)$$

Proof. Using the definitions (6.8), (6.13) and Sylvester's identity,

$$u(L/M+1)-v(L+1/M+1)$$

$$=\frac{\begin{array}{c}C(L+1/M)C(L-1/M+1)C(L+1/M+1)\\-C(L/M)C(L/M+2)C(L+1/M)\end{array}}{C(L/M)C(L/M+1)C(L+1/M+1)}$$

$$=\frac{C(L+1/M)[C(L/M+1)]^2}{C(L/M)C(L/M+1)C(L+1/M+1)}$$

$$=\frac{C(L+1/M)C(L/M+1)}{C(L/M)C(L+1/M+1)}.$$

Similarly, we find $u(L+1/M+1)-v(L+1/M)$ yields the same answer.

Having established these identities, we display them pictorially in Figure 1.

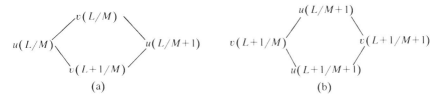

Figure 1. Pictorial representation of the rhombus rules: (a) product rule, (b) addition rule.

All the quantities $u(L/M)$, $v(L/M)$, for $L, M=0,1,2,\ldots$ can be constructed with the aid of these identities (6.16), (6.17), the initializing values $u(L/1)=c_{L-1}/c_L$ given by (6.9), and the artificial initializing values

$$v(L/0)=0, \qquad L=0,1,2,\ldots, \tag{6.18}$$

for the first column and

$$u(0/M)=0, \qquad M=1,2,3,\ldots, \tag{6.19}$$

for the first row. We verify that (6.18) gives the correct initializing values by using (6.17) with $M=0$ together with (6.8) and (6.13). Likewise, (6.19) is verified by using (6.17) with $L=0$ together with (6.8) and (6.13). The quantities $u(L/M)$, $v(L/M)$ are usually displayed in the u-v table [Gragg, 1972], shown in Table 1.

Table 1. The u-v Table

	0		0		0		0		
0		$v(1/1)$		$v(1/2)$		$v(1/3)$		$v(1/4)$	$\cdots$	v_1
	$u(1/1)$		$u(1/2)$		$u(1/3)$		$u(1/4)$			
0		$v(2/1)$		$v(2/2)$		$v(2/3)$		$v(2/4)$	$\cdots$	v_2
	$u(2/1)$		$u(2/2)$		$u(2/3)$		$u(2/4)$			
0		$v(3/1)$		$v(3/2)$		$v(3/3)$		$v(3/4)$	$\cdots$	v_3
	$u(3/1)$		$u(3/2)$		$u(3/3)$		$u(3/4)$			
	$\vdots$		$\vdots$		$\vdots$		$\vdots$			
	u_1		u_2		u_3		u_4			

Summary. The first two columns and the first row initialize the construction of the u-v table, according to (6.9), (6.18), and (6.19). The other entries are constructed by the Q.D. algorithm, expressed by (6.16) and (6.17). The poles $u_1, u_2, u_3,\ldots$ of $f(z)$ are shown as the limiting values of the "u-columns", and the zeros $v_1, v_2, v_3,\ldots$ of $f(z)$ are shown as limiting values of the "v-rows".

In principle, the rate of convergence of the Q.D. algorithm for poles and zeros is geometric. To see this, we refer to analysis of Section 6.2. From (6.2.11), we find that

$$|\det C| = |\det C(L/M)|, \tag{6.20}$$

and from (6.2.22) we note that the dominant part of (6.20) is given by

$$|\det D| = K \prod_{i=1}^{M} |u_i|^{-L}, \tag{6.21}$$

where K is a nonzero constant (independent of L). An inspection of the equations (6.2.11), (6.2.14)–(6.2.22), and (6.2.33) shows that

$$|\det C(L/M)| = K \left| \prod_{i=1}^{M} u_i^{-L} \right| \left\{ 1 + O\left(\left(\frac{|u_M|}{R} \right)^L \right) \right\}. \tag{6.22}$$

Equation (6.22) is Hadamard's formula. It holds under the conditions that $f(z)$ is analytic in the disc $|z| \leq R$ except for precisely M poles, counting multiplicity, in the annulus $0 < |z| < R$.

Let us assume that the poles at u_{M-1}, u_M, and u_{M+1} have distinct moduli:

$$|u_{M-1}| < |u_M| < |u_{M+1}|. \tag{6.23}$$

Substitute (6.22) into (6.8) to show that

$$u(L/M) = u_M \left\{ 1 + O(\theta^L) \right\} \tag{6.24}$$

where

$$\theta = \max \left\{ \left| \frac{u_{M-1}}{u_M} \right|, \left| \frac{u_M}{u_{M+1}} \right| \right\}.$$

The hypothesis (6.23) ensures that $0 < \theta < 1$, and (6.24) shows that the u-columns of the u-v table converge geometrically, in principle.

In practice, the Q.D. algorithm, as expressed in the summary preceding, is unstable. Rounding error accumulates in the u-columns whose limits are theoretically $u_2, u_3, u_4, \ldots$, according to (6.10) and (6.24). To demonstrate this, we consider

$$\frac{v(L+1/M)}{v(L/M)} = \frac{C(L/M+1)}{C(L-1/M+1)} \frac{C(L+1/M-1)}{C(L/M-1)} \frac{C(L-1/M)}{C(L+1/M)}.$$

which is obtained from (6.13). Hence, from (6.10),

$$\lim_{L \to \infty} \frac{v(L+1/M)}{v(L/M)} = \frac{u_M}{u_{M+1}},$$

and we crudely estimate $v(L/M)$ by the formula

$$v(L/M) \approx (u_M/u_{M+1})^L.$$

Consequently, $v(L/M) \to 0$ as $L \to \infty$. We deduce that calculation of the second column $\{v(L/1), \ L=0,1,2,...\}$ necessarily involves substantial rounding error (low relative precision), introduced by cancellation of comparable quantities in the first u-column. This low relative precision is directly transmitted by the multiplication rule (6.16) to each element of the column $\{u(L/2), \ L=0,1,2,...\}$. Hence calculation of u_2 by the basic Q.D. algorithm is necessarily unstable, and a similar argument extends to the poles $u_3, u_4, \ldots$. To remedy this instability, Henrici [1958] proposed the progressive form of the Q.D. algorithm. The order of calculation is changed so as to avoid the necessarily inaccurate arithmetic computations.

PROGRESSIVE Q.D. ALGORITHM.   This algorithm may be used for the computation of the poles of $f(z)$, provided the moduli of these poles are distinct. It is initialized by construction of the coefficients g_i of the power series

$$g(z) = \sum_{i=0}^{\infty} g_i z^i = \left[f(z) \right]^{-1} = \left[\sum_{i=0}^{\infty} c_i z^i \right]^{-1}$$

from the coefficients $\{c_i\}$. The $\{g_i, \ i=1,2,...\}$ are constructed iteratively using the identity

$$g_i = c_0^{-1} \sum_{k=0}^{i-1} g_k c_{i-k}.$$

Hence the quantitities

$$v(1/M) = g_{M-1}/g_M, \qquad M=1,2,...,$$

are constructed, according to (6.14). The other initializing equations (6.9), (6.18), and (6.19) are retained. The progressive form of the Q.D. algorithm requires that the order of calculation using (6.16), (6.17) be as indicated in Table 2. The entries in this schematic section of the u-v table are calculated in numerical order $1,2,3,...,11,12,13,...,21,22,...$ as indicated there.

Table 2. Elements of the u-v table showing schematically the direction of calculation with the progressive form of the Q.D. algorithm.

	0		0		0		0	
0		$v(1/1)$		$v(2/1)$		$v(3/1)$		$v(4/1)$
	$u(1/1)$		(1)		(2)		(3)	
0		(11)		(12)		(13)		
	$u(2/1)$		(21)		(22)			
0		(31)		(32)				
	$u(3/1)$		(41)					
	$\vdots$		$\vdots$					
	u_1		u_2					

Notice that the progressive form of the Q.D. algorithm may be naturally regarded as a practical method for finding the zeros of a meromorphic function $g(z)=\sum_{i=0}^{\infty} g_i z^i$, defined by its power series coefficients g_i, provided that these zeros of $g(z)$ are known to have distinct moduli. For a treatment of the difficult case where the zeros of $g(z)$ have equal modulus, we refer to Henrici [1974, p. 642], or to Rutishauser [1954, p. 35].

Exercise 1. If $f(z)$ is a polynomial, the sequence (6.9) is ill defined. Explain how the Q.D. algorithm may continue to be used in this case.

Exercise 2. Is it a good idea to extrapolate the entries of a u-column of the u-v table using the ε-algorithm?

Exercise 3. Construct the u-v table for $f(z)=\exp(z)$.

CHAPTER 4

Connection with Continued Fractions

4.1 Definitions and the Recurrence Relation

In this chapter we do not aspire to summarize the companion volume of Jones and Thron [1980] which is devoted to the general theory of continued fractions. There is a selected bibliography on continued-fraction theory at the end of this volume. Here, we set out to present a working knowledge of the basic concepts of continued fractions, so that we may give a self-contained account of how continued-fraction theory supplements our understanding of Padé approximation. The discovery of continued fractions in the West seems to have been by Bombelli [1572]; Jones and Thron [1980] and Brezinski give historical surveys. In Section 4.7, we quote the basic convergence theorems for general continued fractions, and refer to the companion volume for the proofs. We are primarily concerned with continued fractions associated with power series, for which the continued fractions happen to be Padé approximants. Indeed, in the next chapter we will see that *S*-fractions are associated with Stieltjes series and that real *J*-fractions are associated with Hamburger series. The convergents of these fractions form simple sequences in the Padé table.

There is no doubt that part of Padé-approximation theory grew out of continued-fraction theory. We choose to regard the Padé table as the fundamental set of rational approximants, and the convergents of various continued fractions derived from power series as particular subsequences of the Padé table. We suggest that which continued-fraction representation is the most useful is often seen most clearly by considering first which sequence from the Padé table has the desired asymptotic behavior or rate of

ENCYCLOPEDIA OF MATHEMATICS and Its Applications, Gian-Carlo Rota (ed.). Vol. 13: George A. Baker, Jr., and Peter R. Graves-Morris, Padé Approximants: Basic Theory, Part I   ISBN 0-201-13512-4

convergence. This view does not mirror the historical development of the subject.

A continued fraction has the general form

$$
b_0 + a_1 \over {b_1 + a_2 \over {b_2 + a_3 \over {b_3 + \cdots}}}
\tag{1.1}
$$

The entries in (1.1), a_i and b_i, are called the elements of the continued fraction. They are usually real or complex numbers. The fraction may be written more compactly as

$$
b_0 + \frac{a_1}{b_1} + \frac{a_2}{b_2} + \frac{a_3}{b_3} + \cdots
\tag{1.2}
$$

with precisely the same meaning as (1.1). By truncating the fractions, we define its convergents, which we denote by ratios A_i / B_i for $i = 0, 1, 2, \ldots$. We find

$$
\frac{A_0}{B_0} = b_0, \qquad \frac{A_1}{B_1} = b_0 + \frac{a_1}{b_1} = \frac{b_0 b_1 + a_1}{b_1},
$$

$$
\frac{A_2}{B_2} = b_0 + \frac{a_1}{b_1 + a_2 / b_2} = \frac{b_0 b_1 b_2 + a_2 b_0 + a_1 b_2}{b_1 b_2 + a_2}.
\tag{1.3}
$$

If the fraction has only a finite number of elements, it takes the form

$$
b_0 + \frac{a_1}{b_1} + \frac{a_2}{b_2} + \cdots + \frac{a_n}{b_n}.
\tag{1.4}
$$

This is equivalent to (1.2) with $a_{n+1} = 0$; (1.4) is called a terminating fraction. The value of a terminating fraction is defined by finite arithmetic. We also note that (1.4) is the $(n+1)$st convergent of (1.2).

In general, a continued fraction is said to converge and have the value v if

$$
\lim_{n \to \infty} \left(\frac{A_n}{B_n} \right) = v.
\tag{1.5}
$$

In other words, provided the limit of the ratios A_n / B_n as $n \to \infty$ exists, it defines a value of the continued fraction. Otherwise, the continued fraction is said to diverge.

The name "series" is used to describe $\Sigma_{i=0}^{\infty} c_i$, meaning a set of numbers $c_0, c_1, c_2, \ldots$ to be added. The word "series" is also used as the value of the sum indicated, provided this value is finite. The same verbal ambiguity arises with continued fractions. Expressions such as (1.1), (1.2), or

$$K\left(\frac{a_n}{b_n}\right) \quad \text{or} \quad \overset{\infty}{\underset{i=1}{\Phi}} \frac{a_i}{b_i}$$

are to be found in the literature. They denote the fact that the pairs (a_1, b_1), (a_2, b_2), $(a_3, b_3), \ldots$ define the continued fraction, or else the value of the fraction if it converges. Once noticed, the ambiguity causes no confusion, and is unimportant in practice.

Part of the definition (1.5) of convergence of a continued fraction refers to the ratios A_n/B_n which are the values of the convergents of the continued fraction. It is possible to construct different continued fractions which have all their convergents equal in value. Such fractions are called equivalent, and, by definition, equivalent fractions all have the same value.

As an example of an equivalence transformation, consider

$$\cfrac{a_1}{b_1 + \cfrac{a_2}{b_2 + \cfrac{a_3}{b_3 + \cdots}}} = \cfrac{(a_1/b_1)}{1 + \cfrac{(a_2/b_1 b_2)}{1 + \cfrac{(a_3/b_2 b_3)}{1 + \cdots}}} . \tag{1.6}$$

By division of the "first" numerator and denominator by b_1, and by division of the "nth" numerator and denominator by b_n for $n = 2, 3, 4, \ldots$, the denominator elements have been reduced to unity. The values of the convergents are unaltered, but the elements of the derived fraction are

$$\left(\frac{a_1}{b_1}, 1\right), \quad \left(\frac{a_2}{b_1 b_2}, 1\right), \quad \left(\frac{a_3}{b_2 b_3}, 1\right), \ldots .$$

Another simple example shows that the numerator elements may be reduced to unity. We find that

$$\frac{a_1}{b_1} + \frac{a_2}{b_2} + \frac{a_3}{b_3} + \cdots = \frac{1}{b_1/a_1} + \frac{1}{b_2 a_1/a_2} + \frac{1}{b_3 a_2/(a_3 a_1)}$$

$$+ \frac{1}{b_4 a_1 a_3/(a_2 a_4)} + \cdots . \tag{1.7}$$

This freedom of representation of the continued fractions using different elements constitutes a group of equivalence transformations. A general member of the group is represented by

$$\frac{e_1 a_1}{e_1 b_1} + \frac{e_1 e_2 a_2}{e_2 b_2} + \frac{e_2 e_3 a_3}{e_3 b_3} + \frac{e_3 e_4 a_4}{e_4 b_4} + \cdots \qquad (1.8)$$

in terms of the parameters $\{e_1, e_2, e_3, \ldots\}$ which are required to be invertible.

The convergents of the continued fractions are ratios A_n / B_n, as is emphasized by (1.3) and (1.5). However, it is useful to define the numerators A_n and denominators B_n separately, but consistently, so that the $(n+1)$th convergent is given by (1.4). The definitions and consistency are expressed by

THEOREM 4.1.1. [Euler, 1737] *For the continued fraction*

$$b_0 + \frac{a_1}{b_1} + \frac{a_2}{b_2} + \frac{a_3}{b_3} + \cdots + \frac{a_m}{b_m} + \frac{a_{m+1}}{b_{m+1}} + \cdots, \qquad (1.9)$$

we define numerators A_i by $A_0 = b_0$, $A_1 = b_1 b_0 + a_1$, and

$$A_i = b_i A_{i-1} + a_i A_{i-2} \qquad for \quad i = 2, 3, 4, \ldots. \qquad (1.10)$$

The denominators are defined by $B_0 = 1$, $B_1 = b_1$, and

$$B_i = b_i B_{i-1} + a_i B_{i-2} \qquad for \quad i = 2, 3, 4, \ldots \qquad (1.11)$$

With this definition, the ratio A_n / B_n is the $(n+1)$th convergent of (1.9).

Proof. By inspection, A_0 / B_0 and A_1 / B_1 are the values given by (1.3). We prove (1.10) and (1.11) by induction. Suppose that they hold for $i = m$. Then the $(m+1)$th convergent of (1.9) is

$$\frac{A_m}{B_m} = \frac{b_m A_{m-1} + a_m A_{m-2}}{b_m B_{m-1} + a_m B_{m-2}}. \qquad (1.12)$$

To obtain the $(m+2)$th convergent of (9) from the $(m+1)$th, we replace b_m by $b_m + a_{m+1}/b_{m+1}$ wherever it appears in the algebraic expression for A_m / B_m. Since b_m does not occur in the algebraic expressions for A_{m-1}, A_{m-2}, B_{m-1} or B_{m-2} defined by (1.10) and (1.11), we find that

$$\frac{A_{m+1}}{B_{m+1}} = \frac{(b_m + a_{m+1}/b_{m+1}) A_{m-1} + a_m A_{m-2}}{(b_m + a_{m+1}/b_{m+1}) B_{m-1} + a_m B_{m-2}}$$

$$= \frac{b_{m+1} A_m + a_{m+1} A_{m-1}}{b_{m+1} B_m + a_{m+1} B_{m-1}}, \qquad (1.13)$$

where the induction hypothesis has been used to obtain (1.13). Clearly, the definitions (1.10) and (1.11) for $i = m + 1$ are consistent with the values of the $(m + 2)$th convergent derived in (1.13).

In this theorem we have derived the most important formula needed for continued-fraction theory: the recurrence relation for the numerators A_i and the denominators B_i defined by (1.10) and (1.11). As an example, we may inspect A_2/B_2 in (1.3) and see that it is given correctly by the recurrence formula. The trivial modification of taking $b_0 = 0$ allows the recurrence to apply to the fraction (1.6). A consequence of Theorem 4.1.1 is that it shows that the following alternative definition of convergence of a continued fraction is entirely equivalent to the previous one.

ALTERNATIVE DEFINITION. The continued fraction

$$b_0 + \frac{a_1}{b_1} + \frac{a_2}{b_2} + \frac{a_3}{b_3} + \cdots$$

is said to converge and have value v if the ratio A_n/B_n of the quantities A_n and B_n defined recursively by (1.10) and (1.11) tends to v as $n \to \infty$.

Exercise 1. Show that

$$\frac{\alpha_1/\beta_1}{1} + \frac{\alpha_2/\beta_2}{1} + \frac{\alpha_3/\beta_3}{1} + \cdots = \frac{\alpha_1}{\beta_1} + \frac{\beta_1 \alpha_2}{\beta_2} + \frac{\beta_2 \alpha_3}{\beta_3} + \cdots.$$

Exercise 2. Show that

$$\frac{a_1}{1} + \frac{a_2}{1} + \frac{a_3}{1} + \cdots = \frac{1}{1/a_1} + \frac{1}{a_1/a_2} + \frac{1}{a_2/(a_1 a_3)} + \frac{1}{a_1 a_3/(a_2 a_4)}$$

$$+ \frac{1}{a_2 a_4/(a_1 a_3 a_5)} + \cdots.$$

Exercise 3. Let A_n/B_n be the nth convergent of

$$b_0 + \frac{a_1}{b_1} + \frac{a_2}{b_2} + \frac{a_3}{b_3} + \cdots$$

as described in the text. Prove that the elements are given in terms of the numerators and denominators by $b_0 = A_0$, $a_1 = A_1 - B_1 A_0$, $b_1 = B_1$ and for $i \geq 2$,

$$a_i = \frac{A_i B_{i-1} - A_{i-1} B_i}{A_{i-2} B_{i-1} - A_{i-1} B_{i-2}}, \qquad b_i = \frac{A_{i-2} B_i - A_i B_{i-2}}{A_{i-2} B_{i-1} - A_{i-1} B_{i-2}}.$$

4.2   Continued Fractions Derived from Maclaurin Series

A formal power series may be manipulated into the form of a continued fraction very easily. In this section, we ignore all questions of convergence. We assume that all the inverses we need exist and that we do not encounter degenerate cases.

The given power series is

$$f(z) = c_0 + c_1 z + c_2 z^2 + \cdots. \tag{2.1}$$

We calculate the reciprocal of the series

$$1 + \frac{c_2 z}{c_1} + \frac{c_3 z^2}{c_1} + \cdots = \left(1 + c_1^{(1)} z + c_2^{(1)} z^2 + \cdots\right)^{-1},$$

which allows the reexpansion

$$c_0 + c_1 z + c_2 z^2 + \cdots = c_0 + \frac{c_1 z}{1 + c_1^{(1)} z + c_2^{(1)} z^2 + \cdots}. \tag{2.2}$$

Next we calculate the reciprocal of the series

$$1 + \frac{c_2^{(1)} z}{c_1^{(1)}} + \frac{c_3^{(1)} z}{c_1^{(1)}} + \cdots = \left(1 + c_1^{(2)} z + c_2^{(2)} z^2 + \cdots\right)^{-1},$$

which allows another reexpansion

$$c_0 + c_1 z + c_2 z^2 + \cdots = c_0 + \cfrac{c_1 z}{1 + \cfrac{c_1^{(1)} z}{1 + c_1^{(2)} z + c_2^{(2)} z^2 + \cdots}}. \tag{2.3}$$

It is clear [Salzer, 1962], that by forming the reciprocal series, we have devised an iterative procedure which allows us to write formally

$$f(z) = c_0 + \frac{c_1 z}{1} + \frac{c_1^{(1)} z}{1} + \frac{c_1^{(2)} z}{1} + \frac{c_1^{(3)} z}{1} + \cdots, \tag{2.4}$$

which corresponds to the series (2.1). The convergents of (2.4) are rational fractions in the variable z. To be quite general, we assume only that the resultant fraction representing the power series takes the form

$$f(z) = b_0 + \frac{a_1 z}{b_1} + \frac{a_2 z}{b_2} + \frac{a_3 z}{b_3} + \cdots \tag{2.5}$$

and (2.4) is just a special case of (2.5). The first few convergents of (2.5) may be easily calculated:

$$\frac{A_0(z)}{B_0(z)}=b_0, \qquad \frac{A_1(z)}{B_1(z)}=\frac{b_0b_1+a_1z}{b_1},$$

$$\frac{A_2(z)}{B_2(z)}=\frac{b_0b_1b_2+(a_2b_0+a_1b_2)z}{b_1b_2+a_2z}. \qquad (2.6)$$

We see that (2.6) is equivalent to (1.3) with the replacement $a_i \rightarrow a_i z$ for all i. Following the analysis of Section 4.1, especially Theorem 4.1.1, we see that the numerators and denominators of (2.5) are generated by

$$A_0(z)=b_0, \qquad A_1(z)=b_0b_1+a_1z,$$

$$A_i(z)=b_iA_{i-1}(z)+a_izA_{i-2}(z), \qquad i=2,3,4,\dots, \qquad (2.7a)$$

and

$$B_0(z)=1, \qquad B_1(z)=b_1,$$

$$B_i(z)=b_iB_{i-1}(z)+a_izB_{i-2}(z), \qquad i=2,3,4,\dots. \qquad (2.7b)$$

The connection between the convergents of the fractions (2.4) and (2.5) and the entries in the Padé table is expressed by

THEOREM 4.2.1. *Provided that c_1 and every coefficient $c_1^{(j)}$ are nonzero, the continued fraction (2.4) has the Maclaurin expansion (2.1). In this case, the Padé approximants of (2.1) are identified with the convergents of the continued fraction by*

$$[M/M]_f(z)=\frac{A_{2M}(z)}{B_{2M}(z)} \quad \text{and} \quad [M+1/M]_f(z)=\frac{A_{2M+1}(z)}{B_{2M+1}(z)} \quad (2.8)$$

for $M=0,1,2,\dots$.

Remarks. No statement is implied by this theorem about the domains of convergence in the z-plane, if any, of either the series (2.1) or the fraction (2.4). If nontrivial domains of convergence exist, they are likely to be different. Even if (2.1) and (2.4) are convergent, the theorem does not directly assert equality of these values.

Proof. Since each expression (2.1), (2.2), (2.3) has the same formal Maclaurin expansion, the first part of the theorem is true by induction. To establish (2.8), we note that

$$\deg\{A_0(z)\}=0, \qquad \deg\{B_0(z)\}=0,$$
$$\deg\{A_1(z)\}\leqslant1, \qquad \deg\{B_1(z)\}=0. \tag{2.9}$$

We prove (2.8) by induction. Suppose that

$$\deg\{A_{2m}(z)\}\leqslant m, \qquad \deg\{B_{2m}(z)\}\leqslant m,$$
$$\deg\{A_{2m+1}(z)\}\leqslant m+1, \qquad \deg\{B_{2m+1}(z)\}\leqslant m, \tag{2.10}$$

for $m=0,1,2,\ldots, M$. Then

$$\deg\{A_{2m+2}(z)\}=\deg\{b_{2m+2}A_{2m+1}(z)+a_{2m+2}zA_{2m}(z)\}\leqslant m+1,$$
$$\deg\{A_{2m+3}(z)\}=\deg\{b_{2m+3}A_{2m+2}(z)+a_{2m+3}zA_{2m+1}(z)\}\leqslant m+2,$$
$$\deg\{B_{2m+2}(z)\}=\deg\{b_{2m+2}B_{2m+1}(z)+a_{2m+2}zB_{2m}(z)\}\leqslant m+1,$$
$$\deg\{B_{2m+3}(z)\}=\deg\{b_{2m+3}B_{2m+2}(z)+a_{2m+3}zB_{2m+1}(z)\}\leqslant m+1.$$

Hence the fractions $A_m(z)/B_m(z)$ have numerators and denominators of the requisite orders for all m, and power series which agree with (2.1) to order z^m inclusive. Consequently the fractions $A_m(z)/B_m(z)$ are the Padé approximants of (2.1) of the orders indicated by (2.8).

Notice that the fractions $\{A_m(z)/B_m(z)\}$ defined in this section occupy a descending staircase sequence in the Padé table, which starts with a horizontal tread, as shown in Figure 1.

As an example of a continued fraction of this type, we may use five terms of the Maclaurin expansion of $\exp(z)$ to show that

$$\exp(z)=1+\frac{z}{1}-\frac{z}{2}+\frac{z}{3}-\frac{z}{2}+\cdots. \tag{2.11}$$

This example shows an advantage of using (2.5) rather than (2.4), because the elements of the fraction may be taken to be integers with this representation. We will derive the general term of (2.11) in Section 4.6. If we consider the reciprocal of (2.11) and replace z by $-z$, we get a different representation:

$$\exp(z)=\frac{1}{1}-\frac{z}{1}+\frac{z}{2}-\frac{z}{3}+\cdots. \tag{2.12}$$

This is of the general type

$$f(z)=\frac{a_1}{b_1}+\frac{a_2z}{b_2}+\frac{a_3z}{b_3}+\cdots. \tag{2.13}$$

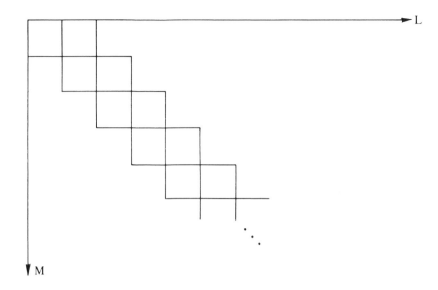

Figure 1. A descending staircase sequence in the Padé table corresponding to convergents of (2.4).

The numerators and denominators of (2.13) are derived from

$$A_1(z)=a_1, \qquad A_2(z)=b_2a_1$$
$$A_{i+1}(z)=b_{i+1}A_i(z)+a_{i+1}zA_{i-1}(z), \qquad i=2,3,4,\dots, \tag{2.14a}$$

and

$$B_1(z)=b_1, \qquad B_2(z)=b_1b_2+a_2z$$
$$B_{i+1}(z)=b_{i+1}B_i(z)+a_{i+1}zB_{i-1}(z), \qquad i=2,3,4,\dots. \tag{2.14b}$$

The Padé approximants which are the convergents of (2.13) are given by

$$[M/M]_f(z)=\frac{A_{2M}(z)}{B_{2M}(z)} \quad \text{and} \quad [M/M+1]_f(z)=\frac{A_{2M+1}(z)}{B_{2M+1}(z)} \tag{2.15}$$

These occupy a descending sequence in the Padé table which begins with a stair, as shown in Figure 2.

The comparison of (2.5) or (2.13) with the sequence of Padé approximants indicates that (2.5) is to be preferred in particular asymptotic regions of the z-plane where $|f(z)|$ is increasing, and (2.13) is to be preferred where $|f(z)|$ is decreasing as $|z|$ increases.

If functions are even or odd, they are degenerate in a rather trivial way, and there is no purpose in making a great issue of this. If the function is

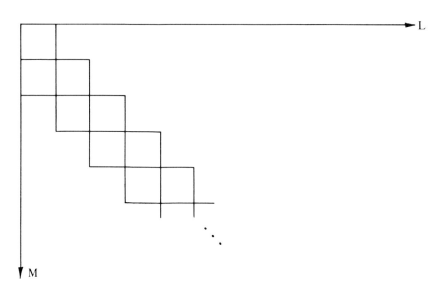

Figure 2. A descending staircase sequence in the Padé table corresponding to convergents of (2.13).

even, either one uses

$$f^{(\text{even})}(z) = \frac{a_1}{b_1} + \frac{a_2 z^2}{b_2} + \frac{a_3 z^2}{b_3} + \cdots$$

or

$$f^{(\text{even})}(z) = b_0 + \frac{a_1 z^2}{b_1} + \frac{a_2 z^2}{b_2} + \cdots.$$

If the given function is odd, normally one chooses

$$f^{(\text{odd})}(z) = \frac{a_1 z}{b_1} + \frac{a_2 z^2}{b_2} + \frac{a_3 z^2}{b_3} + \cdots. \tag{2.16}$$

It is customary to adopt the most convenient form of continued fraction without discussing alternative possible representations. Consideration of the sequence of convergents as they appear in the Padé table with the desired asymptotic properties and consideration of any known degeneracies give a guide to the best continued-fraction representation to use.

Exercise 1.   Prove that the Padé approximants which are the convergents of (2.13) are given by (2.15).

Exercise 2.   Which entries in the Padé table are occupied by the convergents of (2.16)?

4.3 Algebraic and Numerical Methods

In this section, we first consider the task of constructing continued-fraction representations from power series. This is the topic which we considered from a theoretical viewpoint in the previous section. Second, we discuss methods by which the continued fractions which we have constructed, or derived, should be evaluated numerically.

Viskovatov's method [1803] is a handy method for constructing continued-fraction expansions, or equivalently staircase sequences of Padé approximants, from the given power series. It avoids having to reciprocate any series numerically, as was part of the construction of (2.2) and (2.3). It is a labor saving device which is as useful for people as for computers. The method is based on the formal identity

$$\frac{\sum\limits_{r=0}^{\infty} a_r z^r}{\sum\limits_{r=0}^{\infty} b_r z^r} = \frac{a_0}{b_0} + \cfrac{z}{\cfrac{\sum\limits_{r=0}^{\infty} b_r z^r}{\sum\limits_{r=0}^{\infty} (a_{r+1} - a_0 b_{r+1}/b_0) z^r}}. \tag{3.1}$$

Viskovatov's method is best described by an illustrative example.

$$\exp(z) = 1 + z + \frac{z^2}{2} + \frac{z^3}{6} + \frac{z^4}{24} + \cdots$$

$$= 1 + \cfrac{z}{\left(\cfrac{1}{1 + \frac{z}{2} + \frac{z^2}{6} + \frac{z^3}{24} + \cdots}\right)}$$

$$= 1 + \cfrac{z}{1 + \cfrac{z}{\left(\cfrac{1 + \frac{z}{2} + \frac{z^2}{6} + \cdots}{-\frac{1}{2} - \frac{z}{6} - \frac{z^2}{24} + \cdots}\right)}}$$

$$= 1 + \cfrac{z}{1 + \cfrac{z}{-2 + \cfrac{z}{\cfrac{-\frac{1}{2} - \frac{z}{6} - \frac{z^2}{24} + \cdots}{\frac{1}{6} + \frac{z}{12} + \cdots}}}}$$

$$= 1 + \cfrac{z}{1 + \cfrac{z}{-2 + \cfrac{z}{-3 + \cdots}}}. \tag{3.2}$$

Thus the generation of staircase sequences of Padé approximants is absolutely straightforward in the nondegenerate case. The other straightforward method of economical construction of continued fractions from power series is the Q.D. algorithm, which we treat fully in the next section.

The other main question we consider in this section is that of how to evaluate a fraction such as (3.2) numerically. There does not seem to be an agreed best method of computing continued fractions, and so we state three principal methods without making any clear recommendation about which is "best". The problem is, given the elements a_i, b_i and a value of the variable z, to compute

$$f(z) = \frac{a_1 z}{b_1} + \frac{a_2 z}{b_2} + \frac{a_3 z}{b_3} + \cdots, \tag{3.3}$$

which we assume to be a convergent fraction.

FORWARD RECURRENCE METHOD.   The numerators $A_i(z)$ and denominators $B_i(z)$ are calculated from the recurrence relation (2.7). Because of the ease of computing the values of successive convergents, the forward recurrence method is the standard method for computing continued fractions.

It is sometimes the case in practice that the crude approximation

$$\begin{aligned} A_n &\simeq \alpha r_1^n + \beta r_2^n, \\ B_n &\simeq \gamma r_1^n + \delta r_2^n \end{aligned} \tag{3.4}$$

gives a rough and ready estimate of the rate of growth of A_n and B_n. Equation (3.4) is exact in the special case that the coefficients a_i, b_i in (2.7) are constant. If $|r_1| > |r_2|$, the α, γ terms are the dominant components and the β, δ terms become less significant as $n \to \infty$. Such considerations assist in the understanding of the following numerical hazards of the forward recurrence method.

(a) *Floating point overflow of A_n, B_n.* If $|r_1| \gg 1$, floating-point overflow may occur at some point during the actual computation. Rescaling of A_n, A_{n-1}, B_n, and B_{n-1} at the critical point in the obvious way is recommended. Underflow must likewise be anticipated.

(b) *Suppression of required solution by a dominant solution.* If the true solutions A_n, B_n of the exact recurrence relation (2.7) are the subdominant solutions, they may be suppressed numerically. Other solutions, $\tilde{A}_n$, $\tilde{B}_n$, generated by different initial conditions (e.g. $\tilde{A}_0 \neq A_0$) for which $A_n / \tilde{A}_n \to 0$, $B_n / \tilde{B}_n \to 0$, exist in this case. Rounding error introduced in the initial stages always becomes a dominant effect in such calculations. The connection with the concepts of Miller's backward recurrence algorithm is obvious.

(c) *Accumulation of roundoff error in a dominant solution.* If $|A_n|$ and $|B_n|$ are decreasing sequences, rounding error may accumulate using the forward recurrence method. This has been discovered empirically [Jones and Thron, 1974] for the following continued fraction:

$$f(z) = \frac{z}{1} - \frac{z}{1} - \frac{z}{1} - \frac{z}{1} - \cdots \qquad (3.5)$$

evaluated at $z = 1/4$, using fixed *decimal* precision in the floating point operations. (Why would binary operations be different?) The fraction is being evaluated on the edge of its domain of convergence, and so it is a natural prototype for investigation of the buildup of roundoff error.

BACKWARD RECURRENCE METHOD. This method of evaluating the nth convergent of (3.3) starts at the "tail" end of the convergent. We define

$$f_{n+1}^{(n)} = 0,$$
$$f_i^{(n)}(z) = \frac{a_i z}{b_i + f_{i+1}^{(n)}(z)}, \qquad i = n, n-1, n-2, \ldots, 1, \qquad (3.6)$$

and then $f_1^{(n)}(z)$ is the nth convergent of $f(z)$. The major drawback of this method is that of deciding in advance which is the appropriate value of n to choose so that the nth convergent is an adequate approximation to $f(z)$.

SUMMATION FORMULA. The nth convergent of (3.3) is given by the formula

$$\frac{A_n(z)}{B_n(z)} = \sum_{i=1}^{n} \prod_{j=1}^{i} \rho_i, \qquad (3.7)$$

where the quantities $\rho_1, \rho_2, \rho_3, \ldots, \rho_n$ are defined in terms of the elements of the continued fraction by

$$\rho_1 = \frac{a_1 z}{b_1}, \qquad 1 + \rho_2 = \frac{1}{1 + a_2 z / (b_1 b_2)} \qquad (3.8)$$

and recursively by

$$1 + \rho_i - \left(1 + \frac{a_i z (1 + \rho_{i-1})}{b_i b_{i-1}}\right)^{-1}, \qquad i - 3, 4, \ldots. \qquad (3.9)$$

Proof. We construct a summation formula by using the identity

$$\frac{A_n}{B_n} = \left(\frac{A_n}{B_n} - \frac{A_{n-1}}{B_{n-1}}\right) + \left(\frac{A_{n-1}}{B_{n-1}} - \frac{A_{n-2}}{B_{n-2}}\right) + \left(\frac{A_{n-2}}{B_{n-2}} - \frac{A_{n-3}}{B_{n-3}}\right) + \cdots + \frac{A_1}{B_1}$$

$$= \frac{a_1 z}{b_1} + \sum_{i=2}^{n} \frac{(-1)^{i+1}}{B_i B_{i-1}} \prod_{j=1}^{i} (a_i z); \tag{3.10}$$

a formula equivalent to (3.10) is derived as (7.6). A comparison of Equations (3.7), (3.10) leads us to define

$$\rho_i = \frac{-a_i z B_{i-2}}{B_i}, \qquad i = 3, 4, 5, \dots. \tag{3.11}$$

Equation (3.8) defines the first two convergents correctly. To establish (3.9), we use the recurrence relation

$$B_i = b_i B_{i-1} + a_i z B_{i-2}$$

to prove that

$$\frac{B_{i-1}}{B_{i-2}} = \frac{-a_i z}{b_i} \left\{ \frac{1}{\rho_i} + 1 \right\}, \tag{3.12}$$

and the recurrence

$$B_{i-1} = b_{i-1} B_{i-2} + a_{i-1} z B_{i-3}$$

to prove that

$$\frac{B_{i-2}}{B_{i-1}} = \frac{1}{b_{i-1}} \{ 1 + \rho_{i-1} \}. \tag{3.13}$$

Equation (3.9) is proved by multiplying (3.12) and (3.13).

A consequence of (3.8), (3.9) and the equivalence transformation (1.6) is the new representation

$$\frac{A_n(z)}{B_n(z)} = \frac{\rho_1}{1} - \frac{\rho_2}{1 + \rho_2} - \frac{\rho_3}{1 + \rho_3} - \cdots - \frac{\rho_n}{1 + \rho_n}, \tag{3.14}$$

which is a representation of (5.55)

The summation formula expressed by (3.7)–(3.9) has the advantage that it is suitable for iterative computation, which is important if the rate of convergence is not known *a priori*.

For further details of the numerical methods, and especially for error estimates, we refer to Jones and Thron's companion volume [1980], to Blanch [1964], and to Gautschi [1967].

Exercise 1. Prove that the value of the repeating fraction (assumed convergent for $z \leqslant \frac{1}{4}$)

$$f = \frac{z}{1} - \frac{z}{1} - \frac{z}{1} - \frac{z}{1} - \cdots$$

satisfies the equation

$$f = \frac{z}{1-f}.$$

Deduce the value of the fraction. Pringsheim [1910] gives a full discussion of the evaluation of periodic continued fractions.

Exercise 2. Solve the recurrence relation for the numerators and denominators of the fraction

$$F = \frac{p}{2q} + \frac{p}{2q} + \frac{p}{2q} + \cdots.$$

If $q > 0$ and $q^2 + p > 0$, deduce that

$$F = -q + \sqrt{q^2 + p}.$$

4.4 Various Representations of Continued Fractions

We are primarily concerned with continued fractions whose convergents form a sequence of Padé approximants. We will consider continued fractions containing the complex variable z explicitly, so as to maintain the connection with Padé approximation. The formulas remain valid with $z = 1$; this may be more useful for other purposes connected with the general theory of continued fractions. We will first establish the existence of tridiagonal determinantal formulas for the numerators and denominators of a continued fraction.

THEOREM 4.4.1. *The convergents of the continued fraction*

$$f(z) = b_0 + \frac{a_1 z}{b_1} + \frac{a_2 z}{b_2} + \frac{a_3 z}{b_3} + \cdots \tag{4.1}$$

have numerators given, for n=0,1,2,..., by

$$
A_n(z) = \begin{vmatrix}
b_0 & -a_1 z & & & & & & \\
1 & b_1 & -a_2 z & & & 0 & & \\
 & 1 & b_2 & -a_3 z & & & & \\
 & & \ddots & \ddots & \ddots & & & \\
 & & & 1 & b_{n-2} & -a_{n-1}z & & \\
 & 0 & & & 1 & b_{n-1} & -a_n z & \\
 & & & & & 1 & b_n &
\end{vmatrix}, \quad (4.2)
$$

and denominators given, for n=1,2,3,..., by

$$
B_n(z) = \begin{vmatrix}
b_1 & -a_2 z & & & & & & \\
1 & b_2 & -a_3 z & & & 0 & & \\
 & 1 & b_3 & -a_4 z & & & & \\
 & & \ddots & \ddots & \ddots & & & \\
 & & & 1 & b_{n-2} & -a_{n-1}z & & \\
 & 0 & & & 1 & b_{n-1} & -a_n z & \\
 & & & & & 1 & b_n &
\end{vmatrix}. \quad (4.3)
$$

Proof. We expand the determinant (4.2) by its last column. This leads to the equation

$$
A_n(z) = b_n A_{n-1}(z) + a_n z A_{n-2}(z), \quad (4.4)
$$

which is the standard recurrence relation (2.7a). For $n=0$ we have $A_0(z)=b_0$, and for $n=1$ we have $A_1(z)=b_0 b_1 + a_1 z$. From (2.14a) we deduce that the tridiagonal representation (4.2) is valid for the numerators of the convergents of the continued fraction (4.1).

We deduce from (2.14b) that the tridiagonal representation (4.3) is valid for the denominators of the convergents of the continued fraction (4.2).

THEOREM 4.4.2. *The convergents of the continued fraction*

$$
f(z) = \frac{a_1}{b_1} + \frac{a_2 z}{b_2} + \frac{a_3 z}{b_3} + \cdots \quad (4.5)
$$

have numerators, given for $n = 2, 3, \ldots$, by

$$A_n = a_1 \begin{vmatrix} b_2 & -a_3 z & & & \\ 1 & b_3 & -a_4 z & 0 & \\ & \ddots & \ddots & \ddots & \\ & 0 & 1 & b_{n-1} & -a_n 2 \\ & & & 1 & b_n \end{vmatrix} \tag{4.6}$$

and denominators given by (4.3).

Proof. The method of proof is identical to that of Theorem 4.4.1.

The determinants (4.2) and (4.3) express the numerators and denominators of Padé approximants of types $[M/M]$ or $[M+1/M]$, according to the discussion of Section 4.2. Equations (4.2) and (4.3) involve the elements a_i and b_i of the continued fraction (4.1). Next, we reexpress (4.1) in terms of Hankel determinants, so as to relate (4.1), (4.2), and (4.3) more directly to the coefficients c_i of the Maclaurin expansion of $f(z)$.

THEOREM 4.4.3. *The continued fraction* (4.1) *may be expressed as*

$$f(z) = c_0 + \frac{c_1 z}{1} - \frac{c_2 z}{c_1} - \frac{z C(2/2)/C(1/1)}{C(2/1)/C(1/1)} - \frac{z C(3/2)/C(2/1)}{C(2/2)/C(2/1)} - \cdots. \tag{4.7}$$

The initial elements of (4.1) *are*

$$b_0 = c_0, \quad a_1 = c_1, \quad b_1 = 1, \quad a_2 = -c_2, \quad b_2 = c_1. \tag{4.8}$$

The other elements are given, for $M = 1, 2, 3, \ldots$, *by*

$$a_{2M+1} = -\frac{C(M+1/M+1)}{C(M/M)}, \tag{4.9}$$

$$b_{2M+1} = \frac{C(M+1/M)}{C(M/M)}, \tag{4.10}$$

and for $M = 2, 3, 4, \ldots$, *by*

$$a_{2M} = -\frac{C(M+1/M)}{C(M/M-1)}, \tag{4.11}$$

$$b_{2M} = \frac{C(M/M)}{C(M/M-1)}. \tag{4.12}$$

With these values (4.8)–(4.12) *for the elements of the continued fraction* $f(z)$, *the determinantal formulas* (4.2) *and* (4.3) *satisfy for all M*

$$A_{2M}(z)=P^{[M/M]}(z), \qquad B_{2M}(z)=Q^{[M/M]}(z),$$
$$A_{2M+1}(z)=P^{[M+1/M]}(z), \qquad B_{2M+1}(z)=Q^{[M+1/M]}(z),$$

(4.13)

where we assume $B_0(z)\equiv 1$.

Remark. The elements a_i, b_i given in (4.8)–(4.12) are only determined up to equivalence transforms if one only requires that (4.1) should have a given Maclaurin expansion; they are determined uniquely by the conditions (4.13).

Proof. In (4.3), we set $z=0$ and obtain

$$B_n(0)= \prod_{i=1}^{n} b_i.$$

(4.14)

Thus we can arrange for the conventional normalization to hold, namely

$$\prod_{i=1}^{2M} b_i = Q^{[M/M]}(0)=C(M/M)$$

(4.15a)

and

$$\prod_{i=1}^{2M+1} b_i = Q^{[M+1/M]}(0)=C(M+1/M),$$

(4.15b)

by choosing the coefficients b_i according to (4.10) and (4.12). By inspection, we verify that the initial values (4.8) are chosen so as to satisfy

$$A_0(z)=b_0=c_0,$$
$$B_0(z)=1,$$
$$A_1(z)=\begin{vmatrix} b_0 & -a_1z \\ 1 & b_1 \end{vmatrix}=c_0+c_1z,$$
$$B_1(z)=b_1=1.$$

(4.16)

To establish (4.9) and (4.11) for the elements a_i, we use Frobenius identities (3.5.3) and (3.5.7).

The $\boxed{\begin{array}{c|c} M/M & M+1/M \\ \hline & M+1/M+1 \end{array}}$ identity is

$$C(M+1/M)Q^{[M+1/M+1]}=C(M+1/M+1)Q^{[M+1/M]}(z)$$
$$-zC(M+2/M+1)Q^{[M/M]}(z). \quad (4.17)$$

The recurrence relation (2.7b) for the even-order denominator $B_{2M+2}(z)$ is

$$B_{2M+2}(z)=b_{2M+2}B_{2M+1}(z)+a_{2M+2}zB_{2M}(z).$$

By comparing this equation with (4.17), we see that the definition (4.11) of a_{2M+2} ensures that

$$B_{2M+2}(z)=Q^{[M+1/M+1]}(z) \qquad (4.18)$$

provided that all the lower-order denominators are identical.

The $\boxed{M/M-1}$ identity is
$\boxed{M/M} \boxed{M+1/M}$

$$C(M/M)Q^{[M+1/M]}(z)=C(M+1/M)Q^{[M/M]}(z)$$
$$-zC(M+1/M+1)Q^{[M/M-1]}(z). \quad (4.19)$$

The odd-order denominator $B_{2M+1}(z)$ is also generated by (2.7b), which we write as

$$B_{2M+1}(z)=b_{2M+1}B_{2M}(z)+a_{2M+1}zB_{2M-1}(z).$$

By comparing this equation with (4.19), we see that the definition (4.9) of a_{2M+1} ensures that

$$B_{2M+1}(z)=Q^{[M+1/M]}(z) \qquad (4.20)$$

provided that all the lower-order denominators are identical. By combining (4.18), (4.20), and (4.8), the representation (4.7) is proved by induction. The Padé numerators $P^{[M/M]}$ and $P^{[M+1/M]}$ satisfy the same recurrences (4.17), (4.19) as the denominators $Q^{[M/M]}$ and $Q^{[M+1/M]}$, as also do the numerators A_n, (4.5).

Hence (4.13) is established by induction, using the initial conditions expressed by (4.16).

THEOREM 4.4.4. *A power series and a continued fraction may be formally identified by*

$$\sum_{i=0}^{\infty} c_i z^i = \frac{a_1}{b_1} + \frac{a_2 z}{b_2} + \frac{a_3 z}{b_3} + \cdots \qquad (4.21)$$

if $a_1 = c_0$, $a_2 = -c_1$, $b_1 = 1$, $b_2 = c_0$, and, for $M = 1, 2, 3, \ldots$,

$$b_{2M+1} = \frac{C(M/M)}{C(M-1/M)},$$

$$a_{2M+1} = -\frac{C(M/M+1)}{C(M-1/M)},$$

$$b_{2M+2} = \frac{C(M/M+1)}{C(M/M)}, \tag{4.22}$$

$$a_{2M+2} = -\frac{C(M+1/M+1)}{C(M/M)}.$$

The numerators and denominators generated by recurrence are identical to those of Theorem 4.4.2.

Proof. The method of proof is identical to that of Theorem 4.4.3.

COROLLARY. Using an equivalence transformation, we may reexpress (4.21) as

$$\sum_{i=0}^{\infty} c_i z^i = \frac{a_1'}{1} + \frac{a_2' z}{1} + \frac{a_3' z}{1} + \cdots, \tag{4.23}$$

where

$$a_1' = c_0,$$
$$a_2' = -c_1/c_0, \tag{4.24}$$
$$a_3' = -\frac{C(1/2)}{C(1/1)c_0},$$

and for $M = 2, 3, 4, \ldots,$

$$a_{2M}' = -\frac{C(M/M)C(M-2/M-1)}{C(M-1/M)C(M-1/M-1)},$$

$$a_{2M+1}' = -\frac{C(M/M+1)C(M-1/M-1)}{C(M/M)C(M-1/M)}. \tag{4.25}$$

As we see in (5.5.25) and (5.6.33), such a representation and its contraction are of especial importance to Stieltjes series.

In fact, (4.23) is a formal identity, and it represents the connection between a formal power series and a sequence of $[M/M]$ and $[M-1/M]$

Padé approximants. The elements a_i' of (4.23) are relatively easily calculated directly from the coefficients c_i using the Q.D. algorithm; no-one would contemplate using (4.25) for iterative numerical computations in view of the likely ill-conditioning of the Hankel determinants. In the context of numerical computation, it is more usual to write (4.23) as

$$\sum_{i=0}^{\infty} c_i z^i = \frac{c_0}{1} - \frac{q_1^0 z}{1} - \frac{e_1^0 z}{1} - \frac{q_2^0 z}{1} - \frac{e_2^0 z}{1} - \cdots, \tag{4.26}$$

where

$$q_1^0 = c_1/c_0,$$

$$e_1^0 = \frac{C(1/2)}{c_1 c_0}, \tag{4.27}$$

and for $M = 2, 3, 4, \ldots,$

$$q_M^0 = \frac{C(M/M)C(M-2/M-1)}{C(M-1/M-1)C(M-1/M)}, \tag{4.28a}$$

$$e_M^0 = \frac{C(M/M+1)C(M-1/M-1)}{C(M-1/M)C(M/M)}. \tag{4.28b}$$

More generally, we use the expansion

$$\sum_{i=0}^{\infty} c_i z^i = \sum_{i=0}^{J-1} c_i z^i + \frac{c_J z^J}{1} - \frac{q_1^J z}{1} - \frac{e_1^J z}{1} - \frac{q_2^J z}{1} - \frac{e_2^J z}{1} - \cdots. \tag{4.29}$$

Equation (4.26) is a special case of (4.29) with $J=0$. With our usual convention that $L = M + J$, it follows from (4.28) that for $M = 2, 3, 4, \ldots,$ and $J \geq 0$,

$$q_M^J = \frac{C(L/M)C(L-2/M-1)}{C(L-1/M-1)C(L-1/M)} \tag{4.30a}$$

and

$$e_M^J = \frac{C(L/M+1)C(L-1/M-1)}{C(L-1/M)C(L/M)} \tag{4.30b}$$

THEOREM 4.4.5. *The elements of* (4.29) *satisfy, for* $J \geq 0$ *and* $M = 1, 2, 3, \ldots,$

$$e_M^J q_{M+1}^J = e_M^{J+1} q_M^{J+1} \tag{4.31}$$

and

$$q_M^J + e_M^J = e_{M-1}^{J+1} + q_M^{J+1}. \tag{4.32}$$

First proof.   Substituting from (4.30),

$$e_M^J q_{M+1}^J = \frac{C(L-1/M-1)C(L+1/M+1)}{C(L/M)^2} = e_M^{J+1} q_M^{J+1}$$

for $M=2,3,4,\ldots,$ and $J\geqslant 0$. The case for $M=1$ is easily verified explicitly, and one may also treat it using the convention that $C(L/0)\equiv 1$. Likewise one verifies (4.32):

$$e_M^J - q_M^{J+1} = \frac{C(L-1/M-1)\{C(L/M+1)C(L/M-1) \\ \qquad - C(L+1/M)C(L-1/M)\}}{C(L-1/M)C(L/M)C(L/M-1)}$$

$$= \frac{C(L/M)C(L-1/M-1)}{C(L-1/M)C(L/M-1)},$$

$$e_{M-1}^{J+1} - q_M^J = \frac{C(L/M)\{C(L-2/M-1)C(L/M-1) \\ \qquad - C(L-1/M-2)C(L-1/M)\}}{C(L-1/M-1)C(L-1/M)C(L/M-1)}$$

$$= \frac{C(L/M)C(L-1/M-1)}{C(L-1/M)C(L/M-1)} = e_M^J - q_M^{J+1}.$$

Second proof.   We consider the algebraic identity from two sequential continued fraction expansions, namely

$$\sum_{i=J}^{\infty} c_i z^{-i} = \frac{c_J z^{-J}}{1} - \frac{q_1^J z}{1} - \frac{e_1^J z}{1} - \frac{q_2^J z}{1} - \frac{e_2^J z}{1} - \cdots$$

$$= c_J z^{-J} + \frac{c_{J+1} z^{-J+1}}{1} - \frac{q_1^{J+1} z}{1} - \frac{e_1^{J+1} z}{1} - \frac{q_2^{J+1} z}{1} - \cdots. \tag{4.33}$$

We make a contraction, given by (5.3) and explained in the next section, on these fractions, and it follows that

$$c_J z^{-J} + \frac{c_J q_1^J z^{-J+1}}{1-(q_1^J+e_1^J)z} - \frac{e_1^J q_2^J z^{-2}}{1-(q_2^J+e_2^J)z} - \frac{e_2^J q_3^J z^{-2}}{1-(q_3^J+e_3^J)z} - \cdots$$

$$= c_J z^{-J} + \frac{c_{J+1} z^{-J+1}}{1-q_1^{J+1}z} - \frac{q_1^{J+1} e_1^{J+1} z^{-2}}{1-(e_1^{J+1}+q_2^{J+1})z} - \frac{q_2^{J+1} e_2^{J+1} z^{-2}}{1-(e_2^{J+1}+q_3^{J+1})z} - \cdots.$$

$$\tag{4.34}$$

By identifying the coefficients in these expansions, (4.31) and (4.32) are proved.

THE Q.D. ALGORITHM. This algorithm may be used to construct the coefficients q_i^0, e_i^0 in the continued-fraction representation (4.26) of the Padé approximants of a given power series. In this context, the initializing values are taken to be

$$e_0^J = 0, \qquad\qquad J = 1, 2, 3, \ldots,$$
$$q_1^J = c_{J+1}/c_J, \qquad J = 0, 1, 2, \ldots, \qquad\qquad (4.35)$$

as is required by (4.34). Equations (4.31) and (4.32) constitute the body of the algorithm. The elements are normally exhibited in the Q.D. table as shown in Table 1.

Table 1. The Q.D. Table
Its two left columns are specified by (4.35), and the remaining elements are determined by (4.31) and (4.32). The elements along any diagonal are the elements of a continued fraction (4.29).

	q_1^0					
e_0^1		e_1^0				
	q_1^1		q_2^0			
e_0^2		e_1^1		e_2^0		
	q_1^2		q_2^1		q_3^0	
e_0^3		e_1^2		e_2^1		$\ddots$
	q_1^3		q_2^2		$\vdots$	
e_0^4		e_1^3		$\vdots$		
	q_1^4		$\vdots$			
e_0^5		$\vdots$				
	$\vdots$					

[THE Q.D. TABLE OF EXP(z)]. The left-hand column consists of zeros, and the next column contains entries

$$\frac{c_{L+1}}{c_L} = \frac{1}{L+1}$$

(see Table 2). The rules (4.31) and (4.32) connect entries at the vertices of rhombi in the Q.D. table, enabling Table 2 to be completed. We deduce that

Table 2.   Part of the Q.D. Table of $\exp(z)$

1					
0		$-\frac{1}{2}$			
	$\frac{1}{2}$		$\frac{1}{6}$		
0		$-\frac{1}{6}$		$-\frac{1}{6}$	
	$\frac{1}{3}$		$\frac{1}{6}$		$-\frac{1}{10}$
0		$-\frac{1}{12}$		$\frac{1}{10}$	
	$\frac{1}{4}$		$\frac{3}{20}$		
0		$-\frac{1}{20}$			
	$\frac{1}{5}$				
0					

formally

$$\exp(z)=\frac{1}{1}-\frac{z}{1}+\frac{\frac{1}{2}z}{1}-\frac{\frac{1}{6}z}{1}+\frac{\frac{1}{6}z}{1}+\frac{\frac{1}{10}z}{1}+\cdots,$$

which agrees with (6.1a).

One may extend the Q.D. table above its diagonal by defining $q_{i+1}^{-i}=0$ for $i=1,2,3,\ldots$. In this way all the elements of the top row except q_1^0 are defined to be zero, and the rhombus rules allow completion of the table. If the coefficients of the reciprocal series are used by setting $e_{i+1}^i=c_{i+1}'/c_i'$ for $i=0,1,2,\ldots$, in the notation of (1.6.10), and the appropriate order of calculation is followed, then the whole table may be calculated more stably by the progressive form of the Q.D. algorithm described in Section 3.6; otherwise the Q.D. algorithm is notoriously unstable. If $f(z)$ is meromorphic, the "q" columns converge to the reciprocals of the poles of $f(z)$, provided the moduli of the poles are distinct. Similarly the "e" rows converge to the reciprocals of the zeros of $f(z)$, provided they have distinct moduli. This property follows from (4.28) and (3.7.6) et seq. It is instructive to compare this form of the Q.D. algorithm with that described in Section 3.6: they are not identical.

Throughout this section we have assumed that all the inverses we need do in fact exist. It is quite possible that $b_k=0$ for some value of k in (4.1) and so only the first k convergents are defined; convergence of the fraction is meaningless. Likewise the Q.D. algorithm may break down if a zero divisor is encountered. Worse, from a numerical point of view, is the possibility that the computed results are generated by rounding error in a zero or near-zero entry. We refer to Claessens and Wuytack [1979] for an extension of (4.31) and (4.32) to the case of a nonnormal Q.D. table.

Exercise 1.   Prove Theorem 4.4.2.
Exercise 2.   Prove Theorem 4.4.4.
Exercise 3.   Construct the Q.D. table corresponding to the Maclaurin series coefficients of the function $_2F_0(\alpha,1;z)$.

4.5   Different Types of Continued Fractions

4.5.1   *Regular Fractions for Nondegenerate Cases*

The type of continued fractions which are fundamental to the representation of power series are the *regular C-fractions*. These have the form

$$C(z)=b_0+\frac{a_1 z}{1}+\frac{a_2 z}{1}+\frac{a_3 z}{1}+\cdots, \tag{5.1}$$

with $a_i \neq 0$ for $i=1,2,3,\dots$. They may be constructed from a given power series by Viskovatov's method, the Q.D. algorithm, or any other convenient method. They are called C-fractions because they form the given power series, and the regularity condition is that $a_i \neq 0$ for all i. An iterative reexpansion of the convergents of (5.1) shows that the successive convergents correspond to the $[0/0],[1/0],[1/1],[2/1],[2/2],\dots$, sequence of Padé approximants to the given power series as shown in Section 4.2.

If, during the construction of (5.1) from the power series as in (2.1)–(2.4), an a_i is found to be zero, a different representation, such as the general C-fraction (5.11), must be used.

An alternative form of the regular C-fraction has the representation

$$C(z)=\frac{a_0}{1}+\frac{a_1 z}{1}+\frac{a_2 z}{1}+\frac{a_3 z}{1}+\cdots \tag{5.2}$$

with different elements a_i from those in (1); we still require that $a_i \neq 0$ for all i, for regularity. Equation (5.2) corresponds to the $[0/0],[0/1],[1/1],[1/2],[2/2],\dots$ sequence of Padé approximants.

The simple algebraic identity

$$1+\frac{pz}{1+\dfrac{qz}{D}}=1+pz-\frac{pqz^2}{qz+D} \tag{5.3}$$

leads to a *contraction* of the continued fraction. By taking $p_i=a_{2i},\ q_i=a_{2i+1}$, we may contract (5.1) and generate its *associated fraction*:

$$A(z)=b_0+\frac{a_1 z}{1+a_2 z}-\frac{a_2 a_3 z^2}{1+(a_3+a_4)z}-\frac{a_4 a_5 z^2}{1+(a_5+a_6)z}-\cdots. \tag{5.4}$$

The convergents of $A(z)$ are alternate convergents of $C(z)$, and occupy the diagonal of the Padé table in this case. A particular case of the regular C-fraction is the Stieltjes or *S-fraction*, which is

$$s(z)=\frac{a_1}{1}+\frac{a_2 z}{1}+\frac{a_3 z}{1}+\cdots \tag{5.5}$$

with $a_i > 0$, $i = 1, 2, 3, \ldots$. The properties of the convergents and convergence of S-fractions are discussed extensively in the next chapter in the context of Padé approximation of Stieltjes functions. We will see that if the S-fraction converges (e.g. if the divergence condition of Section 4.7 is satisfied), then

$$s(z) = \int_0^\infty \frac{d\phi(u)}{1 + zu}, \qquad |\arg(z)| < \pi, \tag{5.6}$$

where $\phi(t)$ is a bounded and nondecreasing function defined on $0 \leq u < \infty$. Using the variable $\omega = z^{-1}$, (5.6) is frequently expressed in the form

$$S(\omega) = zs(z) = \frac{a_1}{\omega} + \frac{a_2}{1} + \frac{a_3}{\omega} + \frac{a_4}{1} + \cdots \tag{5.7}$$

which is generated by a simple equivalence transformation, and the theory of Sections 5.5 and 5.6 shows that

$$S(\omega) = \int_0^\infty \frac{d\phi(u)}{\omega + u}, \qquad |\arg(\omega)| < \pi. \tag{5.8}$$

Continued fractions of the type

$$J(\omega) = \frac{k_1}{l_1 + \omega} - \frac{k_2}{l_2 + \omega} - \frac{k_3}{l_3 + \omega} - \cdots \tag{5.9}$$

in which $k_i \neq 0$, $i = 1, 2, 3, \ldots$, are called J-fractions. If $k_i > 0$ and l_i are real for $i = 1, 2, 3, \ldots$, then (5.9) is called a real J-fraction [Wall, 1931, 1932a, b, 1948]. Such a fraction may be derived from (5.7) by the contraction formula (5.3), with the identifications

$$k_1 = a_1, \qquad l_1 = a_2,$$

$$\left. \begin{array}{l} k_i = a_{2i-2} a_{2i-1} \\ l_i = a_{2i-1} + a_{2i} \end{array} \right\} \qquad \text{for} \quad i = 2, 3, 4, \ldots .$$

Thus we see that the convergents of (5.9) correspond to alternate convergents of (5.7). We will see in Section 5.6 that any convergent real J-fraction has a representation

$$J(\omega) = \int_{-\infty}^\infty \frac{d\psi(u)}{\omega + u}, \qquad \text{Im}\,\omega \neq 0 \tag{5.10}$$

where $\psi(u)$ is a bounded and nondecreasing function defined on $-\infty < u < \infty$. Equation (5.8) is a special case of (5.10) when $\psi(u)$ is constant on $-\infty < u \leq 0$.

4.5.2 *General Fractions for Degenerate Cases*

Next we consider the situation when Viskovatov's method breaks down and the series cannot be represented as a regular C-fraction. The algorithm leaves us with a representation such as

$$b_0 + \cfrac{a_1 z}{1 + \cfrac{a_2 z}{1 + \cfrac{a_3 z}{1 + \cfrac{\ddots}{1 + \cfrac{a_{n-1} z}{1 + c_2^{(n)} z^2 + c_3^{(n)} z^3 + \cdots}}}}} \tag{5.11}$$

with $c_1^{(n)} = 0$. Clearly, if $c_2^{(n)} \neq 0$, a correct procedure is to allow a numerator equal to $c_2^{(n)} z^2$ and continue to develop the fraction. This procedure defines the general corresponding fraction, or general C-fraction [Leighton and Scott, 1939; Scott and Wall, 1940a, b], represented by

$$C(z) = b_0 + \frac{a_1 z^{\alpha_1}}{1} + \frac{a_2 z^{\alpha_2}}{1} + \frac{a_3 z^{\alpha_3}}{1} + \cdots \tag{5.12}$$

with $a_i \neq 0$ and $\alpha_i \geq 1$ for all i. Such fractions need not occupy a simple staircase sequence in the Padé table. Consider the series

$$f(z) = 1 + z^2 - z^5 + \lambda z^6 + \cdots \tag{5.13}$$

The third convergent of (5.12) corresponding to this function is

$$1 + \cfrac{z^2}{1 + z^3} = \frac{1 + z^2 + z^3}{1 + z^3}, \tag{5.14}$$

whereas the [3/3] Padé approximant of $f(z)$ is

$$[3/3] = 1 + \frac{z^2(1 + \lambda z)}{1 + \lambda z + z^3} = \frac{1 + \lambda z + z^2 + (1 + \lambda) z^3}{1 + \lambda z + z^3}. \tag{5.15}$$

The first and second convergents of $f(z)$ are the [0/0] and [2/0] Padé approximants, but the third convergent (5.14) is not a Padé approximant unless $\lambda = 0$. In the case $\lambda \neq 0$, the fourth convergent of $f(z)$ is

$$1 + \frac{z^2}{1} + \frac{z^3}{1 + \lambda z},$$

which is the [3/3] Padé approximant precisely.

Many examples of functions having natural boundaries are expressed as series expansions $\sum_{i=0}^{\infty} c_i z^i$ with Hadamard gaps, such as (2.2.4). We say that the expansion has a gap (n_i, n_i') if $c_j = 0$ for all j in $n_i \leqslant j \leqslant n_i'$. The series is said to have Hadamard gaps if, for some $\lambda > 0$, the series has an infinity of gaps (n_i, n_i') such that $n_i / n_i' \geqslant 1 + \lambda$. In this context, an interesting example of a general C-fraction is Ramanujan's fraction

$$R(z) = \frac{1}{1} + \frac{z}{1} + \frac{z^2}{1} + \frac{z^3}{1} + \cdots,$$

which has a natural boundary on $|z| = 1$ although its power series does not have the Hadamard gaps.

The orderly relation between continued fractions and the Padé table is restored by P-fractions [Magnus, 1962]. P stands for *principal part plus*, and all convergents of P-fractions lie in the Padé table. Using the variable $\omega = z^{-1}$, a P-fraction is

$$P(z) = c_0 + \frac{1}{b_1(\omega)} + \frac{1}{b_2(\omega)} + \frac{1}{b_3(\omega)} + \cdots, \tag{5.16}$$

where each $b_i(\omega)$ is a polynomial in ω of degree N_i precisely. For example, the third convergent of the P-fraction of $f(z)$ given by (5.13) is

$$1 + \frac{1}{z^{-2}} + \frac{1}{z^{-1} + \lambda}.$$

This is the $[3/3]$ Padé approximant of $f(z)$. In fact the recurrence relations (4.4) for A_i and B_i and the accuracy-through-order conditions verify that the ith convergent of (5.15) is a diagonal $[L_i / L_i]$ Padé approximant, where $L_i = \sum_{j=1}^{i-1} N_j$.

Further, by replacing c_0 in (5.15) with $b_0(\omega)$, another polynomial, P-fractions of functions $f(z)$ having a Laurent series with a finite principal part are directly defined. Magnus defined P-fractions in this way so that every entry in the Padé table may be uniquely associated with a P-fraction of $z^s f(z)$ for some integer s.

4.5.3   Viskovatov's Algorithm for the General Case

In order to generalize Viskovatov's algorithm to degenerate cases, one must make an inspired choice for the representation of the particular $[L/M]$ Padé approximant as a continued fraction. We consider

$$R^{(0)}(z) = p_0(z) + \frac{z^{\alpha_0}}{p_1(z)} + \frac{z^{\alpha_1}}{p_2(z)} + \cdots + \frac{z^{\alpha_{n-1}}}{p_n(z)}, \tag{5.17}$$

which should be compared with the general C-fraction (5.12) and Magnus P-fraction (5.16). In (5.17), each $p_i(z)$ is a polynomial with $p_i(0) \neq 0$ and each α_i is a positive integer: each $p_i(z)$ and α_i are fully defined in (5.21) and (5.22). As usual, we suppose that we are given the series expansion

$$f(z) = \sum_{i=0}^{\infty} c_i z^i \qquad (5.18)$$

and that we are requested to find the $[L/M]$ Padé approximant of (5.18) with $L \geqslant M$. If $L < M$, we should remove the factor z^k from (5.18), where k is the highest power such that z^k factors (5.18), and start with its reciprocal series. We choose to consider an infinite series in (5.18) for convenience of representation; in fact only $L + M + 1$ terms of (5.18) are required, and questions of convergence do not arise. Following Werner [1979, 1980], we define the generalized Viskovatov reduction:

Initialization.   Let

$$l_0 = L, \qquad m_0 = M, \qquad (5.19)$$

and

$$f(z) = \sum_{i=0}^{\infty} c_i z^i = \sum_{i=0}^{\infty} c_i^{(0)} z^i. \qquad (5.20)$$

Iteration.   For $j = 0, 1, 2, \ldots$, proceed recursively. Define α_j to be the least integer for which

$$\alpha_j > l_j - m_j \quad \text{and} \quad c_{\alpha_j}^{(j)} \neq 0. \qquad (5.21)$$

We may then define the polynomial $p_j(z)$ of degree strictly less than α_j by

$$\sum_{i=0}^{\infty} c_i^{(j)} z^i = p_j(z) + z^{\alpha_j} \sum_{i=\alpha_j}^{\infty} c_i^{(j)} z^{i - \alpha_j}. \qquad (5.22)$$

Our aim is to find the $[l_j/m_j]$ Padé approximant of the left-hand side of (5.22). If $\alpha_j > l_j + m_j$, then let $n = j$. In this case, (5.17) is the Padé approximant required, as is proved in Theorem 4.5.1. If

$$l_j < \alpha_j \leqslant l_j + m_j, \qquad (5.23)$$

the $[L/M]$ Padé approximant to the given series (5.18) does not exist, as is proved in Theorem 4.5.1.

Next, we consider the reciprocal series defined by

$$\sum_{i=0}^{\infty} c_i^{(j+1)} z^i = \left[\sum_{i=\alpha_j}^{\infty} c_i^{(j)} z^{i-\alpha_j} \right]^{-1}, \tag{5.24}$$

so that, formally at least,

$$\sum_{i=0}^{\infty} c_i^{(j)} z^i = p_j(z) + \frac{z^{\alpha_j}}{\displaystyle\sum_{i=0}^{\infty} c_i^{(j+1)} z^i}. \tag{5.25}$$

In order to find the $[l_j/m_j]$ Padé approximant of the left-hand side of (5.25), we seek the $[m_j/l_j - \alpha_j]$ Padé approximant of the left-hand side of (5.24). Hence we define

$$l_{j+1} = m_j \quad \text{and} \quad m_{j+1} = l_j - \alpha_j. \tag{5.26}$$

The requirement that $l_{j+1} \geqslant m_{j+1}$ is guaranteed by (5.21), and also $m_{j+1} \geqslant 0$ in view of (5.23). Hence we need to find the $[l_{j+1}/m_{j+1}]$ Padé approximant of the series $\sum_{i=0}^{\infty} c_i^{(j+1)} z^i$ defined by (5.24) and (5.26), and the process becomes recursive.

Having defined the iterative process for the construction of (5.17), we must verify that it fulfills its specification. From (5.26), we note that

$$l_{j+1} + m_{j+1} = l_j + m_j - \alpha_j \tag{5.27}$$

and therefore

$$L + M = \sum_{i=0}^{n-1} \alpha_i + \deg\{ p_n(z)\}. \tag{5.28}$$

where $\deg\{ p(z)\}$ is the degree of $p(z)$. We see that the fraction (5.17) has at most $L + M$ divisors, which occurs if $\alpha_i = 1$ for $i = 0, 1, 2, \ldots, n$, and hence the process terminates.

THEOREM 4.5.1. *The generalized Viskovatov reduction defined in (5.19)–(5.26) is reliable: if the $[L/M]$ Padé approximant exists, the process finds it; if not, (5.23) is satisfied and indicates this fact.*

Proof. Suppose that the $[L/M]$ approximant to (5.20) exists, yet the algorithm fails. Let the approximant have the Maclaurin expansion

$$[L/M]_f(z) = \sum_{i=0}^{\infty} c_i^{(0)'} z^i, \tag{5.29}$$

so that

$$c_i^{(0)\prime} = c_i^{(0)} \qquad \text{for} \quad i = 0, 1, \ldots, L + M. \tag{5.30}$$

Since the reduction process (5.25) for calculating the $[L/M]$ approximant is based on the series-expansion coefficients only, the representation (5.17) derived by the reduction is identical when applied to $\sum_{i=0}^{L+M} c_i z^i$ and $[L/M]$ itself. Initially, with $L \geqslant M$, we find

$$[L/M] \equiv \frac{P^{(0)}(z)}{Q^{(0)}(z)} = p_0(z) + z^{\alpha_0} \frac{T^{(0)}(z)}{Q^{(0)}(z)}. \tag{5.31}$$

This division of $P^{(0)}(z)$ by $Q^{(0)}(z)$ yields $p_0(z)$ as the quotient and $z^{\alpha_0} T^{(0)}(z)$ as the remainder. Unless $p_0(z)$ itself is the $[l_0/m_0]$ Padé approximant of $f(z)$, (5.21) ensures that $T^{(0)}(0) \neq 0$, and so we may use the Maclaurin expansion

$$\frac{Q^{(0)}(z)}{T^{(0)}(z)} = \sum_{i=0}^{\infty} c_i^{(1)} z^i + O(z^{L+M+1-\alpha_0}). \tag{5.32}$$

We suppose that $P_0(z)/Q_0(z)$ in (5.31) is reduced sequentially by the process indicated by (5.19)–(5.26) as far as possible. Let

$$\frac{P^{(j)}(z)}{Q^{(j)}(z)} = p_j(z) + z^{\alpha_j} \frac{T^{(j)}(z)}{Q^{(j)}(z)} \tag{5.33}$$

for $j = 0, 1, \ldots, k$. Hence we reduce the Padé approximant sequentially by

$$\frac{P^{(j)}(z)}{Q^{(j)}(z)} = p_j(z) + \frac{z^{\alpha_j}}{P^{(j+1)}(z)/Q^{(j+1)}(z)}, \tag{5.34}$$

where we define

$$P^{(j+1)}(z) = Q^{(j)}(z) \tag{5.35}$$

and

$$Q^{(j+1)}(z) = T^{(j)}(z). \tag{5.36}$$

Suppose that the process fails at stage k, namely because (5.23) is satisfied with $j = k$. From (5.33),

$$P^{(k)}(z) - Q^{(k)}(z) p_k(z) = z^{\alpha_k} T^{(k)}(z). \tag{5.37}$$

where $l_k > \alpha_k > l_k + m_k$. Because the left-hand side of (5.37) has degree at most l_k, then either $\alpha_k \leqslant l_k$ or else $T^{(k)}(z) \equiv 0$. This is incompatible with a failure at stage k. Consequently the reduction process cannot fail if the $[L/M]$ Padé approximant exists.

The proof of the converse result, that $R^{(0)}(z)$ constructed by (5.17)–(5.26) is in fact the $[L/M]$ Padé approximant, uses the definitions of these equations. From (5.22) and (5.28), we see that

$$p_n(z) = \sum_{i=0}^{\beta} c_i^{(n)} z^i, \tag{5.38}$$

where

$$\beta = L + M - \sum_{i=0}^{n} \alpha_i. \tag{5.39}$$

We define $R^{(n)}(z) = p_n(z)$, and for $j = 0, 1, \ldots, n-1$ we define

$$R^{(j)}(z) = p_j(z) + \frac{z^{\alpha_j}}{p_{j+1}(z)} + \frac{z^{\alpha_{j+1}}}{p_{j+2}(z)} + \cdots + \frac{z^{\alpha_{n-1}}}{p_n(z)}. \tag{5.40}$$

We see directly from (5.40) that

$$R^{(j)}(z) = p_j(z) + \frac{z^{\alpha_j}}{R^{(j+1)}(z)} \tag{5.41}$$

for $j = 0, 1, \ldots, n-1$.

Let us suppose that

$$R^{(k)}(z) = \sum_{i=0}^{\infty} c_i^{(k)} z^i + O\left(z^{L+M+1 - \sum_{i=0}^{k} \alpha_i}\right). \tag{5.42}$$

It then follows from (5.25) and (5.41) that (5.42) is true with $k \to k-1$. Using (5.39), we have proved that

$$R^{(0)}(z) = \sum_{i=0}^{\infty} c_i^{(0)} z^i + O\left(z^{L+M+1}\right) \tag{5.43}$$

by induction. This completes the proof of the theorem.

In the course of the proof we can see the connection between (5.31), (5.33), and the Euclidean algorithm. The connection is direct using the variable $w = z^{-1}$, but it is obvious that each common factor of $P^{(j)}(z), Q^{(j)}(z)$ is also a factor of $T^{(j)}(z)$ for $j = 0, 1, \ldots, n$. Hence we see that we may view the reduction process and Theorem 4.5.1 as a proof, using the Euclidean algorithm, that $P^{(0)}(z)$ and $Q^{(0)}(z)$ have no common factors. Of course, the proof in Theorem 1.4.3 is more direct.

There is also a connection between (5.33) and the recurrence relations (2.14). From (5.33), we find that

$$P^{(j)}(z)=p_j(z)Q^{(j)}(z)+z^{\alpha_j}T^{(j)}(z). \tag{5.44}$$

for $j=0, 1,\ldots, n$. We may suppose that the relative normalization $P_j(z):Q_j(z)$ is fixed by defining

$$P^{(n)}(z)=p_n(z) \quad \text{and} \quad Q^{(n)}(z)=1. \tag{5.45}$$

From (5.35), (5.36), and (5.44), it follows that

$$P^{(j-1)}(z)=p_j(z)P^{(j)}(z)+z^{\alpha_j}P^{(j+1)}(z) \tag{5.46}$$

and

$$Q^{(j-1)}(z)=p_j(z)Q^{(j)}(z)+z^{\alpha_j}Q^{(j+1)}(z) \tag{5.47}$$

which are remarkably similar to (5.2.14) when viewed as a recurrence for $j=n-1, n-2,\ldots, 1$. In view of (5.34), (5.41), and (5.45), the recurrence defines the numerator and denominator of

$$R^{(j)}(z)=\frac{P^{(j)}(z)}{Q^{(j)}(z)}=p_j(z)+\frac{z^{\alpha_j}}{P_{j+1}(z)}+\frac{z^{\alpha_{j+1}}}{P_{j+2}(z)}+\cdots+\frac{z^{\alpha_{n-1}}}{P_n(z)}.$$

However, the recurrence

$$\begin{pmatrix} P^{(j-1)}(z) \\ Q^{(j-1)}(z) \end{pmatrix} = \begin{pmatrix} p_j(z) & z^{\alpha_j} \\ 1 & 0 \end{pmatrix}\begin{pmatrix} P^{(j)}(z) \\ Q^{(j)}(z) \end{pmatrix} \tag{5.48}$$

for $j=n, n-1,\ldots, 1$, initialized by (5.45), is probably the more convenient form.

It is now clear how to define a modified Viskovatov algorithm, based on (5.17)–(5.26) and incorporating a generalization of (3.1) for efficiency. In fact, we may as well consider the problem of finding an $[L/M]$ Padé approximant, with $L\geqslant M$, for

$$f(z)=\frac{g(z)}{d(z)}=\frac{\displaystyle\sum_{i=0}^{\infty} g_i^{(0)}z^i}{\displaystyle\sum_{i=0}^{\infty} d_i^{(0)}z^i}. \tag{5.49}$$

In (5.49), the coefficients $g_i^{(0)}$, $d_i^{(0)}$ for $i=0,1,2,\ldots,L+M$ are given, and $d_0^{(0)}\neq 0$. This problem is solved using bigradients in Section 1.6. We outline a generalized Viskovatov algorithm, emphasizing only the alterations to (5.17)–(5.26) [Bultheel, 1980b].

Initialization.   Define $l_0=L$, $m_0=M$.

Recurrence.   The process is iterative for $j=0,1,2,\ldots$. We construct the polynomial $p_j(z)$ of degree at most l_j-m_j such that

$$\sum_{i=0}^{\infty} g_i^{(j)}z^i -p_j(z) \sum_{i=0}^{\infty} d_i^{(j)}z^i = z^{\alpha_j} \sum_{i=0}^{\infty} h_i^{(j)}z^i \tag{5.50}$$

with $\alpha_j>l_j-m_j$. In fact, we need to construct the coefficients of

$$p_j(z)= \sum_{i=0}^{l_j-m_j} p_i^{(j)}z^i \tag{5.51}$$

and $L+M+1-\sum_{i=0}^{j}\alpha_i$ coefficients of the series

$$H^{(j)}(z)= \sum_{i=0}^{\infty} h_i^{(j)}z^i. \tag{5.52}$$

The construction of the coefficients of (5.51) and (5.52) is performed iteratively, according to the algebra implied by the following analysis. Let

$$S^{(j,0)}(z)= \sum_{i=0}^{\infty} g_i^{(j)}z^i.$$

For $k=0,1,\ldots,l_j-m_j$, define

$$p_k^{(j)} =S^{(j,k)}(0)$$

and

$$S^{(j,k+1)}(z)=\left[S^{(j,k)}(z)-p_k^{(j)} \sum_{i=0}^{\infty} d_i^{(j)}z^i\right]z^{-1}.$$

Hence $p_j(z)$ is specified by (5.51), and for the coefficients of (5.52) we have

$$z^{\alpha_j} \sum_{i=0}^{\infty} h_i^{(j)}z^i =z^{l_j-m_j+1}S^{(j,l_j-m_j+1)}(z). \tag{5.53}$$

We define α_j as the least integer such that (5.53) holds with $h_0^{(j)}\neq 0$, and hence the coefficients of (5.52) are fully specified.

If $\alpha_j > l_j + m_j$, or if (5.53) vanishes exactly, the iteration is terminated with $n = j$.

If $l_j < \alpha_j \le l_j + m_j$, the Padé approximant required is nonexistent. Otherwise, let

$$g_i^{(j+1)} = d_i^{(j)} \quad \text{and} \quad d_i^{(j+1)} = h_i^{(j)}$$

for $i = 0, 1, \ldots, l_j + m_j - \alpha_j$. In this case,

$$\frac{\sum\limits_{i=0}^{\infty} g_i^{(j)} z^i}{\sum\limits_{i=0}^{\infty} d_i^{(j)} z^i} = p_j(z) + \cfrac{z^{\alpha_j}}{\sum\limits_{i=0}^{\infty} g_i^{(j+1)} z^i \Big/ \sum\limits_{i=0}^{\infty} d_i^{(j+1)} z^i} \tag{5.54}$$

By defining $l_{j+1} = m_j$ and $m_{j+1} = l_j - \alpha_j$, the jth stage of the reduction process (5.54) may be iterated. The algorithm is now fully specified.

The generalized Viskovatov algorithm in (5.49)–(5.54) is designed to yield an $[L/M]$ Padé approximant reliably. It follows a path on the Padé table which avoids nonexistent entries in blocks in the table. We conclude that the path that the convergents of the fraction (5.17) follows in the Padé table depends on the values of the coefficients in (5.49) as well as the values of L and M. As a computational algorithm, it can be recommended wherever exact arithmetic is available and the values of $\alpha_0, \alpha_1, \ldots, \alpha_{n-1}$ are unambiguously defined. The techniques of the generalized Viskovatov algorithm can be adapted to rational interpolation using a generalized Thiele fraction (see Part II, Section 1.1, and Werner [1979a, b]) except that the algorithm must incorporate a facility for reordering the interpolation points if necessary [Thacher and Tukey, 1960; Graves-Morris and Hopkins, 1978; Graves-Morris, 1980a]. Similar techniques based on the Euclidean algorithm allow Kronecker's algorithm to be generalized to avoid inessential degeneracies (see Section 2.4; Graves-Morris [1980a]).

4.5.4 Continued Fractions for Special Cases

Euler used a method of writing an equivalent continued fraction for Maclaurin series. This fraction has convergents which reduce to truncated Maclaurin series.

A short calculation with the recurrence relations (4.4) reveals that

$$\sum_{i=0}^{n} c_i z^i = c_0 + \cfrac{c_1 z}{1} \underset{-}{} \cfrac{c_2 z}{c_1 + c_2 z} \underset{-}{} \cfrac{c_1 c_3 z}{c_2 + c_3 z} \underset{-}{} \cdots \underset{-}{} \cfrac{c_n c_{n-2} z}{c_{n-1} + c_n z}. \tag{5.55}$$

Needless to say, the convergence of the continued fraction can be no different from that of the original series.

Thron's general *T*-fraction [Thron, 1948] is a useful modern development. It is given by

$$T(z) = e_0 + d_0 z + \frac{z}{e_1 + d_1 z} + \frac{z}{e_2 + d_2 z} + \frac{z}{e_3 + d_3 z} + \cdots \quad (5.56)$$

with all $e_i \neq 0$. This fraction is designed to be able to match the Maclaurin series of a function, and its asymptotic expansion expressed as a power series in z^{-1}. It is often called a two-point Padé approximant. If all the $e_i = 1$ and all the $d_i > 0$, the *T*-fraction has the integral representation

$$T_1(z) = 1 + d_0 z + z \int_0^\infty \frac{d\phi(t)}{z + t} \quad (5.57)$$

where $\phi(t)$ is bounded and nondecreasing on $[0, \infty)$. These *T*-fractions are further discussed in Part II, Section 1.1. A solution for the rational interpolation problem may always be expressed in the form of a convergent of a continued fraction interpolant. These ideas are explained in Part II, Section 1.1.

Exercise 1.   Verify Equation (5.7).
Exercise 2.   Use the results of Section 4.3, Exercise 1, to verify that

$$\frac{1}{\sqrt{1 + z^2}} = \frac{1}{1} + \frac{z^2}{2} + \frac{z^2}{2} + \frac{z^2}{2} + \cdots \quad \text{(i)}$$

$$= \frac{1}{z} + \frac{1}{2z} + \frac{1}{2z} + \frac{1}{2z} + \cdots \quad \text{(ii)}$$

$$= \frac{1}{1+z} - \frac{z}{1+z} - \frac{z}{2(1+z)} - \frac{z}{1+z} - \frac{z}{2(1+z)} - \cdots \quad \text{(iii)}$$

Expansion (i) is designed to be accurate for small $|z|$, (ii) for large $|z|$, and (iii) for all $|z|$.

Show that

(i) converges for all z except on cuts from $\pm i$ to $\pm i\infty$,
(ii) converges for all z except on a cut from i to $-i$,
(iii) converges for all z except on the semicircle $|z| = 1$, $\mathrm{Re}\, z \leq 0$.

4.6   Examples of Continued Fractions Which are Padé Approximants

We present here some examples of continued fractions which are also Padé approximants. The examples are either staircase or diagonal sequences in the Padé table, obtained from *J*-fractions or *S*-fractions. We do not give continued fractions which are merely Euler's corresponding fractions whose convergents are identical to truncated Maclaurin series. We quote a quite comprehensive set of formulas for the functions which are known to have useful continued-fraction expansions. We conclude the section with their formal algebraic derivation, which consists of showing that the Maclaurin series of each continued fraction is the same as that of the given function. The question of convergence is left to Section 4.7.

Exponential function

$$\exp(z) = \frac{1}{1} - \frac{z}{1} + \frac{z}{2} - \frac{z}{3} + \frac{z}{2} - \cdots + \frac{z}{2} - \frac{z}{2n+1} + \cdots \tag{6.1a}$$

$$= 1 + \frac{z}{1} - \frac{z}{2} + \frac{z}{3} - \frac{z}{2} + \frac{z}{5} - \frac{z}{2} + \cdots - \frac{z}{2} + \frac{z}{2n+1} - \cdots \tag{6.1b}$$

$$= 1 + \frac{z}{1 - z/2} + \frac{z^2/(4 \times 3)}{1} + \frac{z^2(4 \times 15)}{1}$$
$$+ \frac{z^2(4 \times 35)}{1} + \cdots + \frac{z^2/(4(4n^2 - 1))}{1} + \cdots. \tag{6.1c}$$

These expansions converge for all z.

Tangent function

$$\tan z = \frac{z}{1} - \frac{z^2}{3} - \frac{z^2}{5} - \frac{z^2}{7} - \cdots - \frac{z^2}{2n+1} - \cdots. \tag{6.2}$$

This converges for all z except $z = (2n+1)\pi/2$, n integral.

Hyperbolic tangent

$$\tanh z = \frac{z}{1} + \frac{z^2}{3} + \frac{z^2}{5} + \frac{z^2}{7} + \cdots + \frac{z^2}{2n+1} + \cdots. \tag{6.3}$$

This converges for all z except $z = (2n+1)i\pi/2$, n integral.

Binomial function

$$(1+z)^\nu = 1 + \frac{\nu z}{1} + \frac{(1-\nu)z}{2} + \frac{(1+\nu)z}{3} + \frac{(2-\nu)z}{2} + \cdots$$
$$+ \frac{(n-\nu)z}{2} + \frac{(n+\nu)z}{2n+1} + \cdots. \tag{6.4}$$

This converges for all z except $-\infty < z \leq 1$, unless ν is integral, when the continued fraction terminates and the result is exact.

Inverse tangent

$$\tan^{-1} z = \frac{z}{1} + \frac{1 \times z^2}{3} + \frac{4 \times z^2}{5} + \cdots + \frac{n^2 z^2}{2n+1} + \cdots. \tag{6.5}$$

This converges for all z in the z-plane cut from i to $i\infty$ and from $-i$ to $-i\infty$.

Inverse hyperbolic tangent

$$\tanh^{-1}(z)=\tfrac{1}{2}\ln\frac{1+z}{1-z}=\frac{z}{1}-\frac{1\times z^2}{3}-\frac{4\times z^2}{5}-\cdots-\frac{n^2z^2}{2n+1}-\cdots. \quad (6.6)$$

This fraction converges in the whole z-plane cut by $(-\infty,-1]$ and $[1,+\infty)$.

Natural logarithm

$$\ln(1+z)=\frac{z}{1}+\frac{1^2z}{2}+\frac{1^2z}{3}+\frac{2^2z}{4}+\frac{2^2z}{5}+\cdots+\frac{n^2z}{2n}+\frac{n^2z}{2n+1}+\cdots. \quad (6.7)$$

This fraction converges for all z except $-\infty<z\leqslant-1$.

Exponential integral

$$E_n(z)=\int_1^\infty \frac{e^{-zt}}{t^n}\,dt$$

$$=e^{-z}\left(\frac{1}{z+n}-\frac{n}{z+n+2}-\frac{2(n+1)}{z+n+4}-\cdots-\frac{(r+1)(n+r)}{z+n+2r+2}-\cdots\right).$$
$$(6.8)$$

This is valid in the entire z-plane cut along $-\infty<z\leqslant0$; the integral representation is only valid for $\mathrm{Re}\,z>0$. See Exercise 1.

Complementary error function

$$\mathrm{erfc}(z)=1-\mathrm{erf}(z)=\frac{2}{\sqrt{\pi}}\int_z^\infty e^{-t^2}\,dt$$

$$=\frac{e^{-z^2}}{\sqrt{\pi}}\left(\frac{1}{z}+\frac{1/2}{z}+\frac{1}{z}+\frac{3/2}{z}+\frac{2}{z}+\cdots+\frac{n/2}{z}+\cdots\right). \quad (6.9)$$

This converges for $\mathrm{Re}\,z>0$.

Prym's incomplete Gamma function

$$\Gamma(a,z)=\int_z^\infty t^{a-1}e^{-t}\,dt$$

$$=e^{-z}z^a\left(\frac{1}{z+1+a}-\frac{1-a}{z+3-a}-\frac{2(2-a)}{z+5-a}-\cdots-\frac{n(n-a)}{z+2n+1-a}-\cdots\right).$$
$$(6.10)$$

This is valid for all z except in $-\infty<z\leqslant0$. If a is a positive integer, the fraction terminates and so the representation is valid for all z. The connection with (6.8) is that $E_n(z)=z^{n-1}\Gamma(1-n,z)$.

Error function

$$\mathrm{erf}(z) = \frac{2}{\sqrt{\pi}} \int_0^z e^{-t^2}\, dt$$

$$= \frac{2ze^{-z^2}}{\sqrt{\pi}} \left(\frac{1}{1} - \frac{2z^2}{3} + \frac{4z^2}{5} - \frac{6z^2}{7} + \cdots + \frac{4nz^2}{4n+1} - \frac{(4n+2)z^2}{4n+3} + \cdots \right).$$

(6.11)

This converges for all z. However, convergence is not fast for $\mathrm{Re}\, z \gg 2$, and $\mathrm{erfc}(z)$ and its continued fraction (6.10) are more useful in such applications. The relation with Dawson's integral, $e^{-z^2}\int_0^z e^{t^2}\, dt$, is given in (II.1.1.44), and T-fraction representations are given by (II.1.1.45d).

Incomplete Gamma function

$$\gamma(a, z) = \int_0^z t^{a-1} e^{-t}\, dt$$

$$= z^a e^{-z} \left(\frac{1}{a} - \frac{az}{a+1} + \frac{z}{a+2} - \frac{(a+1)z}{a+3} + \frac{2z}{a+4} - \cdots \right.$$

$$\left. + \frac{nz}{a+2n} - \frac{(a+n)z}{a+2n+1} + \cdots \right), \tag{6.12}$$

where a is not a negative integer or zero. If a is a strictly positive integer, this converges for all z. If a is not integral, the continued fraction in (6.12) converges, but $\gamma(a, z)$ is only defined in the z-plane cut by $-\infty < z \leqslant 0$. A T-fraction representation is given by (I.1.1.45a), and an error formula is given by Luke [1975].

Definition of hypergeometric functions. We use the hypergeometric function $_pF_q(a_1, a_2, \ldots, a_p, b_1, b_2, \ldots, b_q; z)$ with p numerator parameters and q denominator parameters. The examples

$$_0F_1(a; z) = 1 + \frac{z}{a} + \frac{z^2}{a(a+1)2!} + \cdots + \frac{z^n}{a(a+1)\cdots(a+n-1)n!} + \cdots,$$

$$_2F_1(a, b; z) = 1 + \frac{abz}{c} + \frac{a(a+1)b(b+1)z^2}{c(c+1)2!} + \cdots$$

$$+ z^n \frac{a(a+1)\cdots(a+n-1)b(b+1)\cdots(b+n-1)}{c(c+1)\cdots(c+n-1)n!} + \cdots,$$

$$_2F_0(a, b; z) = 1 + abz + \frac{a(a+1)b(b+1)z^2}{2!} + \cdots$$

$$+ z^n \frac{a(a+1)\cdots(a+n-1)b(b+1)\cdots(b+n-1)}{n!} + \cdots$$

make the definition clear. The definitions are valid formally provided the denominator parameters are not negative integers or zero. Notice that $_0F_1(a; z)$ is an entire function, $_2F_1(a, g, c; z)$ is analytic in the z-plane cut by $1 \leqslant z < \infty$, and $_2F_0(a, b; z)$ has a purely formal definition, since the radius of convergence of the series is zero. The given expansion is a formal expansion of

$$_2F_0(a, b; z) = \frac{1}{\Gamma(a)} \int_0^\infty \frac{e^{-t}t^{a-1}}{(1-zt)^b}\, dt,$$

which is the proper definition, valid for $\operatorname{Re} a > 0$ and z not on the positive real axis.

$_0F_1$ *hypergeometric-function relation*

$$\frac{_0F_1(a+1; z)}{_0F_1(a; z)} = \frac{a}{a} + \frac{z}{a+1} + \frac{z}{a+2} + \frac{z}{a+3} + \cdots + \frac{z}{a+n} + \cdots. \quad (6.13a)$$

This converges for all z not a zero of $_0F_1(a; z)$. A relation for Bessel functions follows from the formula relating $_0F_1$ hypergeometric functions to Bessel functions,

$$J_\nu(z) = \frac{\left(\frac{1}{2}z\right)^\nu {}_0F_1\left(1+\nu, -\frac{1}{4}z^2\right)}{\Gamma(\nu+1)}.$$

We deduce that

$$\frac{J_\nu(z)}{J_{\nu-1}(z)} = \frac{z}{2\nu} - \frac{z^2}{2(\nu+1)} - \frac{z^2}{2(\nu+2)} - \cdots - \frac{z^2}{2(\nu+n)} - \cdots. \quad (6.13b)$$

Confluent-hypergeometric-function relation

$$\frac{_1F_1(a+1, b+1; z)}{_1F_1(a, b; z)} = \frac{b}{b} - \frac{(b-a)z}{b+1} + \frac{(a+1)z}{b+2} - \frac{(b-a+1)z}{b+3} + \cdots$$

$$+ \frac{(a+n)z}{b+2n} - \frac{(b-a+n)z}{b+2n+1} + \cdots. \quad (6.14)$$

This converges for all z except for the zeros of $_1F_1(a, b; z)$.

$_2F_0$ *hypergeometric-function relation*

$$\frac{_2F_0(a, b+1; z)}{_2F_0(a, b; z)} = \frac{1}{1} - \frac{az}{1} - \frac{(b+1)z}{1} - \frac{(a+1)z}{1} - \frac{(b+2)z}{1} - \cdots$$

$$- \frac{(b+n)z}{1} - \frac{(a+n)z}{1} - \cdots. \quad (6.15)$$

This converges in the cut z plane except in the cut $0 \leqslant z < \infty$.

Hypergeometric-function relation [Gauss, 1813]

$$\frac{{}_2F_1(a,b+1,c+1;z)}{{}_2F_1(a,b,c;z)} = \frac{c}{c} - \frac{a(c-b)z}{c+1} - \frac{(b+1)(c-a+1)z}{c+2} - \cdots$$

$$- \frac{(a+n)(c-b+n)z}{c+2n+1} - \frac{(b+n+1)(c-a+n+1)z}{c+2n+2} - \cdots$$

$$(6.16)$$

This converges in the cut z-plane except on the cut $1 \leqslant z < \infty$, and except for the zeros of ${}_2F_1(a,b,c;z)$.

Other continued-fraction developments which are Padé approximants for special functions are known [Wall, 1945, p. 369]: they mostly involve integrals of hyperbolic and elliptic functions.

The derivation of each of the preceding formulas (6.1)–(6.16) consists of both an algebraic and an analytical part. First we show that the Maclaurin series of the two sides of the equations agree term by term. Since the results (6.1)–(6.12) are corollaries of (6.13), (6.14), and (6.15), we discuss these cases first.

Proof of (6.1).   Take $a=0$ in (6.14). ${}_1F_1(0,b;z)=1$ by definition. This step is used in most of the corollaries. Accordingly, we find that

$$ {}_1F_1(1,b+1;z) = \frac{1}{1} - \frac{z}{b+1} + \frac{1 \times z}{b+2} - \frac{(b+1)z}{b+3} + \cdots $$

$$ + \frac{nz}{b+2n} - \frac{(b+n)z}{b+2n+1} + \cdots. $$

Taking $b=0$, the left-hand side is $\exp(z)$, and (6.1a) follows. To prove (6.1b), write $\exp(z)=\{\exp(-z)\}^{-1}$ and use the representation (6.1a) for $\exp(-z)$. To prove (6.1c), use

$$ \frac{e^y - e^{-y}}{e^y + e^{-y}} = \tanh y. $$

Hence, (6.1c) follows from (6.3) by taking $2y=z$ and using the formula

$$ e^z = 1 - \frac{2}{1 - [\tanh(z/2)]^{-1}}. $$

Proof of (6.2).

$$ \sin z = z \, {}_0F_1(\tfrac{3}{2}; -z^2/4), $$

$$ \cos z = {}_0F_1(\tfrac{1}{2}; -z^2/4), $$

and the result follows from (6.13).

Proof of (6.3).

$$\sinh z = z\ {}_0F_1\!\left(\tfrac{3}{2};\, z^2/4\right)$$
$$\cosh z = {}_0F_1\!\left(\tfrac{1}{2};\, z^2/4\right)$$

and the result follows from (6.13).

Proof of (6.4)–(6.7).

$$(1+z)^\nu = {}_2F_1(-\nu,1,1;\, z),$$
$$\tan^{-1}z = z\ {}_2F_1\!\left(\tfrac{1}{2},1,\tfrac{3}{2};\, -z^2\right),$$
$$\tanh^{-1}z = z\ {}_2F_1\!\left(\tfrac{1}{2},1,\tfrac{3}{2};\, z^2\right),$$
$$\ln(1+z) = z\ {}_2F_1(1,1,2;\, -z),$$

and (6.4)–(6.7) follow from (6.16) with $b=0$.

Proof of (6.8).   Use the representation

$$_2F_0(a,b;\, z') = \frac{1}{\Gamma(a)}\int_0^\infty \frac{e^{-t}t^{a-1}\,dt}{(1-z't)^b}$$

taking $a=1$, $b=n$, $z'=-1/z$. For $\operatorname{Re} z > 0$, $E_n(z) = e^{-z}z^{-1}\,{}_2F_0(1,n;\,-1/z)$, and (6.8) follows from (6.15).

Proof of (6.9).

$$\operatorname{erfc}(z) = \frac{1}{\sqrt{\pi}}\,\Gamma\!\left(\tfrac{1}{2},\, z^2\right) = \frac{1}{\sqrt{\pi}}\int_{z^2}^\infty t^{-1/2}e^{-t}\,dt,$$

and so (6.9) follows from (6.10).

Proof of (6.10).   This is the same as for (6.8).

Proof of (6.11).

$$\operatorname{erf}(z) = \frac{2}{\sqrt{\pi}}\int_0^z e^{-t^2}\,dt$$
$$= \frac{2ze^{-z^2}}{\sqrt{\pi}}\,{}_1F_1\!\left(1,\tfrac{3}{2};\, z^2\right),$$

and so (6.11) follows from (6.14).

Proof of (6.12).

$$\gamma(a, z) = \int_0^z t^{a-1} e^{-t}\, dt$$

$$= z^a a^{-1}\, {}_1F_1(a, 1+a; -z), \qquad \text{by expansion}$$

$$= e^{-z} \int_0^z (z-u)^{a-1} e^u\, du, \qquad \text{by setting } t = z - u$$

$$= z^a e^{-z} a^{-1}\, {}_1F_1(1, 1+a; z), \qquad \text{by expansion.}$$

The continued-fraction expansion then follows from (6.14).

Proof of (6.13).   Series expansion of the hypergeometric function shows that

$$_0F_1(a+1; z) = {}_0F_1(a; z) - \frac{z}{a(a+1)}\, {}_0F_1(a+2; z).$$

Therefore

$$\frac{{}_0F_1(a+1; z)}{{}_0F_1(a; z)} = \cfrac{1}{1 + \cfrac{z}{a(a+1)} \cfrac{{}_0F_1(a+2; z)}{{}_0F_1(a+1; z)}}.$$

This formula is simple to iterate and is used to generate the continued-fraction expansion (6.13a).

Proof of (6.14).   Series expansion of the confluent hypergeometric function shows that

$$_1F_1(a+1, b+1; z) = {}_1F_1(a, b; z) + \frac{z(b-a)}{b(b+1)}\, {}_1F_1(a+2, b+2; z).$$

Therefore

$$\frac{{}_1F_1(a+1, b+1; z)}{{}_1F_1(a, b; z)} = \cfrac{1}{1 - \cfrac{z(b-a)}{b(b+1)} \cfrac{{}_1F_1(a+2, b+2; z)}{{}_1F_1(a+1, b+1; z)}}.$$

Again, this formula is simple to iterate and is used to generate the continued-fraction expansion (6.14).

Proof of (6.15).   Formal operations with the power series similar to the previous operations lead to the formula

$$_2F_0(a, b+1; z) = {}_2F_0(a, b; z) + az\, {}_2F_0(a+1, b+1; z) \qquad (6.17)$$

However, we must use the representation

$$_2F_0(a,b;z)=\frac{1}{\Gamma(a)}\int_0^\infty \frac{e^{-t}t^{a-1}}{(1-zt)^b}\,dt,\tag{6.18}$$

which is valid for $\text{Re}\,a>0$ and z not on the positive real axis, to establish the result (6.17). The identity

$$\frac{t^{a-1}}{(1+zt)^b}=\frac{t^{a-1}}{(1+zt)^{b+1}}+\frac{zt^a}{(1+zt)^{b+1}}$$

leads to equality of the integrands and provides the proof of (6.15) for $\text{Re}\,a>0$. Extension to complex values of a is by analytic continuation. Hence

$$\frac{_2F_0(a,b+1;z)}{_2F_0(a,b;z)}=\frac{1}{1-az\dfrac{_2F_0(a+1,b+1;z)}{_2F_0(a,b+1;z)}}.\tag{6.19}$$

Since $_2F_0(a,b;z)=\,_2F_0(b,a;z)$, it follows from (6.19) that

$$\frac{_2F_0(a+1,b+1;z)}{_2F_0(a,b+1;z)}=\frac{1}{1-(b+1)z\dfrac{_2F_0(a+1,b+2;z)}{_2F_0(a+1,b+1;z)}}.\tag{6.20}$$

Equations (6.19) and (6.20) together provide a formula which may be iterated to yield (6.15).

Proof of (6.16).   The expansion of the hypergeometric function leads to the identity

$$_2F_1(a,b,c;z)=\,_2F_1(a,b+1,c+1;z)-\frac{a(c-b)}{c(c+1)}z\,_2F_1(a+1,b+1,c+2;z).$$

This may be rewritten as

$$\frac{_2F_1(a,b+1,c+1;z)}{_2F_1(a,b,c;z)}=\frac{1}{1-\dfrac{a(c-b)}{c(c+1)}z\dfrac{_2F_1(a+1,b+1,c+2;z)}{_2F_1(a,b+1,c+1;z)}}.\tag{6.21}$$

Since $_2F_1(a,b,c;z)=\,_2F_1(b,a,c;z)$, we may rewrite (6.21), replacing c by

$c+1$, etc., as

$$\frac{{}_2F_1(a+1,b+1,c+2;z)}{{}_2F_1(a,b+1,c+1;z)}=$$

$$\frac{1}{1-z\dfrac{(b+1)(c-a+1)\,{}_2F_1(a+1,b+2,c+3;z)}{(c+1)(c+2)\,{}_2F_1(a+1,b+1,c+2;z)}}. \qquad (6.22)$$

Together (6.21) and (6.22) yield a formula which connects ratios of hypergeometric functions in which the numerator parameters a and b are increased by 1 and the denominator parameter c is increased by 2. The result (6.16) follows by iteration.

To summarize this section, we observe that a variety of familiar functions have continued-fraction expansions given by (6.1)–(6.16). Using the algebraic results (6.17), (6.19)–(6.22), we have proved that the Maclaurin expansion of each function is the same as that of the corresponding continued fraction. In this sense, the results (6.1)–(6.16) are formal equalities. In the next section, we find the domain of values of z for which (6.1)–(6.16) are true equalities.

Exercise 1. Prove that the contracted form of

$$e^z E_n(z)=\frac{1}{z}+\frac{n}{1}+\frac{1}{z}+\frac{n+1}{1}+\frac{2}{z}+\cdots+\frac{n+r}{1}+\frac{r+1}{z}+\cdots$$

is given by (6.8).

Exercise 2. Prove that the contracted form of

$$e^z z^{-a}\Gamma(a,z)=\frac{1}{z}+\frac{1-a}{1}+\frac{1}{z}+\frac{2-a}{1}+\cdots+\frac{n}{z}+\frac{n+1-a}{1}+\cdots$$

is given by (6.10).

4.7 Convergence of Continued Fractions

The derivation given in Section 4.6 of the continued-fraction expansions of the familiar functions of mathematics is purely algebraic. We can justify the usual meaning of the equality signs in Section 4.6 if we can show that

(i) each continued fraction converges to a limit function, and
(ii) this limit function is the same function as the one from which the fraction originated.

We simply state the relevant theorems here, and refer to the companion volume of Jones and Thron for the proofs.

Since each convergent of a continued fraction of the types discussed in Section 4.6 is a rational function of the variable z, one expects to be able to specify conditions which are sufficient to ensure that the limit function is meromorphic in z. Normally, most authors prove theorems by showing that the convergents of a continued fraction [such as (6.14)] converge on bounded domains of the z-plane which do not contain the poles or other singularities, if any, of the limit function. However, some authors prefer to discuss convergence of continued fractions in terms of the chordal metric, which means the same as convergence on the Riemann sphere. This ambiguity rarely leads to confusion, provided the trap is anticipated. We use convergence in the ordinary sense, unless otherwise stated.

We consider a continued fraction in the reduced standard form

$$b_0 + \frac{a_1 z}{1} + \frac{a_2 z}{1} + \frac{a_3 z}{1} + \cdots. \tag{7.1}$$

Naturally enough, convergence criteria for continued fractions are always essentially properties of the "tail" of the fraction; the value of b_0 is totally immaterial. The *divergence condition* is an important criterion for the convergence of (7.1).

THEOREM 4.7.1. *If the continued fraction* (7.1) *converges for any nonzero value of* z, *then either*

$$\sum_{n=1}^{\infty} \left| \frac{a_2 a_4 \cdots a_{2n}}{a_3 a_5 \cdots a_{2n+1}} \right| \quad or \quad \sum_{n=1}^{\infty} \left| \frac{a_3 a_5 \cdots a_{2n-1}}{a_4 a_6 \cdots a_{2n}} \right| \tag{7.2}$$

must diverge.

Remarks. Note that divergence of one of the series (7.2) is a necessary but not a sufficient condition for convergence of (7.1). It must be supplemented by further conditions on the elements a_i if convergence is to be proved, as is done in Theorem 4.7.3.

We do not prove Theorem 4.7.1. Instead we show the scope of such proofs by proving a similar result due to Seidel which is self-contained and very relevant to Stieltjes series.

THEOREM 4.7.2. *Consider the continued fraction*

$$\frac{1}{b_1} + \frac{1}{b_2} + \frac{1}{b_3} + \cdots, \tag{7.3}$$

where $b_i > 0$ *for all* i. *The fraction* (7.3) *converges if and only if* $\sum_{i=1}^{\infty} b_i$ *diverges.*

Proof. We first prove that convergence of (7.3) implies divergence of Σb_i. First we prove an identity which shows that the sequence of convergents of (7.3) may be written as an alternating series. The nth convergent of (7.3) is

$$\frac{A_n}{B_n} = \left(\frac{A_n}{B_n} - \frac{A_{n-1}}{B_{n-1}} \right) + \left(\frac{A_{n-1}}{B_{n-1}} - \frac{A_{n-2}}{B_{n-2}} \right) + \left(\frac{A_{n-2}}{B_{n-2}} - \frac{A_{n-3}}{B_{n-3}} \right) + \cdots + \frac{A_1}{B_1}.$$

$$(7.4)$$

By using the recurrences

$$A_i = b_i A_{i-1} + A_{i-2}, \tag{7.5a}$$

$$B_i = b_i B_{i-1} + B_{i-2}, \tag{7.5b}$$

we find by iteration that

$$A_n B_{n-1} - B_n A_{n-1} = (-1)^n (A_2 B_1 - B_2 A_1) = (-1)^{n+1}.$$

Substituting this result in (7.4), we find that

$$\frac{A_n}{B_n} = \frac{1}{b_1} + \sum_{i=2}^{n} \frac{(-1)^{i+1}}{B_i B_{i-1}}, \qquad n = 2, 3, 4, \ldots. \tag{7.6}$$

Since $B_1 = b_1$, $B_2 = 1 + b_1 b_2$, it follows from (7.5b) that

$$B_i \geqslant m = \min(1, b_1) > 0 \qquad \text{for all} \quad i \geqslant 1.$$

Therefore, each convergent of (7.3) may be expressed by (7.6) as an alternating series.

The alternating-series test states [Ferrar, 1938, p. 47] that if (i) u_n is a decreasing sequence, (ii) $u_n \to 0$ as $n \to \infty$, and (iii) $u_n > 0$ for all n, then the series $\Sigma_{n=0}^{\infty} (-1)^n u_n$ is called an alternating series and it converges.

We show that the condition that Σb_n diverges implies that $A_n / B_n \to f$ as $n \to \infty$. Note that

$$B_i B_{i-1} = (b_i B_{i-1} + B_{i-2}) B_{i-1} \geqslant m^2 b_i + B_{i-1} B_{i-2} \qquad \text{for all } i,$$

and this shows that

$$B_n B_{n-1} \geqslant m^2 \sum_{i=2}^{n} b_i.$$

Hence condition (ii) of the alternating-series test is valid, and we deduce that $A_n / B_n \to f$ as $n \to \infty$.

If Σb_n is not divergent, let

$$\sum_{n=1}^{\infty} b_n = B.$$

We may easily prove by induction that

$$B_n < (1+b_1)(1+b_2)\cdots(1+b_n)$$

using (7.5b), and hence

$$\ln B_n < B \text{ and } B_n < e^B \qquad \text{for all } n.$$

Consequently, the terms of (7.6) do not decrease in modulus, and so the ratios A_n/B_n do not converge to any limiting value.

An interesting application of Theorem 4.7.2 is that we may show directly that the Stieltjes fraction (5.5.24) converges on the positive real z-axis. The connection between Theorems 4.7.1 and 4.7.2 becomes evident from an equivalence transformation, as we state in Exercise 1.

Our next theorem provides a sufficient condition for the convergence of a continued fraction.

THEOREM 4.7.3 (Parabola theorem) [Scott and Wall, 1940b; Thron, 1974]. *The continued fraction*

$$b_0 + \frac{a_1 z}{1} + \frac{a_2 z}{1} + \frac{a_3 z}{1} + \cdots \tag{7.7}$$

converges provided that

(i) α *may be found in the range* $-\pi/2 < \alpha < \pi/2$, *and* n_0 *exists such that the elements of* (7.7) *satisfy*

$$|a_n z| - \mathrm{Re}\left(a_n z e^{-2i\alpha}\right) \leq \tfrac{1}{2}\cos^2 \alpha \tag{7.8}$$

for all $n > n_0$; *and*
(ii) *the divergence condition* (7.2) *is satisfied.*

COROLLARY. If it so happens that the sequence $\{a_n\}$ has a nonzero limit, let $a_n \to a$. We choose to take $\alpha = 0$. The key condition (7.8) is then satisfied if we impose a constraint on z, namely

$$|z| - \mathrm{Re}\{z e^{i\arg(a)}\} < \frac{1}{2|a|}.$$

This defines a parabolic domain $\mathcal{P}$ shown in Fig. 1. The parabola has its focus at $z=0$ and its axis running through $z=-1/(4a)$. The geometric

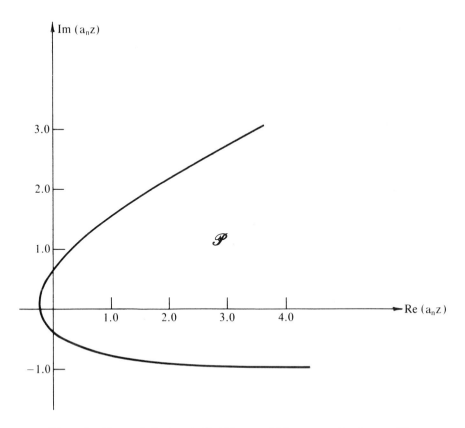

Figure 1.   The parabolic domain $\mathscr{P}$ of Theorem 4.7.3 corresponding to $\alpha = \pi/12$.

interpretation of (7.8) is that z must be nearer the origin than the directrix of the parabola. The hypothesis that $a_n \to a$, $a \neq 0$, is also sufficient to satisfy the divergence condition, and so this extra hypothesis allows a simple corollary of the parabola theorem.

THEOREM 4.7.4 (Cardioid theorem) [Paydon and Wall, 1942; Dennis and Wall, 1945; Thron, 1974]. *Provided that n_0 and k may be found such that*

(a) $k > 0$
(b) $|a_n| - \operatorname{Re} a_n \leqslant 1/(2k)$ *for all* $n > n_0$, *and*
(c) *the divergence condition (7.2) is satisfied, then the fraction (7.7) converges for all z in the cardioid*

$$|z| < k[1 + \cos(\arg(z))].\tag{7.9}$$

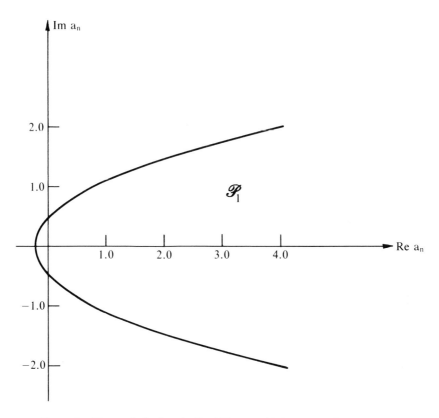

Figure 2. The parabolic domain $\mathscr{P}_1$ of Theorem 4.7.4 corresponding to $k = 1.07$.

Interpretation. Condition (b) of the theorem requires that all the partial numerator coefficients a_n of the continued fraction lie in a parabolic domain $\mathscr{P}_1$, shown in Figure 2. If the conditions (a), (b), and (c) of the theorem are satisfied, convergence of the fraction is assured in the cardioid shown in Figure 3. Note that a larger value of k gives a more restrictive parabolic constraint and a larger cardioid domain of convergence for the fraction.

THEOREM 4.7.5. *If the sequence of convergents of (7.7) converges uniformly in $|z| < R$ with $R > 0$, to a limit function $f(z)$, then $f(z)$ is analytic in $|z| < R$ and its power series generates the fraction (7.7).*

Remark. This theorem is a consequence of Theorem 4.7.3. and is proved using Weierstrass's theorem [Titchmarsh, 1939, p. 95; Copson, 1948, p. 97].

THEOREM 4.7.6 [Van Vleck, 1904]. *If $a_n \to 0$ as $n \to \infty$, the fraction (7.7) converges to a meromorphic function of z. If $a_n \to a$ as $n \to \infty$ with $a \neq 0$, then (7.7) converges to a function $f(z)$ which is meromorphic in the cut z-plane. The*

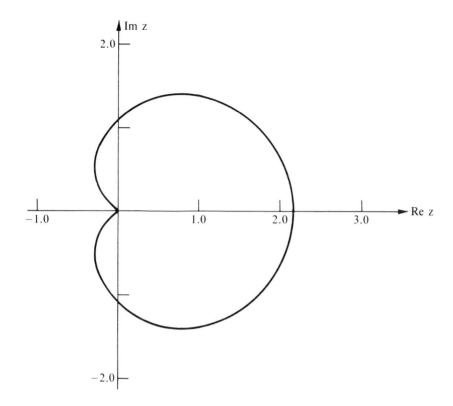

Figure 3. The cardioid domain of Theorem 4.7.4 corresponding to $k = 1.07$.

cut is placed in the shadow of $-(4a)^{-1}$ from the origin, as shown in Figure 4. In each case, convergence is uniform on any compact set containing no poles of the limit function, and the limit function has the continued fraction expansion (7.7).

Next, we will use some of these theorems to prove the quoted results of the previous section. Van Vleck's theorem is used to prove (6.13), (6.14) and (6.16); (6.15) is proved by using the cardioid theorem.

Proof of (6.13). We have shown that both left- and right-hand side of the equation

$$\frac{{}_0F_1(a+1;z)}{{}_0F_1(a;z)} = \frac{1}{1} + \frac{a_1 z}{1} + \frac{a_2 z}{1} + \cdots + \frac{a_n z}{1} + \cdots \qquad (7.10)$$

with $a_n = \{(a+n-1)(a+n)\}^{-1}$ have the same formal power-series expansion. Since $a_n \to 0$, Theorem 4.7.6 states that (7.10) is an identity for all z not a zero of ${}_0F_1(a;z)$.

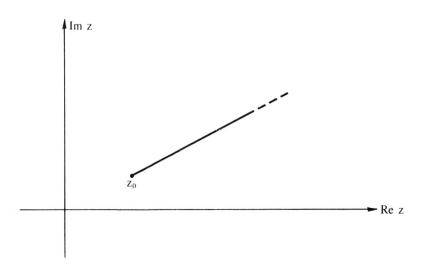

Figure 4.   The cut from $z_0 = -(4a)^{-1}$ to ∞ in the complex z-plane for Theorem 4.7.6.

Proof of (6.14).   We have shown that both left- and right-hand side of the equation

$$\frac{{}_1F_1(a+1, b+1; z)}{{}_1F_1(a, b; z)} = \frac{1}{1} + \frac{a_1 z}{1} + \frac{a_2 z}{1} + \cdots + \frac{a_k z}{1} + \cdots \quad (7.11)$$

with

$$a_{2n} = (a+n)\{(b+2n-1)(b+2n)\}^{-1}, \qquad n=1,2,3,\ldots$$

and

$$a_{2n+1} = (a-b-n)\{(b+2n)(b+2n+1)\}^{-1}, \qquad n=1,2,3,\ldots$$

have the same formal power-series expansion. Since $a_n \to 0$, Theorem 4.7.6 asserts that the right-hand side of (7.11) is convergent and that (7.11) is an identity for all z not a zero of ${}_1F_1(a, b; z)$.

PARTIAL PROOF OF (6.15).   We have shown that both left- and right-hand side of the identity

$$\frac{{}_2F_0(a, b+1; z)}{{}_2F_0(a, b; z)} = \frac{1}{1} + \frac{a(-z)}{1} + \frac{(b+1)(-z)}{1}$$

$$+ \frac{(a+1)(-z)}{1} + \cdots + \frac{-a_n z}{1} + \cdots$$

with

$$a_{2n} = b + n, \qquad n = 1, 2, 3, \ldots,$$
$$a_{2n+1} = a + n, \qquad n = 0, 1, 2, \ldots, \tag{7.12}$$

have the same formal expansion. By noting that the ratio of successive numerator coefficients $a_k / a_{k+1} \rightarrow 1$, we may show that the divergence condition is satisfied, which is one necessary condition for the convergence of (7.12). It is more convenient to use the variable $z' = -z$, so that (7.12) becomes formally

$$\frac{{}_2 F_0(a, b+1; -z')}{{}_2 F_0(a, b; -z')} = \frac{1}{1} + \frac{a_1 z'}{1} + \frac{a_2 z'^2}{1} + \cdots + \frac{a_k z'^k}{1} + \cdots. \tag{7.13}$$

This equation has the status of a formal algebraic identity, and we seek to show that it represents an identity between function values in the cut z'-plane.

The sequence a_n is shown in Figure 5. We see that for any $k \geqslant 0$, $n_0 = n_0(k)$ exists such that

$$|a_n| - \operatorname{Re} a_n \leqslant \frac{1}{2k} \qquad \text{for all} \quad n > n_0.$$

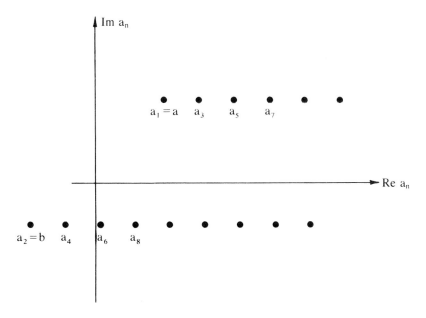

Figure 5. The numerator elements a_n in the complex plane defined by (7.12).

Hence the continued fraction (7.13) converges for all z' in the cardioid

$$|z'| < k[1 + \cos(\arg z')].$$

Since this is true for any $k > 0$, the continued fraction converges for all z' except on $-\infty < z' < 0$, which corresponds to the positive real $z = $ axis.

To establish equality between left- and right-hand side of (7.13), we note that for the special case of $b = 0$, $a > 0$, we have a strict Stieltjes series in the variable z', and the theory of Section 5.5 is applicable. Using Carleman's theorem, (7.13) is established directly as a true equality for $|\arg(z')| < \pi$. To extend this argument to the cases in question, we refer to Wall [1945].

Proof of (6.16).   We have shown that both left- and right-hand side of the equation

$$\frac{{}_2F_1(a, b+1; c+1; z)}{{}_2F_1(a, b; c; z)} = \frac{1}{1} + \frac{a_1 z}{1} + \frac{a_2 z}{1} + \cdots + \frac{a_n z}{1} + \cdots \quad (7.14)$$

with

$$a_{2n} = \frac{(b+n)(a-c-n)}{(c+2n-1)(c+2n)}$$

and

$$a_{2n+1} = \frac{(a+n)(b-c-n)}{(c+2n)(c+2n+1)}$$

have the same formal expansion. Since $a_n \to -\frac{1}{4}$, Theorem 4.7.6 asserts that (7.14) is an identity valid in the z-plane cut along $1 \leq z < \infty$. This is, of course, the usual domain of definition of a ${}_2F_1$ hypergeometric function. Hence the fraction (7.14) converges for all z not on the cut and not a zero of ${}_2F_1(a, b; c; z)$.

Exercise 1.   Consider the fraction

$$b_0 + \frac{z}{b_1} + \frac{z}{b_2} + \frac{z}{b_3} + \cdots$$

with $b_i > 0$ for all i. Show that the divergence condition (7.2) is precisely the condition that $\sum_{i=1}^{\infty} b_i$ diverges.

Exercise 2. Prove that the condition that $a_n/a_{n+1} \to c$ as $n \to \infty$ is sufficient to satisfy the divergence condition.

Exercise 3. Prove that the fraction

$$\frac{(1\times 2)^\nu}{1} + \frac{(2\times 3)^\nu}{1} + \cdots + \frac{j(j+1)^\nu}{1} + \cdots$$

converges for $0 \leqslant \nu \leqslant 1$ and diverges for $\nu > 1$.

CHAPTER 5

Stieltjes Series and Pólya Series

5.1 Introduction to Stieltjes Series

DEFINITION. A Stieltjes function is defined by the Stieltjes-integral* representation

$$f(z) = \int_0^\infty \frac{d\phi(u)}{1+zu}, \tag{1.1}$$

where $\phi(u)$ is a bounded, nondecreasing function (taking infinitely many different values) on $0 \leqslant u < \infty$ and with finite real-valued moments given by

$$f_j = \int_0^\infty u^j d\phi(u), \qquad j = 0, 1, 2, \ldots. \tag{1.2}$$

From (1.1), it follows immediately that $f(z)$ is a real symmetric function, defined in the cut z-plane with the cut along the negative real axis as shown in Figure 1; real symmetric functions are defined by (1.6).

A formal expansion of (1) always provides a series expansion of $f(z)$, called a Stieltjes series, and given by

$$f(z) = \sum_{j=0}^\infty f_j(-z)^j \tag{1.3}$$

The series is called formal because it may not converge for any z (except $z = 0$); nevertheless it is a useful representation of the function $f(z)$, if

---

*A good explanation of Stieltjes integrals is given in Perron [1957, Vol. 2, p. 180] or Rudin [1976, Chapter 6].

ENCYCLOPEDIA OF MATHEMATICS and Its Applications, Gian-Carlo Rota (ed.). Vol. 13: George A. Baker, Jr., and Peter R. Graves-Morris, Padé Approximants: Basic Theory, Part I   ISBN 0-201-13512-4

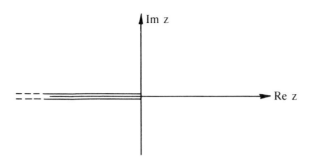

Figure 1.   The cut z-plane in which $f(z)$ is defined by (1.1).

properly reinterpreted, as we will show in this chapter. It is easier to use the positive definite coefficients $\{f_j\}$ in the expansion (1.3) rather than our standard notation, $f(z) = \sum_{j=0}^{\infty} c_j z^j$, because the determinantal inequalities of Theorem 5.1.2 take on a simpler form.

The phase "taking infinitely many different values" in the definition of a Stieltjes series is made part of the definition so as to exclude the following special case. If $\phi(u)$ takes on a finite number of values, say $m+1$ distinct values, then $\phi(u)$ is piecewise constant on $m+1$ intervals covering the range $0 \leqslant u < \infty$. Suppose that

$$\phi(u) = 0 \quad \text{on} \quad 0 \leqslant u < u_1,$$

$$\phi(u) = \phi_i \quad \text{on} \quad u_i < u < u_{i+1} \qquad \text{for} \quad i = 1, 2, \ldots, m-1,$$

$$\phi(u) = \phi_m \quad \text{on} \quad u > u_m.$$

Then $d\phi(u) = 0$ except in neighborhoods of $u = u_i$, $i = 1, 2, \ldots, m$, and so

$$f(z) = \int_0^{\infty} \frac{d\phi(u)}{1+zu_i} = \sum_{i=1}^{m} \frac{1}{1+zu_i} \int_{\text{nhd. of } u_i} d\phi(u)$$

$$= \sum_{i=1}^{m} \frac{\phi(u_i+) - \phi(u_i-)}{1+zu_i}$$

$$= \sum_{i=1}^{m} \frac{\lambda_i}{1+zu_i} \qquad \text{with} \quad \lambda_i > 0 \quad \text{for} \quad i = 1, 2, \ldots, m. \qquad (1.4)$$

Hence $f(z)$ is a rational function of z with m simple poles at $z = -u_i^{-1}$ on the negative real axis and with positive definite residues. Furthermore, all Padé approximants of $f(z)$ with $L \geqslant m-1$ and $M \geqslant m$ are exact. Thus, the case when $\phi(u)$ takes a finite number of values only is a special case, and it is usually excluded, by definition, from being a Stieltjes series.

If $\phi(u)$ is constant on $\lambda \leqslant u < \infty$, then

$$f(z) = \int_0^\lambda \frac{d\phi(u)}{1+zu}. \tag{1.5}$$

In this case, $f(z)$ is defined in the cut z-plane, cut along $-\infty < z < -\lambda^{-1}$. The power-series expansion of $f(z)$ is given by (1.3) is then convergent in the disk

$$|z| < \lambda^{-1}$$

shown in Figure 2.

Whether or not the formal series $\sum_{j=0}^\infty f_j(-z)^j$ of $f(z)$ has a zero radius of convergence, the Padé approximants of the series are vital for its analysis and are useful for its numerical evaluation, as we will see in this chapter. We can prove convergence of the Padé approximants largely because we can prove that the poles of the Padé approximants lie on the cuts of the Stieltjes function. Stieltjes functions are real symmetric functions defined in the cut plane, with the negative real axis as the cut.

A function is defined to be real symmetric if it takes complex conjugate values when the variable is complex-conjugated [Titchmarsh, 1939, p. 155].

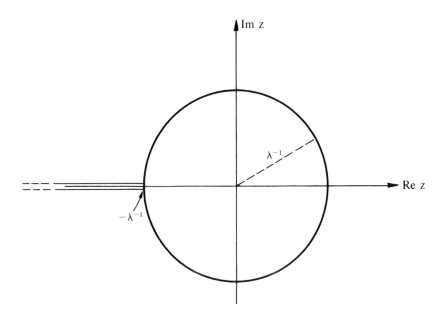

Figure 2.   The cut z-plane in which $f(z)$ is defined by (1.5).

This condition is that

$$f(z^*) = [f(z)]^*. \tag{1.6}$$

An important and immediate consequence of this applies to a function $f(z)$ which is analytic at a point $z = x_0$ on the real axis, so that the expansion

$$f(z) = \sum_{i=0}^{\infty} d_i (z - x_0)^i$$

is convergent in some small disk enclosing $z = x_0$. The coefficients d_i, $i = 0, 1, 2, \ldots$, are real if and only if $f(z)$ is real symmetric, which justifies the name "real symmetric".

Stieltjes functions are real symmetric and, as can be shown from (1.1), have a negative imaginary part on the cut in the sense that

$$\mathrm{Im}\, f(x + i\varepsilon) = - \mathrm{Im}\, f(x - i\varepsilon) = \frac{\pi}{x} \phi'\left(-\frac{1}{x}\right)$$

$$\text{if} \quad x = \mathrm{Re}\, z < 0, \tag{1.7}$$

provided the implied limit ($\varepsilon \to 0$) exists (cf. Lemma 3 in Section 5.6). We distinguish the three cases (i) $\phi(u)$ is differentiable, (ii) $\phi(u)$ is continuous but not differentiable, and (iii) $\phi(u)$ is discontinuous at a point u [Riesz and Nagy, 1955].

Before embarking on the proofs of the properties of Stieltjes series, let us consider an illustrative example a Stieltjes series. The function is

$$f(z) = \frac{1}{z} \ln(1 + z)$$

$$= 1 - \tfrac{1}{2} z + \tfrac{1}{3} z^2 + \cdots \qquad \text{for} \quad |z| < 1.$$

Its coefficients are given by

$$f_j = \frac{1}{j+1}$$

Hence the density function defined by

$$\phi(u) = u, \quad \phi'(u) = 1 \qquad \text{on} \quad 0 \leqslant u \leqslant 1,$$
$$\phi(u) = 1, \quad \phi'(u) = 0 \qquad \text{on} \quad 1 < u < \infty \tag{1.8}$$

ensures that

$$f_j = \int_0^{\infty} u^j \, d\phi(u) = \frac{1}{j+1}. \tag{1.9}$$

Further,

$$f(z) = \int_0^1 \frac{du}{1+zu} = \int_0^\infty \frac{d\phi(u)}{1+zu},$$

so that $f(z)$ is a Stieltjes series according to (1.1), (1.2), where $\phi(u)$ is a nondecreasing function taking on infinitely many values [given by (1.8)] and all the moments are well defined by (1.9). The real symmetry is evident from the real coefficients in the expansion, and also

$$\operatorname{Im} f(z+i\varepsilon) = \frac{\pi}{x} \qquad \text{for all} \quad x - \operatorname{Re} z < -1,$$

illustrating (1.7).

The following theorem shows the effect on a Stieltjes series of deletion of the first J terms of the series.

THEOREM 5.1.1. *Let $f(z)$ be a Stieltjes series with a formal expansion and representation*

$$f(z) = \sum_{j=0}^\infty f_j(-z)^j, \qquad f(z) = \int_0^\infty \frac{d\phi(u)}{1+zu}.$$

Then $g(z)$ given by the formal expansion

$$g(z) = (-z)^{-J-1} \sum_{j=J+1}^\infty f_j(-z)^j \tag{1.10}$$

is also a Stieltjes series represented by

$$g(z) = \int_0^\infty \frac{u^{J+1} d\phi(u)}{1+zu}. \tag{1.11}$$

Proof. The results follow immediately from the definitions (1.1)–(1.3).

Stieltjes series may be recognized by virtue of the determinantal conditions satisfied by the power-series coefficients f_j. We use the following definition, which is more convenient than using the determinants $C(L/M)$ defined in (1.4.8) in this context:

$$D(m,n) = \begin{vmatrix} f_m & f_{m+1} & \cdots & f_{m+n} \\ f_{m+1} & f_{m+2} & \cdots & f_{m+n+1} \\ \vdots & \vdots & & \vdots \\ f_{m+n} & f_{m+n+1} & \cdots & f_{m+2n} \end{vmatrix}. \tag{1.12}$$

These definitions are related by the identities (see exercise 1)

$$D(L-M+1, M-1)=C(L/M) \qquad \text{if } L-M \text{ is odd,}$$
$$D(L-M+1, M-1)=(-)^M C(L/M) \quad \text{if } L-M \text{ is even.}$$

THEOREM 5.1.2. *A necessary condition for $f(z)$ to be a Stieltjes series is that all the determinants $D(m, n)$ with $m \geqslant 0$, $n \geqslant 0$, are positive.*

Remark. The condition is also sufficient, provided that $f(z)$ is uniquely determined by its power series in principle. The proof of this result follows later.

Proof. We use the properties of Stieltjes series

$$f(z)=\int_0^\infty \frac{d\phi(u)}{1+uz}, \qquad f_j=\int_0^\infty u^j d\phi(u),$$

and we must prove that the coefficients f_j satisfy $D(m, n)>0$. Notice that each coefficient f_m is defined to be positive, and so $D(m,0)>0$. Let us define

$$G(x_0, x_1,\ldots, x_n)=\int_0^\infty u^m(x_0+x_1 u+ \cdots +x_n u^n)^2 d\phi(u)$$

$$=\int_0^\infty \left(\sum_{i=0}^n \sum_{j=0}^n u^{m+i+j} x_i x_j \right) d\phi(u)$$

$$= \sum_{i=0}^n \sum_{j=0}^n f_{i+j+m} x_i x_j, \tag{1.13}$$

which is seen to be a real positive quadratic form in the $n+1$ real variables $x_0, x_1,\ldots, x_n$. Therefore $G(x_0,\ldots, x_n)$ has a minimum value on the hypersphere

$$S(x_0,\ldots x_n) \equiv \sum_{i=0}^n x_i^2 \equiv x_0^2+x_1^2 + \cdots +x_n^2 = 1. \tag{1.14}$$

The values of the $\{x_i\}$ at this minimum are given by Lagrange's undetermined-multiplier method, following variation of the n independent coordinates. The equations are

$$\frac{\partial G}{\partial x_i} -\lambda \frac{\partial S}{\partial x_i} =0, \qquad i=0,1,\ldots, n, \tag{1.15}$$

and $S(\mathbf{x})=1$.

Hence from (1.15),

$$\sum_{j=0}^{n} f_{i+j+m} x_j - \lambda x_i = 0 \qquad \text{for} \quad i = 0, 1, \ldots n. \tag{1.16}$$

which is a set of $n+1$ homogeneous equations for $x_0 \ldots x_n$ with a consistency condition that λ must be an eigenvalue of the real and symmetric matrix

$$H = \begin{pmatrix} f_m & f_{m+1} & \cdots & f_{m+n} \\ f_{m+1} & f_{m+2} & \cdots & f_{m+n+1} \\ \vdots & & & \vdots \\ f_{m+n} & f_{m+n+1} & \cdots & f_{m+2n} \end{pmatrix}. \tag{1.17}$$

Let $\mathbf{x}^{(k)}$ be the eigenvector associated with the eigenvalue $\lambda^{(k)}$. Then we find from (1.13) that

$$G(\mathbf{x}) = \mathbf{x}^T H \mathbf{x},$$

and from (1.14) that

$$G(\mathbf{x}^{(k)}) = \mathbf{x}^{(k)T} H \mathbf{x}^{(k)} = \lambda^{(k)}.$$

If any eigenvector $\mathbf{x}^{(k)}$ of H had a strictly negative eigenvalue $\lambda^{(k)}$, then the quadratic form G of (1.13) would be negative, which is impossible. Hence all the eigenvalues of H are nonnegative, det H is the product of these eigenvalues, and so $D(m, n) \geq 0$ for all $m \geq 0$, $n > 0$.

This argument completes the proof except for the special case in which it might happen that $D(m, n) = 0$. This is associated with the existence of a nontrivial solution of the homogeneous linear equations

$$H\mathbf{x} = 0,$$

and therefore

$$G(\mathbf{x}) = \mathbf{x}^T H \mathbf{x} = 0 \qquad \text{for} \quad \mathbf{x}^T \cdot \mathbf{x} = 1,$$

From (1.13), we discover a nontrivial polynomial $p_n(u)$ for which

$$\int_0^\infty u^m [p_n(u)]^2 d\phi(u) = 0.$$

This can only occur if $d\phi(u) = 0$ except at the zeros of $p_n(u)$, contradicting the hypothesis that $\phi(u)$ takes on infinitely many values, which is a requirement for $f(z)$ to be a Stieltjes series.

COROLLARY 1. All $[L/M]$ Padé approximants to Stieltjes series exist and are nondegenerate, if $L \geq M - 1$.

The proof follows immediately from the basic representation (1.1.8). The sequences of Padé approximants to Stieltjes series of the next section are characterized by $J = L - M = \text{const}$, with the natural condition that $J \geqslant -1$.

COROLLARY 2. For any given Stieltjes series, there exists a regular C-fraction with the same formal expansion denoted by

$$\sum_{i=0}^{\infty} c_i z^i = \frac{a_1' z}{1} + \frac{a_2' z}{1} + \frac{a_3' z}{1} + \cdots \tag{1.18}$$

with $a_i' > 0$ for all $i \geqslant 1$. This justifies the nomenclature (4.5.5) for the S-fraction, and demonstrates the identification with Stieltjes series.

Proof. In Chapter 4, we saw from (4.2.13) and (4.4.5) the possibility of expressing a formal power series as a regular C-fraction. Theorem 5.1.2 enables the signs of the elements of the fraction to be determined. We find that $a_{2M+2} > 0$, $a_{2M+1} < 0$ and

$$\text{sign}\,(b_{2M+2}) = \text{sign}\,(b_{2M+1}) = (-1)^M.$$

Elementary equivalence transformations of the form (4.1.6) may be used to show that the fraction has the representation (1.18) with positive definite elements $\{a_i', \, i = 1, 2, \dots\}$. For example, note that $a_1' = c_0 > 0$, $a_2' = -c_1/c_0 > 0$, etc.

As general historical references for Stieltjes series (and not just for this section) we cite Stieltjes [1889, 1894], Tchebycheff [1858] and Van Vleck [1903]. An excellent review is given by Perron [1957], and material related to continued fractions is treated by Wall [1948].

Exercises 1. Use the definitions (1.3), (1.12), (1.1.1), and (1.4.8) to prove that

$$D(m, n) = (-1)^{m(n+1)} C(m + n/n + 1).$$

Exercise 2. Check the statements of Corollary 2.
Exercise 3. Use the definitions (1.2), (1.12) to prove that

$$D(m, n) = \int_0^{\infty} d\phi(u_0) \int_0^{\infty} d\phi(u_1) \cdots \int_0^{\infty} d\phi(u_n)$$

$$\times u_0^m u_1^{m+1} \cdots u_n^{m+n} \begin{vmatrix} 1 & u_0 & \cdots & u_0^n \\ 1 & u_1 & \cdots & u_1^n \\ \vdots & \vdots & & \vdots \\ 1 & u_n & \cdots & u_n^n \end{vmatrix}.$$

By permuting the variables and summing, deduce that

$$(n+1)!D(m,n) = \int_0^\infty d\phi(u_0) \int_0^\infty d\phi(u_1) \ldots \int_0^\infty d\phi(u_n)$$

$$\times u_0^m u_1^m \cdots u_n^m \begin{vmatrix} 1 & u_0 & \cdots & u_0^n \\ 1 & u_1 & \cdots & u_1^n \\ \vdots & \vdots & & \vdots \\ 1 & u_n & \cdots & u_n^n \end{vmatrix}^2 .$$

Hence, deduce that

$$D(m,n) = \frac{1}{(n+1)!} \int_0^\infty u_0^m \, d\phi(u_0) \int_0^\infty u_1^m \, d\phi(u_1) \ldots \int_0^\infty u_n^m \, d\phi(u_n)$$

$$\times \prod_{i=0}^{n} \prod_{j=0}^{i-1} (u_i - u_j)^2$$

and that $D(m,n) > 0$ [Bessis, 1979].

5.2   Convergence of Stieltjes Series

The convergence properties of Padé approximants of Stieltjes series hinge on the fact that all the poles of the $[M+J/M]$ Padé approximants (with $J \geqslant -1$) to Stieltjes series lie on the negative real axis and have positive residues. As a special exception (1.4) shows that if $f(z)$ is a Stieltjes series, *except* that $\phi(u)$ has precisely m points of increase, then $f(z)$ is a rational function with poles on the negative real axis with positive residues. In this case, all the $[M+J/M]$ Padé approximants to $f(z)$ with $J \geqslant -1$ and $M \geqslant m$ are identical to $f(z)$. For the case of genuine Stieltjes series, we have the following important theorem.

THEOREM 5.2.1. *If $f(z)$ is a Stieltjes series, then the poles of the $[M+J/M]$ Padé approximants (with $J \geqslant -1$) to $f(z)$ are simple poles which lie on the negative real axis and have positive residues.*

Remark. The proof uses the determinantal inequalities of Theorem 5.1.2 for the coefficients f_j, and does not require the fundamental representation of $\{f_j\}$ given by (1.2) and based on the existence of $\phi(u)$.

Proof. For $J \geqslant -1$, we define $\Delta_0^{(J)}(x) = 1$, and then for $M = 1, 2, 3, \ldots,$

$$\Delta_M^{(J)}(x) = \begin{vmatrix} f_{1+J} + xf_{2+J} & f_{2+J} + xf_{3+J} & \cdots & f_{M+J} + xf_{M+J+1} \\ f_{2+J} + xf_{3+J} & f_{3+J} + xf_{4+J} & \cdots & f_{M+J+1} + xf_{M+J+2} \\ \vdots & \vdots & & \vdots \\ f_{M+J} + xf_{M+J+1} & f_{M+J+1} + xf_{M+J+2} & \cdots & f_{2M+J-1} + xf_{2M+J} \end{vmatrix}.$$

(2.1)

Apart from a sign, these quantities are compact expressions for the denominators of a diagonal sequence of Padé approximants (see Section 1.3). They are related by

$$Q^{[M+J/M]}(x) = (-1)^{M(J+1)} \Delta_M^{(J)}(x). \tag{2.2}$$

Notice that $\Delta_M^{(J)}(x)$ is a real-valued function of the real variable x. When we consider sequences $\{\Delta_M^{(J)}(x), M = 0, 1, 2, \ldots\}$ with J fixed, we often omit the superscript (J). We first show that the functions

$$\Delta_0(x), \ \Delta_1(x), \ \Delta_2(x), \ldots \tag{2.3}$$

form a Sturm sequence. This means that if, for some x, $\Delta_j(x) = 0$, then $\Delta_{j-1}(x)$ and $\Delta_{j+1}(x)$ have opposite signs. We apply Sylvester's determinant identity to $A = \Delta_{j+1}(x)$ given by (2.1), (2.2). Using subscripts to denote the deleted rows and columns, Sylvester's identity is

$$A \times A_{M-1, M; M-1, M} = A_{M; M} A_{M-1; M-1} - A_{M; M-1} A_{M-1; M}.$$

By symmetry,

$$A_{M; M-1} A_{M-1; M} = (A_{M; M-1})^2 \geqslant 0.$$

We identify

$$A_{M-1, M; M-1, M} = \Delta_{j-1}(x) \quad \text{and} \quad A_{M, M} = \Delta_j(x),$$

and therefore

$$\Delta_{j-1}(x) \Delta_{j+1}(x) \leqslant 0 \qquad \text{if} \quad \Delta_j(x) = 0. \tag{2.4}$$

Also, (2.3) is valid if we identify $\Delta_0(x) = 1$ and take $j = 1$. We exclude the possibility that

$$\Delta_{j-1}(x) \Delta_{j+1}(x) = 0 \quad \text{and} \quad \Delta_j(x) = 0$$

by using the Frobenius $(*\,_**)$ identity, which would imply that $\Delta_j(x)=0$ for all j if any two successive Δ's vanish at the same value of x. Therefore (2.3) is a Sturm sequence.

To locate the roots of $\Delta_M(x)$, which are the poles of the $[M+J/M]$ Padé approximant, we consider the Sturm sequence (23) and recall that $f_i>0$. The first few members of the sequence are

$$\Delta_0(x)=1,$$
$$\Delta_1(x)=f_{1+J}+xf_{2+J},$$
$$\Delta_2(x)=\Delta_1(x)(f_{3+J}+xf_{4+J})-(f_{2+J}+xf_{3+J})^2.$$

From the representation (1), we have the general properties that

$$\left.\begin{array}{l}\Delta_j(0)>0,\\[4pt]\Delta_j(-\infty)=(-)^j\cdot\infty\end{array}\right\}\quad\text{for all}\quad j\qquad\qquad(2.5)$$

We define $x_{j,k}$ to be the kth zero of $\Delta_j(x)$, as is shown in Figure 1 and according to the following interlacing scheme:

$$\Delta_1(x)=0\qquad\text{at}\quad x=x_{1,1}=-\frac{f_{1+J}}{f_{2+J}}.$$

From (2.3),

$$\Delta_2(x_{1,1})<0$$

and from (2.4)

$$\Delta_2(0)>0;$$

therefore

$$\Delta_2(x)=0\qquad\text{at}\quad x=x_{2,1}\in(x_{1,1},0).$$

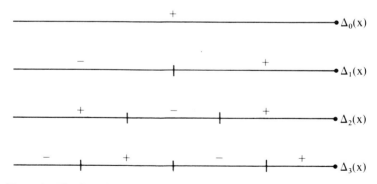

Figure 1. The sign of the first four Padé denominators of Stieltjes series for $x\leqslant0$.

From (2.3) and (2.5)

$$\Delta_2(x_{1,1})<0 \quad \text{and} \quad \Delta_2(-\infty)=+\infty;$$

therefore $\Delta_2(x)=0$ at $x=x_{2,2}\in(-\infty, x_{1,1})$.

The complete argument for the interlacing of the zeros of $\Delta_j(x)$ follows by induction in an obvious way. In short, $\Delta_j(x)=0$ has roots $x=x_{j,k}$, for $k=1,2,\ldots, j$ which lie on the negative real axis and are distinct, because each zero of $\Delta_{j-1}(x)$ lies in an interval $(x_{j,k+1}, x_{j,k})$. Hence the poles of the Padé approximants are distinct and lie on the negative real axis. To prove that the residues are positive, we use the identity (3.5.18),

$$[M+J+1/M+1]-[M+J/M]=\frac{(-x)^{2M+J+1}[D(1+J, M)]^2}{\Delta_{M+1}(x)\Delta_M(x)}. \quad (2.6)$$

Writing $P^{[M+J/M]}(x)=\Gamma_M(x)$ for short, it follows from (2.6) that

$$\Gamma_{M+1}(x)\Delta_M(x)-\Gamma_M(x)\Delta_{M+1}(x)>0 \quad \text{for} \quad x<0.$$

If $\Delta_{M+1}(x)=0$, $\text{sign}[\Gamma_{M+1}(x)]=\text{sign}[\Delta_M(x)]$ and at the particular root $x=x_{M+1,k}$

$$\text{sign}[\Gamma_{M+1}(x_{M+1,k})]=(-)^{k+1}.$$

Because $\Delta_{M+1}(x)$ is a polynomial of degree $M+1$, $\Delta'_{M+1}(x)$ has precisely one zero in each interval $(x_{M+1,k}, x_{M+1,k+1})$. Therefore, since $\Delta_{M+1}(0)>0$,

$$\frac{d}{dx}\Delta_{M+1}(x)\bigg|_{x=x_{M+1,k}}=(-)^{k+1},$$

and so

$$\frac{\Gamma_{M+1}(x)}{\Delta'_{M+1}(x)}\bigg|_{\Delta_{M+1}(x)=0}>0.$$

Thus the residues of the poles of the Padé approximants are positive, and the theorem is proved.

We now proceed to consider properties of the diagonal and first subdiagonal sequence of Padé approximants to Stieltjes series for $x>0$.

THEOREM 5.2.2. *Let* $\sum_{j=0}^{\infty} f_j(-z)^j$ *be a Stieltjes series. For* z *real and positive, its Padé approximants obey the following inequalities:*

$$[M/M+1] > [M-1/M] \qquad (*_*), \qquad\qquad (2.7)$$

$$[M+1/M+1] < [M/M] \qquad (*_*), \qquad\qquad (2.8)$$

$$[M-1/M] > [M/M-1] \qquad (_*{}^*), \qquad\qquad (2.9)$$

$$[M/M] < [M+1/M-1] \qquad (_*{}^*), \qquad\qquad (2.10)$$

$$[M/M] > [M-1/M] \qquad (**), \qquad\qquad (2.11)$$

$$[M/M]' > [M-1/M]' \qquad (**). \qquad\qquad (2.12)$$

Remark. The proof uses properties based on the determinantal inequalities for the coefficients f_j which are used to construct the Padé approximants. Like the other theorems of this section, it does not assume the representation (1.2) which characterizes the basic function.

Proof. From (3.5.18), the $(*_*)$ identity is

$$[M+J+1/M+1] - [M+J/M] = \frac{x^{2M+J+1}[D(1+J, M)]^2}{Q^{[M+J+1/M+1]}(x)Q^{[M+J/M]}(x)}.$$
$$(2.13)$$

From the proof of Theorem 5.2.1, the denominator functions $\Delta_M^{(J)}(x)$ (and $\Delta_{M+1}^{(J)}(x)$) are positive for $x \geqslant 0$. By virtue of (2.2), (2.7) and (2.8) follow from (2.13) with $J = -1$ and $J = 0$ respectively.

From (3.5.20), the $(_*{}^*)$ identity is

$$[M+J/M] - [M+J+1/M-1] = \frac{x^{2M+J+1}[D(2+J, M-1)]^2}{Q^{[M+J/M]}(x)Q^{[M+J+1/M-1]}(x)}$$
$$(2.14)$$

Hence (2.9) and (2.10) follow from (2.14) with $J = -1$ and $J = 0$ respectively.

From (3.5.17), the $(**)$ identity becomes

$$[M/M] - [M-1/M] = \frac{x^{2M}D(0, M)D(1, M-1)}{\Delta_M^{(0)}(x)\Delta_M^{(-1)}(x)}.$$
$$(2.15)$$

(2.11) follows from (2.15). The coefficients of every power of x in $\Delta_M^{(0)}(x)$ and $\Delta_M^{(-1)}(x)$ are positive; see Exercise 1. We define

$$R(x) \equiv x^{-2M}\Delta_M^{(0)}(x)\Delta_M^{(-1)}(x)$$

$$= \sum_{i=1}^{2M} r_i x^{-i}$$

and note that each $r_i > 0$, which implies that

$$R'(x) < 0 \qquad \text{for} \quad x > 0.$$

Hence $[M/M]' - [M-1/M]' > 0$ for $x > 0$.

COROLLARY. With the hypotheses of the theorem,

$$(-)^{J+1}\{[M+J+1/M+1] - [M+J/M]\} \geqslant 0,$$
$$(-)^{J+1}\{[M+J/M] - [M+J+1/M-1]\} \geqslant 0,$$

and these also hold when differentiated once.

Proof. The proof follows that of the theorem in every respect.

The essential result of Theorem 5.2.2 which is carried further is that the $[M-1/M]$ Padé approximants, for fixed $x > 0$, form a strictly increasing sequence. This leads to a convergence theorem after some preliminary theorems.

These inequalities [(2.11), for example] can be extended into the complex plane. There they become nesting inclusion regions for the higher-order approximants. They can be derived by exploiting Theorem 5.5.9 and allied results. The inclusion regions are convex, and lens-shaped. The two vertices of the boundaries are the $[M/M]$ and $[M-1/M]$ approximants. The new lens-shaped region touches the boundary of the old in only two places, as shown in Figure 2. A more thorough discussion is given in EPA and by Henrici and Pfluger [1966], Gargantini and Henrici [1967], and Baker [1969].

THEOREM 5.2.3. *The sequence of $[M-1/M]$ Padé approximants to a Stieltjes series is uniformly bounded as $M \to \infty$ in the domain $\mathcal{D}(\Delta)$. $\mathcal{D}(\Delta)$ is a bounded region of the complex z-plane which is at least at a distance Δ from the cut $-\infty < z \leqslant 0$ along the negative real axis.*

Proof. From Theorem 5.2.1, we may write

$$[M-1/M] = \sum_{i=1}^{M} \frac{\beta_i}{1+\gamma_i z}, \qquad (2.16)$$

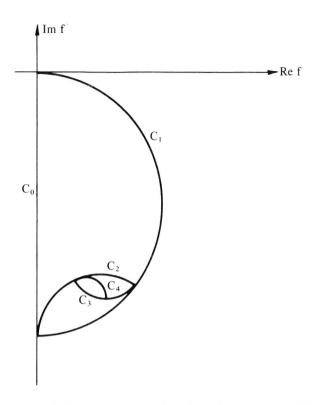

Figure 2. Nesting inclusion regions for the values of a staircase sequence of Padé approximants of a Stieltjes series.

where $\beta_i > 0$, $\gamma_i > 0$ for $i = 1, 2, \ldots, M$. Each of the Padé approximants may be bounded by

$$|[M-1/M]| \leqslant \sum_{i=1}^{M} \frac{\beta_i}{|1+\gamma_i z|}, \qquad (2.17)$$

and therefore

$$|[M-1/M]| \leqslant \sum_{i=1}^{M} \beta_i \qquad \text{if} \quad \text{Re}\, z \geqslant 0 \qquad (2.18)$$

The right-hand side of (2.18) is interpreted by taking $z=0$ in (2.16), so that

$$|[M-1/M]|_{z=0} = f_0 = \sum_{i=1}^{M} \beta_i, \qquad (2.19)$$

and thus $[M-1/M](z)$ is uniformly bounded for $\text{Re}\, z \geqslant 0$. If $\text{Re}\, z < 0$, because $\mathcal{D}(\Delta)$ is bounded we may assume that $\mathcal{D}(\Delta) \subset \{z, |z| < R_{\max}\}$ and

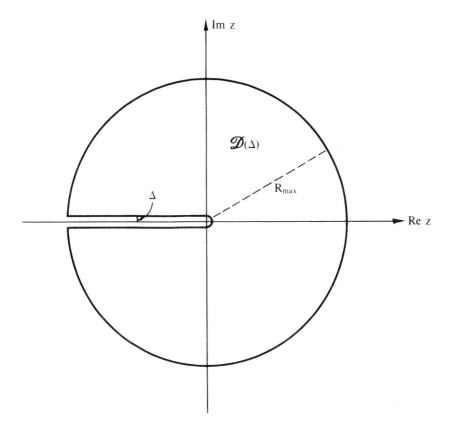

Figure 3. The bounded domain $\mathcal{D}(\Delta)$.

then

$$|[M-1/M]| \leqslant \sum_{i=1}^{M} \left| \frac{\beta_i}{1+\gamma_i z} \right|$$

$$\leqslant \left(\sum_{i=1}^{M} \beta_i \right) \sup_{\gamma>0} \left| \frac{1}{1+\gamma z} \right|$$

$$= f_0 \left[\inf_{\gamma>0} \sqrt{1+2\gamma \operatorname{Re}(z)+\gamma^2|z|^2} \right]^{-1}.$$

An elementary calculation shows that the minimum is achieved for $\gamma = -(\operatorname{Re} z)/|z|^2$, and also that

$$|[M-1/M]| \leqslant \frac{f_0|z|}{|\operatorname{Im} z|} \leqslant f_0 \frac{R_{\max}}{\Delta}. \tag{2.20}$$

Hence the theorem is proved

COROLLARY. *All paradiagonal sequences $[M+J/M]$ of Padé approximants (with $J \geqslant -1$) to Stieltjes series are uniformly bounded in the domain $\mathcal{D}(\Delta)$.*

Proof. For $J \geqslant 0$, the polynomial $\tilde{f}(z) = \Sigma_{i=0}^{J} f_i(-z)^i$ is uniformly bounded on $\mathcal{D}(\Delta)$ and may be treated separately. Then $(-z)^{-J-1}[f(z) - \tilde{f}(z)]$ may be expressed in the form (2.16). The rest of the proof is straightforward.

We have now established that the sequence of $[M-1/M]$ Padé approximants to a Stieltjes series is strictly increasing and bounded at any given point on the positive real axis, and so it is convergent. In fact, we are building up to a much stronger result than pointwise convergence, and this requires the concept of equicontinuity of a sequence [Courant and Hilbert, 1953, Chapter 2].

DEFINITION. *A sequence of functions $f_m(z)$, $m = 0, 1, \ldots$, defined on a domain $\mathcal{D}$ is equicontinuous if, given any $\varepsilon > 0$, there exists $\delta > 0$, depending only on ε, such that*

$$|f_m(z_1) - f_m(z_2)| < \varepsilon \qquad \text{for all} \quad m = 0, 1, \ldots, \tag{2.21}$$

for any pairs of points $z_1, z_2 \in \mathcal{D}$ satisfying $|z_1 - z_2| < \delta$.

The significant part of the latter definition is that $\delta = \delta(\varepsilon)$ does not depend on m, z_1 or z_2. Thus equicontinuity embraces the properties of uniform continuity of each member of the sequence with independence of which member of the sequence is selected.

THEOREM 5.2.4. *The sequence of $[M-1/M]$ Padé approximants to a Stieltjes series is equicontinuous on $\mathcal{D}(\Delta)$.*

Proof. From Theorem 5.2.1, we may write

$$[M-1/M] = \sum_{i=1}^{M} \frac{\beta_i}{1+\gamma_i z}, \qquad \beta_i > 0, \ \gamma_i > 0 \qquad \text{for} \quad i = 1, 2, \ldots, M.$$

Therefore

$$|[M-1/M](z_1) - [M-1/M](z_2)| \leqslant |z_1 - z_2| \sum_{i=1}^{M} \frac{\beta_i \gamma_i}{|1+\gamma_i z_1||1+\gamma_i z_2|}$$

From (2.16), we find that $\Sigma_{i=1}^{M} \beta_i \gamma_i = f_1$. By inspection of (2.20), we deduce that

$$|[M-1/M](z_1) - [M-1/M](z_2)| \leqslant |z_1 - z_2| f_1 \frac{R_{\max}^2}{\Delta^2}$$

provided $z_1, z_2 \in \mathcal{D}(\Delta)$. Hence the sequence of $[M-1/M]$ Padé approximants is equicontinuous on $\mathcal{D}(\Delta)$.

COROLLARY. All paradiagonal sequences $[M+J/M]$ of Padé approximants (with $J \geqslant -1$) to Stieltjes series are equicontinuous on $\mathcal{D}(\Delta)$.

Proof. Completely parallel to that of the theorem.

The property of equicontinuity established is exploited by Arzela's theorem:

THEOREM 5.2.5. *For any set of functions which are uniformly bounded and equicontinuous on $\mathcal{D}$, there exists a subsequence which converges uniformly to a continuous function defined on $\mathcal{D}$.*

Proof. Take a countable, dense set of points in $\mathcal{D}$. This is easily done by using a countable set of rationals $\{r_j\}$, which is dense on $(-\infty, \infty)$ and forming the countable set $\{z_{jk} = r_j + ir_k\}$. The subset of this contained in $\mathcal{D}$ is the point set required.

Let P_J be the set of the first J points, so that $P_J = \{z_i, i=1,2,\ldots,J\}$ is a subset of $\mathcal{D}$.

Let the given equicontinuous functions form a sequence S,

$$S = \{f_n(z), n=0,1,2,\ldots\}.$$

Because S is a sequence of functions which is uniformly bounded on $\mathcal{D}$, we may define S_1 to be a subsequence of S convergent at z_1, S_2 to be a subsequence of S_1 convergent at z_2, etc. Then, by construction, S_i is an infinite subsequence of S which converges at $z_1, z_2, \ldots, z_i$.

Given any $\varepsilon > 0$, define $\delta = \delta(\varepsilon)$ (2.20), so that

$$|f_n(z) - f_n(z')| < \varepsilon \qquad \text{for} \quad n=0,1,2,\ldots$$

whenever $|z - z'| < \delta$ and $z, z' \in \mathcal{D}$. δ is independent of n by the equicontinuity hypothesis.

Choose J sufficiently large so that the δ-neighborhoods of all the points P_J cover $\mathcal{D}$. This condition means that

$$\mathcal{D} \subset \bigcup_{j=1}^{J} \{z, |z_j - z| < \delta\},$$

and this choice is possible because the z_j are dense in $\mathcal{D}$. Then, for the chosen ε and any given $z \in \mathcal{D}$, z_j exists with $j \leqslant J$ and $|z_j - z| < \delta$. Because the sequence S_J converges at z_j, N exists such that

$$|f_m(z_j)-f_n(z_j)|<\varepsilon$$

for all $f_m, f_n \in S_J$, $m, n > N$, and $j = 1, 2 \ldots J$. By equicontinuity

$$|f_m(z_j)-f_m(z)|<\varepsilon,$$
$$|f_n(z_j)-f_n(z)|<\varepsilon,$$

and hence

$$|f_m(z)-f_n(z)|<3\varepsilon \qquad \text{for all} \quad f_m, f_n \in S_J, \quad m, n > N.$$

By choosing the subsequence of functions to be the Jth element of S_J, $J = 1, 2, \ldots$, we find a subsequence of the given sequence which satisfies Cauchy's condition for uniform convergence, and prove Arzela's theorem.

We have proved in Theorem 5.2.3 that the sequence of $[M+J/M]$ Padé approximants to a Stieltjes are uniformly bounded on $\mathcal{D}(\Delta)$, and Theorem 5.2.4 establishes that the sequence is equicontinuous on $\mathcal{D}(\Delta)$. Thus Arzela's theorem asserts that a subsequence of $[M+J/M]$ Padé approximants to a Stieltjes series converges uniformly to a continuous limit function $f^{(J)}(z)$ on $\mathcal{D}(\Delta)$.

Next we need a familiar theorem on uniformly convergent sequences of analytic functions:

WEIERSTRASS'S THEOREM [Titchmarsh, 1939, p. 95]. *Let each member of a sequence of functions* $g_1(z), g_2(z), g_3(z), \ldots$ *be analytic in a domain* $\mathcal{D}_1$ *and converge to a limit function* $g(z)$ *in any domain* $\mathcal{D}_2$ *in the interior of* $\mathcal{D}_1$. *Then* $g(z)$ *is analytic in* $\mathcal{D}_2$.

We apply Weierstrass's theorem directly to assert that $f^{(J)}(z)$ is analytic in $\mathcal{D}(2\Delta)$. But Δ was chosen (see Theorem 5.2.3) as an arbitrary small positive number, and can be replaced by $\frac{1}{2}\Delta$ without further implications. Thus we may deduce that for arbitrary positive Δ, $f^{(J)}(z)$ is analytic in $\mathcal{D}(\Delta)$.

Following Theorem 5.2.3, we noted that the entire sequence converges pointwise on the positive real axis, and so this pointwise limit is the real function $f^{(J)}(x)$. Since $f^{(J)}(x)$ is analytic on an interval of the real axis, the analytic continuation of $f^{(J)}(x)$ to $\mathcal{D}(\Delta)$, the domain of analyticity, is unique. Thus we have proved

THEOREM 5.2.6. *The sequence of* $[M+J/M]$ *Padé approximants of a Stieltjes series* (*with* $J \geqslant -1$) *converges uniformly on* $\mathcal{D}(\Delta)$, *as shown in Figure 2, to a real symmetric function* $f^{(J)}(z)$, *analytic on* $\mathcal{D}(\Delta)$.

Having established convergence of the paradiagonal sequences (with $J \geqslant -1$), the obvious questions are what the limit function $f^{(J)}(z)$ is and how convergence is achieved. We answer the second question first by showing that the power-series coefficients of $(-z)^j$ of the expansions of the $[M+J/M]$ Padé approximants approach the coefficients f_j from below.

THEOREM 5.2.7. *Let* $f(z)$ *be a Stieltjes series given by the formal power series*

$$f(z) = \sum_{i=0}^{\infty} f_i(-z)^i.$$

Its $[L/M]$ *Padé approximant has the power-series expansion*

$$[L/M] = \sum_{i=0}^{\infty} f_i^{[L/M]}(-z)^i. \tag{2.22}$$

Then, for all i and $L \geqslant M-1$,

$$0 \leqslant f_i^{[L/M]} \leqslant f_i \tag{2.23}$$

Proof. Each $[L/M]$ Padé approximant of $f(z)$ with $L \geqslant M-1$ exists, so that (2.22) is a well-defined series with a nontrivial circle of convergence for fixed L and M. If $i \leqslant L+M$, the Padé equations and (1.2) require that

$$0 \leqslant f_i^{[L/M]} = f_i.$$

Otherwise, for $i > L+M$, consider

$$f_i - f_i^{[L/M]} = \left(f_i^{[L+1/M+1]} - f_i^{[L/M]} \right)$$
$$+ \left(f_i^{[L+2/M+2]} - f_i^{[L+1/M+1]} \right)$$
$$+ \cdots$$
$$+ \left(f_i - f_i^{[L+K/M+K]} \right). \tag{2.24}$$

Provided $2K \geqslant i-L-M$, the last term vanishes by virtue of the Padé equations. Now consider the expansion of (2.6),

$$[M+J+1/M+1] - [M+J/M] = \frac{(-z)^{2M+J+1}[D(1+J, M)]^2}{\Delta_{M+1}(z)\Delta_M(z)}.$$

$$\tag{2.25}$$

At $z=0$, $\Delta_{M+1}(z)$ and $\Delta_M(z)$ are positive. The zeros of $\Delta_{M+1}(z)$ and $\Delta_M(z)$ occur at negative values. Thus we may write

$$\left[\Delta_M(z)\right]^{-1}=\alpha_0^{-1}\prod_{i=1}^{M}\left(1+\alpha_i z\right)^{-1} \qquad \text{with} \quad \alpha_0,\alpha_1,\ldots,\alpha_M>0$$

$$=\alpha_0^{-1}\prod_{i=1}^{M}\left[1-\alpha_i(-z)\right]^{-1}$$

$$=\alpha_0^{-1}\prod_{j=1}^{M}\left[1+\alpha_j(-z)+\alpha_j^2(-z)^2+\cdots\right]$$

$$=\alpha_0'+\alpha_1'(-z)+\alpha_2'(-z)^2+\cdots$$

$$\text{with}\quad \alpha_0',\alpha_1',\alpha_2',\ldots>0,$$

and similarly

$$\left[\Delta_M(z)\Delta_{M+1}(z)\right]^{-1}=\alpha_0''+\alpha_1''(-z)+\alpha_2''(-z)^2+\cdots$$

$$\text{with}\quad \alpha_0'',\alpha_1'',\alpha_2'',\cdots>0$$

This expansion, and therefore the expansion of (2.25), is a power series in $(-z)$ with positive coefficients. Hence every bracket of (2.24) is positive, and

$$f_i-f_i^{[L/M]}\geqslant 0,$$

proving the theorem.

As a general reference, we cite EPA, Chapter 15, and we also refer to Baker [1970], Common [1968], Wynn [1968], and Brezinski [1977, p. 82].

Exercise Use the proof of Theorem 5.2.1 to show that the coefficients of each power of x in the polynomials $\Delta_M^{(0)}(x)$ and $\Delta_M^{(-1)}(x)$ are positive.

5.3   Moment Problems and Orthogonal Polynomials

The principal question left open in the previous section is what the $[M+J/M]$ Padé approximants of Stieltjes series converge to. This question is part of a wider question: to what extent do the coefficients f_j determine the measure $d\phi(u)$? The latter is a moment problem. If it is true that for some positive measure $d\phi(u)$ defined on $-\infty<u<\infty$ the coefficients

$$f_j=\int_{-\infty}^{\infty}u^j\,d\phi(u), \qquad j=0,1,2,\ldots \tag{3.1}$$

are finite and well defined, then it is natural to call f_j the moments associated with the measure. The phraseology has a historical setting in which $\phi'(u)$ is a density per unit length, necessarily positive, of a linear mass distribution, such as a beam. If the beam has variable density, $\phi'(u)$ is not constant. If the beam has a weight attached at u_1, $\phi(u)$ has a positive jump discontinuity at u_1. The integration limits in (3.1) are set by the length of the beam, outside of which $d\phi(u)=0$. The mathematical questions which emerge from this physical setting have the name of moment problems. Given the values of all the moments $f_j, j=0,1,2,\ldots$, the problems are

(i) *Existence.* Does a positive measure $d\phi(u)$ exist to allow the representation of $\{f_j\}$ by (3.1)?

(ii) *Determinacy.* Is $d\phi(u)$ uniquely determined?

(iii) *Nature.* Are the $\{f_j\}$ Stieltjes moments? Are they Hamburger moments?

The answers to these problems depend on various conditions on the moments. We will explain the solutions as well as the problems in their various settings. Stieltjes gave a clear answer to some of the outstanding problems in the form of the following theorem.

THEOREM 5.3.1. *If the coefficients* $\{f_j, j=0,1,2\ldots\}$ *satisfy the determinantal conditions* $D(0,n)>0$, $D(1,n)>0$ *for all* $n=0,1,2,\ldots$, *then a Stieltjes measure* $d\phi(u)$ *exists for which*

$$f_j = \int_0^\infty u^j\, d\phi(u). \tag{3.2}$$

Further,

$$f(z) = \int_0^\infty \frac{d\phi(u)}{1+zu} \tag{3.3}$$

is a Stieltjes series with the given coefficients $\{f_j\}$ *in its formal expansion.*

Proof. An inductive proof based on Sylvester's identity establishes that $D(m,n)>0$ for all $m,n>0$ given that $D(0,n)>0$ and $D(1,n)>0$ for all n.

Consider the sequence of $[M-1/M]$ Padé approximants to the formal power series

$$\sum_{j=0}^\infty f_j(-z)^j. \tag{3.4}$$

Theorems 5.2.1, 2, 3, 4, 6, and 7 all concern Padé approximants and are based entirely on properties of the coefficients f_j — in fact, the properties that the determinants $D(m, n)$ are positive. Consequently, Theorem 5.2.6 is based on valid hypotheses, and the $[M-1/M]$ Padé approximants of (3.4) converge uniformly to an analytic function $f^{(-1)}(z)$ in $\mathcal{D}(\Delta)$ shown in Figure 3 of Section 5.2. We deduce from Theorem 5.2.1 that

$$[M-1/M] = \sum_{i=1}^{M} \frac{\beta_i}{1+\gamma_i z} \quad \text{with} \quad \beta_i, \gamma_i > 0 \qquad \text{for} \quad i = 1, 2, \ldots, M.$$

$$= \sum_{i=1}^{M} \frac{\beta_i(1+\gamma_i z^*)}{[1+\gamma_i \operatorname{Re} z]^2 + [\operatorname{Im} z]^2}. \tag{3.5}$$

From (3.5),

$$\operatorname{Im}[M-1/M](z) < 0 \qquad \text{if} \quad \operatorname{Im} z > 0, \tag{3.6}$$

$$\operatorname{Im}[M-1/M](z) > 0 \qquad \text{if} \quad \operatorname{Im} z < 0, \tag{3.7}$$

$$\operatorname{Re}[M-1/M](z) = \operatorname{Re}\{[M-1/M](z)\}^*, \tag{3.8}$$

$$\operatorname{Im}[M-1/M](z) = -\operatorname{Im}\{[M-1/M](z)\}^*. \tag{3.9}$$

Let us apply Cauchy's theorem to $[M-1/M]$ using a contour C which is the boundary of $\mathcal{D}(\Delta)$, and inside which $[M-1/M](z)$ is analytic. This contour is shown in Figure 1, and Cauchy's theorem states that

$$[M-1/M](z) = \frac{1}{2\pi i} \int_C \frac{[M-1/M](\omega)\,d\omega}{\omega - z}.$$

We may now take the limit as $R' \to \infty$, noting that the contribution from the large circle tends to zero, and obtain, using (3.8) and (3.9),

$$[M-1/M](z) = \frac{1}{\pi} \int_{-\infty}^{0} \frac{\operatorname{Im}[M-1/M](\omega + i\Delta)}{\omega - z}\,d\omega \tag{3.10}$$

For given $\Delta > 0$, we may let $M \to \infty$, giving

$$\lim_{M \to \infty} [M-1/M](z) = \frac{1}{\pi} \int_{-\infty}^{0} \frac{\operatorname{Im} f^{(-1)}(\omega + i\Delta)}{\omega - z}\,d\omega$$

$$= -\frac{1}{\pi} \int_{0}^{\infty} \operatorname{Im} f^{(-1)}\left(\frac{-1}{t+i\varepsilon}\right) \frac{dt}{t(1+zt)}. \tag{3.11}$$

For any open interval (u, v) of the negative real axis, we may choose

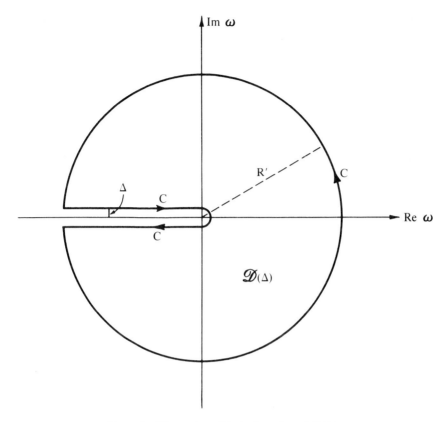

Figure 1.   The contour C is the boundary of $\mathscr{D}(\Delta)$.

$\Delta \ll v-u,\ \varepsilon \ll v-u$ and consider

$$\phi(v-)-\phi(u+)=\int_{u}^{v}d\phi(u)$$

$$=\lim_{\varepsilon \downarrow 0}\int_{u}^{v}-\frac{1}{\pi t}\operatorname{Im} f^{(-1)}\left(\frac{-1}{t+i\varepsilon}\right)dt \qquad (3.12)$$

provided the limit is well defined. The details are given in Section 5.6. Equation (3.12) provides a construction of a Stieltjes measure so that (3.11) may be written as

$$\lim_{M \to \infty}[M-1/M](z)=f^{(-1)}(z)=\int_{0}^{\infty}\frac{d\phi(u)}{1+zu} \qquad (3.13)$$

for z not on the negative real axis, $-\infty<z\leqslant 0$. We have avoided any discussion of the value of $\phi(u)$ at a jump discontinuity, because this value is

usually of no importance, and we say that $\phi(u)$ is substantially determined by (3.12). Equally, we may take $\phi(0)=0$ without loss of generality. The theorem is now proved.

It is important to realize that this theorem answers the questions of existence and nature, but not the question of uniqueness. Proofs of uniqueness are presently based on further hypotheses, as we discuss in Sections 5.4 and 5.5.

Let us now consider a change of variable $w=-z^{-1}$ which reveals a remarkable connection between Padé approximants and orthogonal polynomials. It is, in fact, the historical approach to orthogonal polynomials, and rightly so.

We consider the basic initial representation

$$f(z)=\int_0^\infty \frac{d\phi(u)}{1+zu} \tag{3.14}$$

and rearrange it and define $F(w)$ by

$$F(w)=\int_0^\infty \frac{d\phi(u)}{w-u}=\frac{1}{w}f\left(-\frac{1}{w}\right), \tag{3.15}$$

which has the formal expansion about $w=\infty$,

$$F(w)=\sum_{i=0}^\infty \frac{f_i}{w^{i+1}}=\frac{f_0}{w}+\frac{f_1}{w^2}+\cdots. \tag{3.16}$$

If we construct a set of polynomials $\{\pi_m(u), \ m=0,1,2\cdots\}$ orthogonal over $d\phi(u)$, we fundamentally require that

$$\int_0^\infty \pi_m(u)u^k\,d\phi(u)=0 \qquad \text{for} \quad k=0,1,2,\ldots,m-1. \tag{3.17}$$

This equation is tantamount to the orthogonality condition

$$\int_0^\infty \pi_m(u)\pi_k(u)\,d\phi(u)=0 \qquad \text{for} \quad k=0,1,2,\ldots,m-1, \tag{3.18}$$

and we have taken the usual assumption that each $\pi_k(u)$ is a polynomial of degree precisely equal to k for granted.

By writing

$$\pi_m(u)=u^m \sum_{i=0}^m \beta_i^{(m)}u^{-i}, \tag{3.19}$$

the equation (3.17) becomes

$$\sum_{i=0}^{m} \beta_i^{(m)} f_{m+k-i} = 0, \qquad k=0,1,\ldots,m-1.$$

With the identification $f_j = (-1)^j c_j$ and by taking $\beta_0^{(m)} = 1$, this set of linear equations is seen to be the Padé equations (1.1.6), with the solution $b_i = (-1)^i \beta_i^{(m)}$ for $i=0,1,\ldots,m$. Hence (3.19) becomes

$$\pi_m(u) = u^m B^{[m-1/m]}(-1/u).$$

Since $C(m-1/m) \neq 0$ for Stieltjes series, we prefer to use the equivalent but conventionally normalized orthogonal polynomials given by $\pi_0(u) = 1$ and

$$\pi_m(u) = u^m Q^{[m-1/m]}(-1/u), \qquad m=1,2,3,\ldots. \tag{3.20}$$

These polynomials $\pi_m(u)$, defined by (3.20), satisfy the orthogonality condition (3.18). A natural observation at this point is that the $(*\,*\,*)$ identity (Section 3.5) for three consecutive $Q^{[m-1/m]}(z)$ denominators immediately is interpreted as the recurrence relation for the orthogonal polynomials.

As a corollary to this development, we observe that the set

$$\pi_m^{(J)}(u) = u^m Q^{[m+J/m]}(-u^{-1}), \qquad m=0,1,2,\ldots, \tag{3.21}$$

are orthogonal polynomials over $u^{J+1} d\phi(u)$ for any $J \geqslant -1$.

Next, we proceed with the converse development. We assume the usual properties of the orthogonal polynomials $\{\pi_m(u), n=0,1,2,\ldots\}$ and define polynomials (3.30) whose ratio is the Padé approximant of $f(z)$ defined by (3.14).

To determine the numerator of the Padé approximant to $f(z)$, or of $F(w)$ expanded about $w=\infty$, recall (3.15),

$$F(w) = \frac{1}{w} f\left(\frac{-1}{w}\right) = \int_0^\infty \frac{d\phi(u)}{w-u}, \qquad f(z) = wF(w) = -\frac{1}{z} F\left(-\frac{1}{z}\right),$$

which leads us to expect

$$[m-1/m]_f(z) = \frac{P^{[m-1/m]}(z)}{Q^{[m-1/m]}(z)} = \frac{w^m P^{[m-1/m]}(-1/w)}{w^m Q^{[m-1/m]}(-1/w)} = \frac{w \rho_m(w)}{\pi_m(w)},$$

$$\tag{3.22}$$

where $\rho_m(w)$ is a polynomial of degree $m-1$. Thus we are led to consider

$$\pi_m(w)F(w)=\int_0^\infty \frac{d\phi(u)}{w-u}\pi_m(w)$$

$$=\int_0^\infty \frac{\pi_m(w)-\pi_m(u)}{w-u}d\phi(u)+\int_0^\infty \frac{\pi_m(u)\,d\phi(u)}{w-u}. \quad (3.23)$$

Equation (3.23) splits into two parts, and we find

$$\pi_m(w)F(w)-\rho_m(w)+\varepsilon(w) \qquad (3.24)$$

where we will find that $\varepsilon(w)$ plays the role of an error. The first part is

$$\rho_m(w)\equiv\int_0^\infty \frac{\pi_m(w)-\pi_m(u)}{w-u}d\phi(u) \qquad (3.25)$$

$$=\int_0^\infty \{\text{polynomial in } w, u \text{ of degree } m-1\}\,d\phi(u)$$

$$=\text{polynomial in } w \text{ of degree } m-1.$$

A glimpse forward to (3.30) explains why it is convenient to use a subscript m for a polynomial of degree $m-1$ in this instance. The second part of (3.24) is

$$\varepsilon(w)\equiv\int_0^\infty \frac{\pi_m(u)}{w-u}d\phi(u)$$

$$=\frac{1}{w}\int_0^\infty \pi_m(u)(1-u/w)^{-1}d\phi(u)$$

$$=\frac{1}{w}\int_0^\infty \left\{\left[1+\frac{u}{w}+\cdots+\left(\frac{u}{w}\right)^{m-1}\right]\right.$$

$$\left.+\left(\frac{u}{w}\right)^m\left(1-\frac{u}{w}\right)^{-1}\right\}\pi_m(u)\,d\phi(u)$$

$$=\frac{1}{w^{m+1}}\int_0^\infty \frac{u^m\,d\phi(u)}{1-u/w}\pi_m(u), \qquad (3.26)$$

where the orthogonality property (3.17) has been used. From (3.23),

$$F(w)=\frac{\rho_m(w)}{\pi_m(w)}+\frac{\varepsilon(w)}{\pi_m(w)}. \qquad (3.27)$$

Using the O-notation in the sense of formal series operations, we find from

(3.20) and (3.26) that

$$\varepsilon(w)=O(w^{-m-1}), \qquad \pi_m(w)=O(w^m),$$

and hence from (3.26)

$$F(w)=\frac{\rho_m(w)}{\pi_m(w)}+O(w^{-2m-1}). \tag{3.28}$$

Recalling from (3.25) and (3.20) that $\rho_m(w)$ and $\pi_m(w)$ are polynomials of degrees $m-1$ and m respectively, we have proved that

$$f(z)=wF(w)=\frac{w\rho_m(w)}{\pi_m(w)}+O(w^{-2m}), \tag{3.29}$$

where

$$\frac{w\rho_m(w)}{\pi_m(w)}=\frac{w^{-m+1}\rho_m(w)}{w^{-m}\pi_m(w)}=\frac{(-z)^{m-1}\rho_m(-1/z)}{(-z)^m\pi_m(-1/z)}.$$

Hence, following (3.20), we define

$$\tilde{P}^{[m-1/m]}(z)=(-z)^{m-1}\rho_m(-z^{-1}),$$
$$\tilde{Q}^{[m-1/m]}(z)=(-z)^m\pi_m(-z^{-1}), \tag{3.30}$$

with ρ_m, π_m defined by (3.19) and (3.25). This proves that

$$[m-1/m]_f(z)=\frac{\tilde{P}^{[m-1/m]}(z)}{\tilde{Q}^{[m-1/m]}(z)}. \tag{3.31}$$

Hence we see that, except for normalization, $\tilde{P}^{[m-1/m]}(z)$ and $\tilde{Q}^{[m-1/m]}(z)$ are the numerator and denominator polynomials of Padé approximants. Furthermore, for (3.15), (3.24), (3.26), (3.30), and (3.31) we have the explicit error formula

$$f(z)-[m-1/m]=\frac{(-z)^m}{\pi_m(-z^{-1})}\int_0^\infty\frac{u^m\pi_m(u)\,d\phi(u)}{1+zu}.$$

Other formulas of this kind are given in Part II, Section 3.1.

   Equations (3.14)–(3.31) show a different approach to the construction of Padé approximants. As a bonus, (3.25) indicates that the polynomials $\rho_m(w)$ satisfy exactly the same recurrence relation as $\pi_m(w)$, the proof needing no more than the orthogonality property (3.17).

For general reviews of the scope of this section, we refer to Allen et al. [1975], and Karlson and von Sydow [1976], who show that much of theory of Sections 5.1–5 can be derived using orthogonality methods.

Exercise 1. The Laguerre polynomials $L_n(u)$ satisfy an orthogonality condition of the type (3.18). Deduce that the associated Stieltjes function is given by

$$f(z)=z^{-1}e^{1/z}E_1(1/z),$$

where $E_1(z)$ is defined by (4.6.8) in $|\arg(z)|<\pi$. Use (3.21) and (3.25) to find the $[m-1/m]$ Padé approximant of $f(z)$.

Exercise 2. Only the first $2M+1$ moments, $f_0, f_1, \ldots, f_{2M}$ of a Stieltjes density (1.2) are given. Use the formal expansions

$$f(z)= \sum_{j=0}^{\infty} f_j(-z)^j \quad \text{and} \quad [M/M]_f= \sum_{j=0}^{\infty} f_j^{[M/M]}(-z)^j$$

to prove that the unknown moments are bounded by

$$f_j \geqslant f_j^{[M/M]} \qquad \text{for} \quad j \geqslant 2M+1.$$

What bound would you use if $2M+2$ moments were given?

Exercise 3. Prove the orthogonality property (3.17) using the method of Exercise 3 of Section 5.1.

5.4 Stieltjes Series Convergent in $|z|<R$

If the Stieltjes series discussed in the previous sections have a nonzero radius of convergence, which is to say that they are analytic in a neighborhood of the origin, then convergence theorems are easily proved and the moment problem is determinate. This section is devoted to the results ensuing from the hypothesis that $f(z)$ is a Stieltjes series with a nonzero radius of convergence, and they may be contrasted with the results of Section 5.5.

The property that the poles of the Padé approximants of $f(x)$ are on the cut of $f(z)$ is retained. In fact, we have the stronger result:

THEOREM 5.4.1. *Let* $f(z)$ *be a Stieltjes series convergent in* $|z|<R$. *Then the poles of the* $[M+J/M]$ *Padé approximant, with* $J \geqslant -1$, *lie on the real axis in the interval* $-\infty<z<-R$.

Method 1. Suppose the contrary. From Theorem 5.2.1, there is a pole at $z=z_0$ with $-R \leqslant z_0 < 0$ of the $[M_1+J_1/M_1]$ Padé approximant, with $J_1 \geqslant -1$. The interlacing property implies that every $[M+J_1/M]$ Padé approximant with $M \geqslant M_1$ has a pole in the interval $(z_0,0)$, and let the limit of the poles nearest to $z=0$ be at $z=z_1$. Then from Theorem 5.2.7,

$$R^{-1}\limsup_{m\to\infty}(f_m)^{1/m}\geqslant R^{-1}\limsup_{m\to\infty}(f_m^{[L/M]})^{1/m}=|z_1|^{-1}.$$

Hence, $|z_1|\geqslant R$, contradicting the hypothesis. Therefore, the poles of the Padé approximant lie on the open interval $(-\infty,-R)$.

Method 2. If $f(z)$ is analytic in $|z|<R$, its Stieltjes-integral representation can have no singularities in this circle, and so becomes

$$f(z)=\int_0^{R^{-1}}\frac{d\phi(u)}{1+zu}. \tag{4.1}$$

We may also take $d\phi(u)=0$ on $R^{-1}<u<\infty$.

The poles of the $[m+J/m](z)$ Padé approximant occur at zeros of $\pi_m(u)$, where $z=-u^{-1}$ and $\pi_m(u)$ is a polynomial satisfying the orthogonality conditions (3.17):

$$\int_0^\infty u^j\pi_m(u)u^{J+1}d\phi(u)=0 \qquad\text{for}\quad j=0,1,2,\dots,n-1.$$

Suppose that m_1 zeros of $\pi_m(u)$ do not lie in $0<u<R^{-1}$, but elsewhere in the complex u-plane at $u=u_1,u_2,\dots,u_{m_1}$. Then $\pi_m(u)$ has the representation

$$\pi_m(u)=\kappa\prod_{i=1}^{m_1}(u-u_i)\prod_{j=m_1+1}^{m}(u-u_j),$$

where κ is a normalization constant. Consider

$$I=\int_0^{R^{-1}}c\left\{\prod_{j=m_1+1}^{m}(u-u_j)\right\}\pi_m(u)u^{J+1}d\phi(u).$$

The integral of I is strictly positive, having no sign changes at the points $u_{m_1+1},\dots,u_m$, where it vanishes. But, by orthogonality, $I=0$. Thus $m_1=0$, all zeros of $\pi_m(u)$ lie in $(0,R^{-1})$, and all poles of the $[M+J/M]$ Padé approximant lie on the cut of $f(z)$.

The next theorem concerns the limit functions $f^{(J)}(z)$ of the paradiagonal sequence $[M+J/M]$ of Padé approximants of a Stieltjes series $f(z)$, which are shown to be identical to $f(z)$. Again, we give two methods of proof, based on the integral representation and on orthogonal polynomials.

THEOREM 5.4.2.   *Let $f^{(J)}(z)$ be the limit functions of $[M+J/M]$ Padé approximants with $J\geqslant-1$, to a Stieltjes series $f(z)$, which are analytic in $\mathcal{D}(\Delta)$. If $f(z)$ is analytic in $|z|<R$, then $f^{(J)}(z)=f(z)$ for all $J\geqslant-1$.*

Method 1.   The hypothesis of the theorem implies that

$$f(z) = \int_0^{R^{-1}} \frac{d\phi(u)}{1+zu}$$

as in (4.1). The analytic structure is shown in Figure 1.

From Theorem 5.2.6, we see that $f^{(J)}(z), f(z)$ have an identical Maclaurin expansion. Since $f(z)$ is analytic in $|z| < R$, $f^{(J)}(z)$ is identical to $f(z)$ for any $J \geqslant -1$, and the theorem is proved.

Method 2.   The explicit error formula of (3.27) is used. First we consider the case of $J = -1$ and make the usual changes of variable

$$w = -z^{-1}, \qquad F(w) = -zf(z).$$

Equation (3.26) and (3.27) give

$$\left| F(w) - \frac{\rho_m(w)}{\pi_m(w)} \right| = \left| \frac{\varepsilon(w)}{\pi_m(w)} \right|$$

$$\leqslant \frac{1}{|w|^{m+1}} \int_0^{R^{-1}} \frac{u^m\, d\phi(u)}{|1 - u/w|} \left| \frac{\pi_m(u)}{\pi_m(w)} \right|.$$

From Theorem 5.4.1, the zeros of $\pi_m(u)$ occur at $u = u_i$, $i = 1, 2, \ldots, m$, and

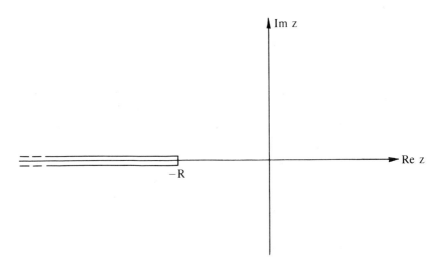

Figure 1.   The domain of analyticity of $f(z)$.

lie in $(0, R^{-1})$. Therefore

$$\left| F(w) - \frac{\rho_m(w)}{\pi_m(w)} \right| \leqslant \frac{1}{|wR|^m} \int_0^{R^{-1}} \frac{d\phi(u)}{|w-u|} \prod_{i=1}^m \left| \frac{u-u_i}{w-u_i} \right|.$$

Let k be an arbitrary constant greater than 1; then provided $|w| > kR^{-1}$,

$$\left| F(w) - \frac{\rho_m(w)}{\pi_m(w)} \right| \leqslant (wR)^{-m}(k-1)^{-m-1} R \int_0^R {}^1 d\phi(u). \qquad (4.2)$$

This proves convergence of the sequence of $[m-1/m]$ Padé approximants in $|w| > 2R^{-1}$ to $f(z)$. If $J > -1$, the first $J+1$ terms of the series must be treated explicitly, so that the problem of $[M+J/M]$ Padé approximants of $f(z)$ reduces to that of $[M-1/M]$ Padé approximants to $[f(z) - \sum_{j=0}^J (-z)^j f_j]/z^{J+1}$.

The second method shows that the rate of convergence is geometrical, as one would expect from the order notation. The sharper result of Theorem 5.4.4 is obtainable by using Schwarz lemma in conjunction with the first method, but the sharpest result (see page 193) follows from a refinement of method 2.

THEOREM 5.4.3 (Schwarz's lemma). *If $f(z)$ is analytic in $|z| < R$ and continuous on $|z| \leqslant R$, and further*

$$f(0) = f'(0) = \cdots = f^{(n)}(0) = 0$$

and

$$\max_{|z|=R} |f(z)| = M,$$

then

$$|f(z)| \leqslant M \left| \frac{z}{R} \right|^{n+1} \qquad \text{if} \quad |z| \leqslant R.$$

Proof. Apply the maximum-modulus theorem to

$$g(z) = z^{-(n+1)} f(z).$$

$g(z)$ is analytic in $|z| \leqslant R$, and so $|g(z)| < R^{-n-1} M$ on $|z| = R$. Therefore the maximum-modulus theorem asserts that

$$|g(z)| \leqslant R^{-(n+1)} M \qquad \text{for} \quad |z| \leqslant R$$

and

$$|f(z)| \leq \left|\frac{z}{R}\right|^{n+1} M \qquad \text{for} \quad |z| \leq R,$$

proving the theorem.

The convergence of Padé approximant to Stieltjes series is much faster than that suggested by the formulas of Theorem 5.4.2 for $\text{Re}\, z > 0$. The following theorem establishes an interesting result about the rate of convergence, and gives a clear idea of what can be expected in general, at best, from the Padé method. This is the first theorem of the chapter to extend beyond paradiagonal sequences, and is possible because convergence to a common limit is established. We are concerned with convergence in the bounded domain shown in Figure 2, and called $\mathcal{D}^{+}(\Delta)$ because convergence is proved in a domain larger than $\mathcal{D}(\Delta)$. $\mathcal{D}^{+}(\Delta)$ is defined by $|z| < R_{max}$, but such that all points of $\mathcal{D}^{+}(\Delta)$ are at least a distance Δ from $-\infty < z < -R$.

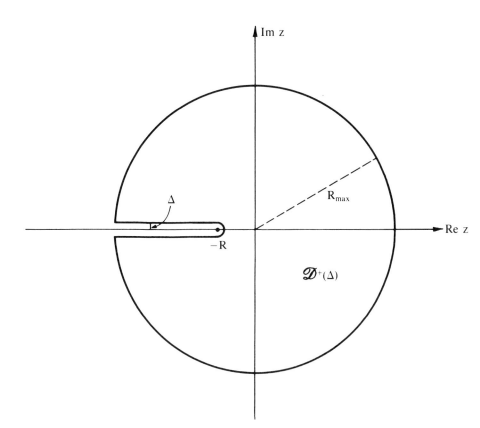

Figure 2. The domain $\mathcal{D}^{+}(\Delta)$.

THEOREM 5.4.4 [Baker, EPA p. 220]. *Let $f(z)$ be a Stieltjes series with radius of convergence $R > 0$. Let $\{P_k(z)\}$ be any sequence of $[M_k + J_k/M_k]$ Padé approximants with $J_k \geqslant -1$ of $f(z)$. Let $\rho = R - \Delta > 0$. Then convergence in $\mathcal{D}^+(\Delta)$ is given by*

$$|P_k(z) - f(z)| \leqslant \left|\frac{z}{\rho}\right|^{J_k} \left|\frac{\sqrt{\rho + z} - \sqrt{\rho}}{\sqrt{\rho + z} + \sqrt{\rho}}\right|^{2M_k} \cdot \text{constant}. \qquad (4.3)$$

Comment 1.   For paradiagonal sequences, $[M + J/M]$, with $J_k = J$ fixed, convergence follows at a rate

$$|P_k(z) - f(z)| \leqslant \left|\frac{\sqrt{\rho + z} - \sqrt{\rho}}{\sqrt{\rho + z} + \sqrt{\rho}}\right|^{2M} \cdot \text{constant}. \qquad (4.4)$$

Comment 2.   For any ray sequence of $[L/M]$ Padé approximants, with $L = \lambda M$ and $\lambda \geqslant 1$, convergence follows at a rate

$$|P_k(z) - f(z)| \leqslant \left|\frac{\sqrt{\rho + z} - \sqrt{\rho}}{\sqrt{\rho + z} + \sqrt{\rho}}\right|^{2M} \left|\frac{z}{\rho}\right|^{(\lambda - 1)M} \cdot \text{constant}. \qquad (4.5)$$

Proof. Let us consider $J \geqslant 0$, and $[M + J/M]$ Padé approximants to $f(z)$. Then, from Section 5.2 , the results

$$[M + J/M] - \sum_{i=0}^{J} f_i(-z)^i = \sum_{i=1}^{M} \frac{\beta_i}{1 + \gamma_i z} \qquad (4.6)$$

and

$$\left|[M + J/M] - \sum_{i=0}^{J} f_i(-z)^i\right| \leqslant \text{constant} \qquad (4.7)$$

follow for $z \in \mathcal{D}(\Delta)$ and all $M, J \geqslant 0$. Clearly (4.7) extends to $\mathcal{D}^+(\Delta)$ with a minor modification of the proof. Using Schwarz's lemma, (4.7) may be altered to

$$\left|[M + J/M] - \sum_{i=0}^{J} f_i(-z)^i\right| < \left|\frac{z}{\rho}\right|^{J+1} \cdot \text{constant}$$

for $|z| < \rho$ and so also for $z \in \mathcal{D}^+(\Delta)$ and all $M, J \geqslant 0$. Further, for $z \in \mathcal{D}^+(\Delta)$,

$$\left|\sum_{i=0}^{J} f_i(-z)^i - f(z)\right| < \left|\frac{z}{\rho}\right|^{J+1} \cdot \text{constant},$$

and hence

$$\lvert [M+J/M]-f(z)\rvert < \left\lvert \frac{z}{\rho} \right\rvert^{J+1} \cdot \text{constant}. \qquad (4.8)$$

Consider the special mapping

$$z = \frac{4\rho w}{(1-w)^2}, \qquad w = \frac{\sqrt{\rho+z}\ -\sqrt{\rho}}{\sqrt{\rho+z}\ +\sqrt{\rho}}. \qquad (4.9)$$

The unit circle in the w-plane is given parametrically by $w = e^{i\theta}$, $0 \leqslant \theta < 2\pi$. This becomes

$$\sqrt{\rho+z}\ -\sqrt{\rho} = e^{i\theta}\left(\sqrt{\rho+z}\ +\sqrt{\rho}\right),$$

or

$$\sqrt{\rho+z} = (e^{i\theta}+1)(1-e^{i\theta})^{-1}\sqrt{\rho},$$

and by squaring,

$$z = -2\rho(1-\cos\theta)^{-1},$$

which is the parametric equation of $-\infty < z \leqslant \rho$. The mapping is origin preserving and conformal except at $w=1$. Let us reconsider (4.8) in the form

$$\left\lvert \frac{[M+J/M](z(w))-f(z(w))}{z(w)^{J+1}} \right\rvert < \rho^{-J-1}c,$$

which is valid for all $z \in \mathfrak{D}^+(\Delta)$, where c is a positive constant independent of $z(w)$, J, and M. For $z \to 0$, the left-hand side is of order z^{2M}, i.e., of order w^{2M}, and so the Schwarz lemma sharpens this result to

$$\lvert [M+J/M]z(w)) - f(z(w)) < \left(\frac{z}{\rho}\right)^{J+1} w^{2M}c,$$

which is finally written as

$$\lvert [M+J/M]-f(z)\rvert < \left(\frac{z}{\rho}\right)^{J+1} \left\lvert \frac{\sqrt{\rho+z}\ -\sqrt{\rho}}{\sqrt{\rho+z}\ +\sqrt{\rho}} \right\rvert^{2M} c, \qquad (4.10)$$

which is also valid for $z \in \mathfrak{D}^{-1}(\Delta)$. The case of $J=-1$ is simpler than the general case of $J \geqslant 0$, because subtraction of the first $J+1$ terms is unnecessary. The proof is otherwise unaltered, and the theorem is proved.

An interesting realization of the implications of this theorem is the application to $f(z) = z^{-1} \ln(1+z)$. This is a Stieltjes series cut along $-\infty < z \leq -1$. We consider the ray sequence of $[4M/M]$ Padé approximants. At a point z, (4.10) implies that the ray sequence converges provided

$$|z|^{3/2} |\sqrt{1+z} - 1| < |\sqrt{1+z} + 1|. \tag{4.11}$$

(4.11) defines a heart-shaped region shown in Figure 3. For this particular function, divergence of the Padé approximant is given by the poles and zeros. The poles are located on the cut, where divergence is obligatory, and so are the interlacing zeros, but the other zeros map out a heart-shaped region of empirical divergence. Those shown in Figure 3 are taken from the $[20/5]$, $[16/4]$, $[12/3]$, $[8/2]$, and $[4/1]$ Padé approximants. It is obvious that the proven region of convergence is smaller than the empirical region, but it is remarkable that the shape is correct and the scales are very comparable. In fact, the zeros of the numerators of a superdiagonal ray sequence of Padé approximants of $z^{-1} \ln(1+z)$ delineate a pierced-heart shaped boundary of convergence, within which convergence has been proved for the same ray sequence of Padé approximants of any Stieltjes series with unit radius of convergence [Graves-Morris, 1981].

Our next theorem about Stieltjes series with a nonzero radius of convergence R is that the moment problem is determinate. This means that the coefficients f_j have the unique representation

$$f_j = \int_0^{R^{-1}} u^j \, d\phi(u). \tag{4.12}$$

This is a corollary of the following simple theorem.

THEOREM 5.4.5. *The moment problem for a finite interval is determinate.*

Proof. Suppose $d\phi_1(u)$ and $d\phi_2(u)$ are two different measures for which

$$f_j = \int_a^b u^j \, d\phi_1(u) = \int_a^b u^j \, d\phi_2(u), \qquad j = 0, 1, 2, \ldots.$$

Then $\int_a^b p(u) \, d(\phi_1 - \phi_2) = 0$ for every polynomial $p(u)$. Using Weierstrass's approximation theorem, $\int_a^b \psi(u) \, d(\phi_1 - \phi_2) = 0$ for every continuous function $\psi(u)$. Since $\phi_1(u) - \phi_2(u)$ is of bounded variation, $\phi_1(u) - \phi_2(u) -$ constant, except possibly at the common points of discontinuity of ϕ_1 and ϕ_2.

In summary, if $f(z)$ is a Stieltjes series with a nonzero radius of convergence, we have found that $[M+J/M]$ paradiagonal sequences with $J \geq -1$ converge uniquely and determine a unique solution of the moment problem.

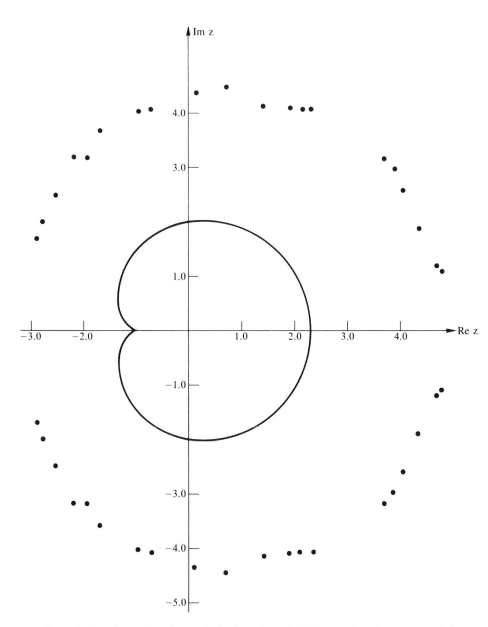

Figure 3. The heart shaped curve is the boundary of (4.11). The dots denote zeros of the [$4N/N$] Padé approximants, $N = 1, 2, 3, 4, 5$.

5.4.1   Hausdorff Moment Problem

If the Stieltjes series defined in (4.1) has radius of convergence $R = 1$, the moment problem is called the Hausdorff moment problem. In fact, a simple change of scale of the variable z in (4.1) allows us to assume that $R = 1$ without loss of generality and to consider Stieltjes series defined by

$$f(z) - \int_0^1 \frac{d\phi(u)}{1 + zu} = \sum_{j=0}^{\infty} f_j(-z)^j, \tag{4.13a}$$

where

$$f_j = \int_0^1 u^j d\phi(u), \qquad j = 0, 1, 2, \ldots, \tag{4.13b}$$

and $\phi(u)$ is a bounded nondecreasing function defined on $0 \leqslant u \leqslant 1$. The moments defined by (4.13b) are called Hausdorff moments, and the Hausdorff moment problem consists of constructing $\phi(u)$ from the given sequence $f_0, f_1, f_2, \ldots$. This we have done in (3.12) using Padé approximants. To make the connection between the Hausdorff moment problem and totally monotone sequences, we define Δ, the forward difference operator, by

$$\Delta f_j = f_{j+1} - f_j, \qquad j = 0, 1, 2, \ldots,$$

and higher differences are defined similarly (see Section 3.1).

DEFINITION. A sequence $\{f_j\}$ is *totally monotone* if

$$(-)^k \Delta^k f_j \geqslant 0 \qquad \text{for all} \quad j, k \geqslant 0. \tag{4.14}$$

This definition immediately implies that a totally monotone sequence $\{f_j\}$ is a positive decreasing sequence.

THEOREM 5.4.6. *The sequence defined by* (4.13) *is totally monotone.*

Proof. By inspection of (4.13), we see that

$$(-)^k \Delta^k f_j = \int_0^1 (1 - u)^k u^j d\phi(u) \geqslant 0$$

for all $j, k \geqslant 0$, proving the result.

The converse of this result—that if a sequence $\{f_j\}$ is totally monotone, then it has a Stieltjes integral representation (4.13)—is also true. The best proof does not use Padé approximants; we refer the reader to the proof in the books by Wall [1948, p. 267] and Widder [1972, p. 109].

For further details, we refer to Brezinski [1978a], Gragg [1968], and Wynn [1966b].

5.4.2   *Integer Moment Problem*

If the Stieltjes moments defined by

$$f_j = \int_0^\Lambda u^j \, d\phi(u), \qquad j=0,1,2,\dots, \tag{4.15}$$

are known to be integers, construction of the Stieltjes function

$$f(z) = \int_0^\Lambda \frac{d\phi(u)}{1+zu}$$

is an integer moment problem [Barnsley et al., 1979]. We consider the simplest cases, $\Lambda=1$ and $\Lambda=2$, as examples.

Example 1 ($\Lambda=1$).   In this case, it follows from (4.15) that the moments $\{f_j, j=0,1,2,\dots\}$ form a positive decreasing sequence. Therefore f_∞ exists such that $f_j \to f_\infty$, and f_∞ is an integer. Hence n exists such that $f_j = f_\infty$ for all $j \geqslant n$, and we find that

$$f_n - f_{n-1} = \int_0^1 (1-u)u^n \, d\phi(u) = 0,$$

$$d\phi(u) = \left[A\delta(u) + B\delta(u-1) \right] du,$$

and

$$f_j = f_1 \qquad \text{for all} \quad j \geqslant 1.$$

The function $f(z)$ is thus given exactly by its $[1/1]$ Padé approximant as

$$f(z) = A + \frac{B}{1+z}.$$

Example 2 ($\Lambda=2$).   In this case, we use the moments (4.15) to construct the sequence of integers

$$m_k = \int_0^2 u^k (2-u)^k \, d\phi(u), \qquad k=0,1,2,\dots,$$

which is a positive decreasing sequence. Therefore m_∞ exists such that

$m_k \to m_\infty$, and so n exists such that $m_j = m_\infty$ for all $j \geq n$. We find that

$$m_n - m_{n+1} = \int_0^2 u^k(2-u)^k(1-u)^2 \, d\phi(u) = 0,$$

$$d\phi(u) = \left[A\delta(u) + B\delta(u-1) + C\delta(u-2)\right] du,$$

and

$$f_j = f_1(2 - 2^{j-1}) + f_2(2^{j-1} - 1) \qquad \text{for all} \quad j \geq 1.$$

Thus it happens again in this example that $f(z)$ is given exactly by one of its Padé approximants. Specifically, we find that

$$f(z) = \frac{A + B + C + (3A + 2B + C)z + 2Az^2}{1 + 3z + 2z^2}.$$

Notice that the possible positions of the poles of $f(z)$ are determined uniquely by the specification $\Lambda = 2$ for an integer moment problem.

Solutions of the integer moment problem expressed by (4.15) are known for $\Lambda \leq 4$. For any $\Lambda < 4$, the solution is a finite-order Padé approximant of $f(z)$, implying that $f(z)$ is rational. When $\Lambda = 4$, a new type of solution

$$f(z) = m(1 + 4z)^{-1/2}, \qquad m \text{ integral},$$

becomes possible. It is also known that the general problem expressed by (4.15) can be reduced to the case of one with $\Lambda \leq 6$ [Barnsley et al., 1979].

We conclude this section by noting that the general question raised in Section 5.3 of the nature of a given sequence of moments is not always answered fully by categorizing them as Stieltjes moments or Hamburger moments. A more complete answer would include further specification of the support of the measure $d\phi(u)$.

Exercise Given that $\{f_j\}$ is totally monotone, prove from first principles that the sequence $\{(-)^k\Delta^k f_j, j = 0, 1, 2, \ldots\}$ is totally monotone for $k \geq 0$. Deduce that $f_n = 0$ for some $n > 0$ implies that $f_n = 0$ for all $n > 0$.

5.5 Stieltjes Series with Zero Radius of Convergence

Let us start with an example of a function $f(z)$ which has a Maclaurin series with zero radius of convergence. Consider, for $a > 0$,

$$f(z) = {}_2F_0(a, 1, -z) = \frac{1}{\Gamma(a)} \int_0^\infty \frac{e^{-u}u^{a-1}}{1 + zu} \, du. \qquad (5.1)$$

$f(z)$ is a Stieltjes function expressed in the standard form

$$f(z)=\int_0^\infty \frac{d\phi(u)}{1+zu},$$ (5.2)

where

$$\phi(u)=\frac{1}{\Gamma(a)}\int_0^u e^{-t}t^{a-1}\,dt.$$ (5.3)

We see that

$$f(z)=\sum_{j=0}^\infty f_j(-z)^j=1-az+a(a+1)z^2+\cdots,$$ (5.4)

where

$$f_j=a(a+1)\cdots(a+j-1)=\int_0^\infty u^j\,d\phi(u).$$ (5.5)

The difficulty with the series (5.4) is that it converges for no values of z except $z=0$, because the individual terms do not tend to zero. However, we know from Section 5.3 that the paradiagonal sequence of $[M+J/M]$ Padé approximants to the formal expansion (5.4) is convergent in the bounded domain $\mathcal{D}(\Delta)$ which does not include the origin. In this section, we describe how Padé approximants are useful for reconstructing functions from power series with zero radius of convergence. We restrict our attention to Stieltjes series, for which the convergence theorems can be established.

A basic precept of Padé approximation is that there is a function $f(z)$ which is determined by its Maclaurin expansion. In general, it is not true that an arbitrary function is determined by its Maclaurin expansion. This is demonstrated by the following example.

Example 1.

$$g(x)=\exp(-1/x),\qquad 0\leqslant x\leqslant\infty.$$

The function $g(x)$ of the real variable x is well defined, and so are all its (right-handed) derivatives for $x\geqslant 0$, and in this sense

$$\left.\frac{d^j g(x)}{dx^j}\right|_{x=0}=0\qquad\text{for }j=0,1,2,\dots.$$

Thus we see that any given function $f(x)$ with a Maclaurin expansion about

$x=0$ has the same expansion as $f(x)+g(x)$. Clearly, conditions must be imposed to define the class of functions which are uniquely determined by their power-series expansions. If we assume that $f(z)$ is analytic at $z=0$, which implies that the Maclaurin expansion of $f(z)$ has nonzero radius of convergence, then $f(z)$ is determined by analytic continuation, and may be uniquely defined in complex plane cut by the Mittag–Leffler star. Analyticity at the origin of $f(z)$ was a foundation of the development of the previous section, but it is an unnecessarily strong hypothesis for our present purposes. We must consider a weaker condition which will enable $f(z)$ to be determined by its Maclaurin expansion.

The second difficulty which we encounter is that the representation (5.2) is not necessarily unique even for Stieltjes series in the case where they have zero radius of convergence. This raises the question of determinacy mentioned in section 5.3. The following examples establish this nonuniqueness by showing that a nontrivial measure $d\phi_0(u)$ exists, corresponding to a function $\phi_0(u)$ of bounded variation, for which

$$\int_0^\infty u^k \, d\phi_0(u)=0, \qquad k=0,1,2,\dots .$$

Example 2 (Rennison).

$$d\phi_0(u)= \sum_{n=0}^\infty \frac{(-)^n \pi^{2n+1}}{(2n+1)!} \delta(u-2^{2n+1}) \, du.$$

This distribution corresponds to a piecewise continuous $\phi_0(u)$ with jumps of oscillating sign. $\phi_0(u)$ is of bounded variation, and

$$\int_0^\infty u^k \, d\phi_0(u)=\sin(2^k\pi)=0 \qquad \text{for} \quad k=0,1,2,\dots .$$

Example 3 (Stieltjes).

$$d\phi_0(u)=u^{-\ln u}\sin(2\pi \ln u)\,du,$$

This distribution corresponds to a continuous $\phi_0(u)$ of bounded variation. The substitution

$$\ln u=t+\frac{n+1}{2}$$

may be used to show that

$$\int_0^\infty u^n \, d\phi_0(u)=\pm \exp\left[\left(\frac{n+1}{2}\right)^2\right]\int_{-\infty}^\infty e^{-t^2}\sin 2\pi t \, dt=0.$$

Thus we are led to impose certain extra conditions on the moments $\{f_j\}$ to ensure that the constructed function $f(x)$ and its generating measure $d\phi(u)$ are unique. These conditions are obviously weaker than the conditions which render $f(z)$ analytic at $z=0$.

Our starting point is a series expansion

$$\sum_{j=0}^{\infty} c_j z^j = \sum_{j=0}^{\infty} f_j(-z)^j. \tag{5.6}$$

The usual definition of convergence of the power series (5.6) is that $\sum_{j=0}^{\infty} c_j z^j$ converges to $f(z_0)$ at $z=z_0$ if

$$\lim_{n \to \infty} \sum_{j=0}^{n} c_j z_0^j = f(z_0). \tag{5.7}$$

The existence of the limit implied by (5.7) in turn implies that the power series (5.6) is convergent for all z such that $|z|<|z_0|$, and a circle of convergence is established. This familiar definition is to be contrasted with the definition of asymptotic convergence needed in this section.

DEFINITION. A power series $\sum_{j=0}^{\infty} c_j z^j$ is *asymptotically convergent* to $f(z)$ if

$$\lim_{z \to 0} \left| \left[f(z) - \sum_{j=0}^{n} c_j z^j \right] z^{-n} \right| = 0 \tag{5.8}$$

for $\arg(z) \in \mathcal{Q}$ and for each $n=0,1,2,\ldots$. $\mathcal{Q}$ is an angular interval, such as $-\alpha < \arg(z) < \beta$, specifying a wedge domain of asymptotic convergence at $z=0$, as shown in Figure 1 [Erdélyi, 1956, p. 22].

Asymptotic convergence is denoted by the special symbol $\simeq$. If (5.8) is satisfied, we write

$$f(z) \simeq \sum_{j=0}^{\infty} c_j z^j, \qquad \alpha < \arg(z) < \beta.$$

Example 4. We quote the result that Euler's series

$$E(z) = {}_2F_0(1,1,-z) \simeq 1 - (1!)z + (2!)z^2 - (3!)z^3 + \cdots \tag{5.9}$$

is an asymptotic expansion of the function given below in (5.11) and defined by this integral representation in $-\pi < \arg(z) < \pi$.

We next quote a powerful theorem of Carleman [1926], which enables us not only to prove the result of Example 4 but to establish the existence of a

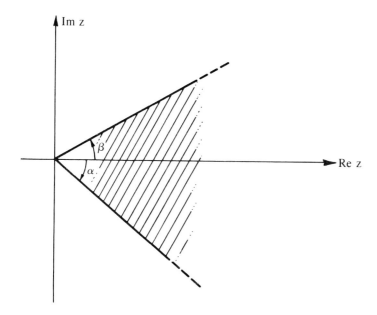

Figure 1. A wedge domain of asymptotic convergence.

unique function $f(z)$ to be associated with certain power series with zero radius of convergence.

CARLEMAN'S CRITERION.   Let $\{f_j\}$ satisfy the determinantal conditions (of Theorem 5.1.2) for Stieltjes series, and also the condition that

$$\sum_{j=1}^{\infty} (f_j)^{-1/(2j)} \text{ diverges.} \qquad (5.10)$$

Then there exists a unique Stieltjes function $f(z)$, analytic in $\mathrm{Re}\, z > 0$, such that the asymptotic equality

$$\sum_{j=0}^{\infty} f_j(-z)^j \simeq f(z)$$

holds in $|\arg(z)| < \pi$. (In fact, this equality also holds in any disk $\mathcal{C}$ such as is shown in Figure 2 of Section 5.6.)

Returning to Example 4, which is (5.4) with $a = 1$, we see that Euler's series (5.9) is a Stieltjes series necessarily satisfying the determinantal conditions of Theorem 5.1.2. From Stirling's formula,

$$f_j = j! \simeq \left(\frac{j}{e}\right)^j \sqrt{2\pi j} \qquad \text{as} \quad j \to +\infty,$$

and

$$(f_j)^{-1/(2j)} \simeq \left(\frac{e}{j}\right)^{1/2} (2\pi j)^{-1/(4j)} \qquad \text{as} \quad j \to +\infty.$$

Since $(2\pi j)^{-1/(4j)} \to 1$, $\sum^\infty f_j^{-1/(2j)}$ diverges, and Carleman's theorem asserts the uniqueness of the Stieltjes function

$$E(z) = {}_2F_0(1,1;-z) = \int_0^\infty \frac{e^{-u}}{1+zu} du \tag{5.11}$$

defined in $-\pi < \arg(z) < \pi$, with the asymptotic expansion (5.9). $E(z)$ has its branch cut along the negative real z-axis.

THEOREM 5.5.1. *Let* $f(z) = \sum_{j=0}^\infty f_j(-z)^j$ *be a series satisfying Stieltjes determinantal conditions and Carleman's criterion. Then all paradiagonal sequences of* $[M+J/M]$ *Padé approximants with* $J \geq -1$ *converge to* $f(z)$ *in the domain* $\mathcal{D}(\Delta)$.

Proof. Carleman's criterion asserts the existence of at most one Stieltjes function $f(z)$, analytic in $\operatorname{Re} z > 0$, with right-handed derivatives at the origin specified by the asymptotic equality

$$f(z) \simeq \sum_{j=0}^\infty f_j(-z)^j.$$

If $J \geq 0$, we use the device of considering the first $J+1$ terms of $f(z)$ explicitly (as in the corollary to Theorem 5.2.3), and so reduce the problem to that with $J = -1$. Theorem 5.3.1 asserts the existence of a limit function $f^{(-1)}(z)$ of the sequence of $[M-1/M]$ Padé approximants, that $f^{(-1)}(z)$ is analytic in $\mathcal{D}(\Delta)$, and from (3.3), $f^{(-1)}(z) \simeq \sum_{j=0}^\infty f_j(-z)^j$. Thus $f(z) = f^{(-1)}(z)$ in $\operatorname{Re} z > 0$ and so throughout $\mathcal{D}(\Delta)$.

An interesting application of the techniques developed so far in this section occurs in the theory of Stirling's formula. In this context, we need Binet's second formula, which is

$$\ln \Gamma(z) = (z - \tfrac{1}{2}) \ln z - z + \tfrac{1}{2} \ln 2\pi + J(z), \tag{5.12}$$

valid for $\operatorname{Re} z > 0$, where

$$J(z) = 2 \int_0^\infty \frac{\tan^{-1}(t/z)\,dt}{\exp(2\pi t) - 1} \tag{5.13a}$$

$$= -\frac{1}{\pi} \int_0^\infty \ln(1 - e^{-2\pi t}) \frac{z\,dt}{z^2 + t^2} \tag{5.13b}$$

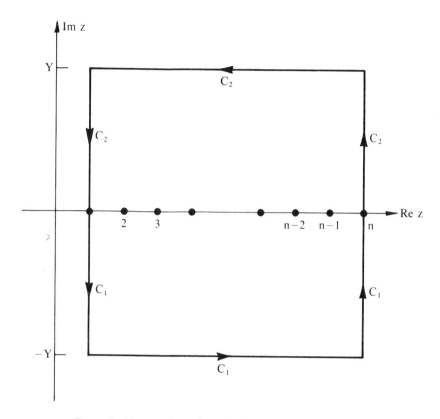

Figure 2. The complex z-plane showing the contours C_1 and C_2.

Proof [Ford, 1960, Chap 1; Hardy, 1956, p 339].   Define

$$R(n) = \sum_{j=1}^{n} \ln j - \tfrac{1}{2} \ln n - \int_{1}^{n} \ln x \, dx. \qquad (5.14)$$

We derive a formula for $R(n)$ by considering the principal-value contour integral

$$\oint_{C_1 + C_2} \frac{\cot \pi z}{2i} \ln z \, dz = \sum_{j=1}^{n} \ln j - \tfrac{1}{2} \ln n \qquad (5.15)$$

over the contour of Figure 1, which has been evaluated using the residue theorem. Because $\ln z$ is a real symmetric function,

$$\int_{1}^{n} \ln x \, dx = \tfrac{1}{2} \int_{C_1} \ln z \, dz - \tfrac{1}{2} \int_{C_2} \ln z \, dz. \qquad (5.16)$$

Substituting (5.15) and (5.16) into (5.14), we find that

$$R(n) = \int_{C_1} \psi_1 \ln z \, dz + \int_{C_2} \psi_2 \ln z \, dz,$$

where

$$\psi_1(z) = \frac{\cot \pi z}{2i} - \frac{1}{2} = \frac{1}{\exp(2\pi i z) - 1}$$

and

$$\psi_2(z) = \frac{\cot \pi z}{2i} + \frac{1}{2} = \frac{1}{1 - \exp(-2\pi i z)}.$$

By extending the contours to $Y = \infty$, we deduce that

$$R(n) = 2 \int_0^\infty \frac{\tan^{-1}(t/n)}{\exp(2\pi t) - 1} \, dt - 2 \int_0^\infty \frac{\tan^{-1} t}{\exp(2\pi t) - 1} \, dt. \qquad (5.17)$$

By direct integration of (5.14), we know that

$$R(n) = \ln(n!) - \tfrac{1}{2} \ln n - [x \ln x - x]_1^n. \qquad (5.18)$$

Using (5.17), (5.18), and the definition (5.13a), we deduce that

$$\ln \Gamma(n) = (n - \tfrac{1}{2}) \ln n - n + J(n) + C,$$

where C is a constant, independent of n. The value of this constant C is given by Stirling's formula (see Titchmarsh [1939, p.150]), since $J(+\infty) = 0$, and so (5.12) is established for all positive integers $z = 1, 2, 3, \dots$. Carlson's uniqueness theorem (see Titchmarsh [1939, p.186]) states that any function $f(z)$ which is analytic for $\mathrm{Re}\, z > 0$, is bounded by $\exp(k|z|)$ with $k < \pi$ for $\mathrm{Re}\, z \geq 0$, and satisfies $f(z) = 0$ for $z = 0, 1, 2, \dots$ is zero identically: $f(z) = 0$ for $\mathrm{Re}\, z > 0$. Hence (5.12) is established.

We exhibit the connection with Stirling's formula by writing (5.12) as

$$\Gamma(z) = z^z e^{-z} \sqrt{2\pi z} \exp\{J(z)\} \qquad \text{for } \mathrm{Re}\, z > 0,$$

whereas Stirling's formula [Titchmarsh, 1939] is

$$\Gamma(z) = z^z e^{-z} \sqrt{2\pi z} \{1 + O(z^{-1})\} \qquad \text{for } |\arg(z)| < \pi, \ |z| \to \infty.$$

Using (5.13a) and the expansion

$$\tan^{-1}\left(\frac{t}{z}\right) = \frac{t}{z} - \frac{1}{3}\left(\frac{t}{z}\right)^3 + \frac{1}{5}\left(\frac{t}{z}\right)^5 - \cdots + \frac{(-)^{n+1} t^{2n-1}}{(2n-1) z^{2n-1}} + \frac{(-)^n}{z^{n-1}} \int_0^t \frac{u^{2n} \, du}{u^2 + z^2},$$

we see that $J(z)$ has the asymptotic expansion

$$J(z) \simeq \sum_{j=0}^{\infty} (-)^j f_j z^{-(2j+1)}, \qquad (5.19)$$

where

$$f_j = \frac{-1}{\pi} \int_0^{\infty} u^{2j} \ln(1 - e^{-2\pi u}) \, du, \qquad j = 0, 1, 2, \dots. \qquad (5.20)$$

Notice also that $J(z)$ is analytic in $\operatorname{Re} z > 0$. This suggests a change of variable in (5.13) and (5.19). Define $y = z^{-2}$ and $K(y) = zJ(z)$. Then

$$K(y) = \sum_{j=0}^{\infty} f_j(-y)^j$$

$$= -\frac{1}{2\pi} \int_0^{\infty} \frac{\ln(1 - e^{-2\pi\sqrt{u}})}{\sqrt{u}} \frac{du}{1 + yu}. \qquad (5.21)$$

Equation (5.21) shows that $K(y)$ is a Stieltjes series with zero radius of convergence. From (5.20),

$$f_j = \frac{(2j)!}{2^{2j+1}\pi^{2j+2}} \sum_{r=1}^{\infty} \frac{1}{r^{2j+2}}, \qquad j = 0, 1, 2, \dots. \qquad (5.22)$$

It is easy to deduce that $\sum_{j=1}^{\infty} f_j^{-1/(2j)}$ diverges and that Carleman's criterion is satisfied. Hence $K(y)$ is uniquely determined by a convergent sequence of $[M-1/M]$ Padé approximants and by its continued-fraction expansion in $|\arg(y)| < \pi$, corresponding to $\operatorname{Re} z > 0$. From Abramowitz and Stegun [1964, Chapter 23] and Wall [1948, p. 364], we see that the actual asymptotic expansion of $J(z)$ is given in terms of Bernoulli numbers by

$$J(z) \simeq \sum_{j=0}^{\infty} \frac{B_{2j+2}}{(2j+1)(2j+2)} z^{-(2j+1)}$$

$$= \frac{z^{-1}}{12} - \frac{z^{-3}}{360} + \frac{z^{-5}}{1260} - \frac{z^{-7}}{1680} + \frac{z^{-9}}{1188} - \cdots.$$

Our conclusion is that the continued-fraction expansion, see Char [1980],

$$J(z) = \frac{\frac{1}{12}}{z} + \frac{\frac{1}{30}}{z} + \frac{\frac{53}{210}}{z} + \frac{\frac{195}{371}}{z} + \cdots \qquad (5.23)$$

converges in $\operatorname{Re} z > 0$. Convergence of this continued fraction is rapid until the eighth-order convergent is reached, when the rate of convergence becomes poor; this is understood in the context of the theory of inclusion regions (see Henrici and Pfluger [1966] or EPA for further details).

This example has led us to consider the connection between the S-fraction (4.5.3) and Stieltjes series defined in (1.1).

We have stated, in Theorem 5.5.1, sufficient conditions under which the sequences of $[M-1/M]$ and $[M/M]$ Padé approximants of a Stieltjes series

$$f(z) = \int_0^\infty \frac{d\phi(u)}{1+zu} = c_0 + c_1 z + c_2 z^2 + \cdots$$

converge to $f(z)$. These two sequences of approximants can also be expressed as the sequence of convergents of the continued fraction

$$f_c(z) = \frac{a_1}{1} + \frac{a_2 z}{1} + \frac{a_3 z}{1} + \cdots, \tag{5.24}$$

where

$$a_1 = c_0 = f_0, \tag{5.25}$$

$$a_2 = \frac{-c_1}{c_0} = \frac{f_1}{f_0},$$

$$a_3 = -\frac{C(1/2)}{c_0 c_1} = \frac{D(0,1)}{D(0,0)D(1,0)},$$

$$a_{2M} = -\frac{C(M/M)C(M-2/M-1)}{C(M-1/M-1)C(M-1/M)} = \frac{D(1,M-1)D(0,M-2)}{D(1,M-2)D(0,M-1)},$$

$$a_{2M+1} = -\frac{C(M/M+1)C(M-1/M-1)}{C(M-1/M)C(M/M)} = \frac{D(0,M)D(1,M-2)}{D(0,M-1)D(1,M-1)},$$

using the results of (4.4.25). Notice that $a_i > 0$ for $i = 1,2,3,\ldots$, so that we have proved the following result.

THEOREM 5.5.2. *If $f(z)$ is a Stieltjes function, the continued fraction (5.24) derived from the corresponding Stieltjes series is an S-fraction.*

The converse result follows from a theorem derived in Perron's book [1957, p. 208]. We quote it in the form that if $zf_c(z)$, derived from (5.24), is a convergent S-fraction, then its convergents form a sequence which converge to a function $f(z)$ which is a Stieltjes series, defined by (1.1), in which $\phi(u)$ is essentially unique. We do not prove this result, but note that (5.25) shows that the property that each $a_i > 0$ is sufficient to ensure that the determinants $D(m, n)$ are positive for all $m, n \geqslant 0$. Also we note that the hypothesis that the S-fraction converges is essential to establish the uniqueness property of the corresponding $\phi(u)$.

To conclude this section, we mention that there are various kinds of conditions on a real function $S(\omega)$, defined on $0 \leqslant \omega \leqslant \infty$, which are necessary and sufficient conditions to ensure that $S(\omega)$ has a Stieltjes representation of the form (4.5.8). As an example, we quote one result:

THEOREM 5.5.3. *Necessary and sufficient conditions for $S(\omega)$ to have the integral representation*

$$S(\omega) = \int_0^\infty \frac{d\phi(t)}{\omega + t} \tag{5.26}$$

with $\phi(t)$ nondecreasing and bounded are that $S(\omega) \geqslant 0$ and

$$(-1)^{k-1} \left(\frac{d}{d\omega}\right)^{2k-1} \{\omega^k S(\omega)\} \geqslant 0, \quad k = 1, 2, \ldots \quad on \quad 0 < \omega < \infty,$$

and that a finite limit of $\omega S(\omega)$ exists as $\omega \to +\infty$.

The proof of this theorem is given in Widder [1972, p. 364], and we stress that this is just one of several similar results which characterize Stieltjes functions [Widder, 1972, Chapter 8]. We omit any details because neither the statements of the theorems nor the proofs involve Padé approximation.

From the basic representation (1.1) of a Stieltjes function $f(z)$, it follows that Stieltjes functions form a subclass of the class of completely monotonic functions defined on $[0, \infty)$. As such, they are directly related to a subclass of the class of absolutely continuous functions [Widder, 1972, Chapter 4].

Exercise $f(z)$ is a Stieltjes series, and $g(z)$ is defined by

$$f(z) = \frac{f(0)}{1 + zg(z)}.$$

Use Hadamard's formula given in Theorem 1.6.1 to establish that the coefficients defined by

$$g(z) = \sum_{j=0}^\infty g_j(-z)^j$$

satisfy the inequalities

$$D_g(m,n) \equiv \begin{vmatrix} g_m & g_{m+1} & \cdots & g_{m+n} \\ g_{m+1} & g_{m+2} & \cdots & g_{m+n+1} \\ \vdots & \vdots & & \vdots \\ g_{m+n} & g_{m+n+1} & \cdots & g_{m+2n} \end{vmatrix} > 0$$

for all $m, n \geq 0$. How else may this result be proved?

5.6   Hamburger Series and the Hamburger Moment Problem

A Hamburger function is defined to be a function with an integral representation

$$f(z) = \int_{-\infty}^{\infty} \frac{d\phi(u)}{1+uz} \tag{6.1}$$

where the moments

$$f_j = \int_{-\infty}^{\infty} u^j d\phi(u), \qquad j=0,1,2,\ldots, \tag{6.2}$$

are finite and $\phi(u)$ is increasing [Hamburger, 1920, 1921].
A Hamburger series is defined to be a series

$$\sum_{j-0}^{\infty} c_j z^j = \sum_{j-0}^{\infty} f_j(-z)^j$$

with moments f_j defined by (6.2). This is the series derived by a formal expansion of (6.1). Just as before, with Stieltjes series, we exclude the case where $\phi(u)$ is piecewise constant with a finite number of jump discontinuities and consequently $f(z)$ is a rational function. The characteristic feature of Hamburger series is the full range $(-\infty, \infty)$ of integration. The inverse problem, which is the determination of $f(z)$ from the moments, is called the Hamburger moment problem. Hamburger series, functions, moments, etc. are sometimes called extended Stieltjes series, functions, moments, etc.

The conditions for Hamburger series are weaker than those for Stieltjes series, and so the Hamburger moments satisfy fewer conditions than the Stieltjes moments. As before (1.12), we define

$$D(m,n) = \begin{vmatrix} f_m & f_{m+1} & \cdots & f_{m+n} \\ f_{m+1} & f_{m+2} & \cdots & f_{m+n+1} \\ \vdots & \vdots & & \vdots \\ f_{m+n} & f_{m+n+1} & \cdots & f_{m+2n} \end{vmatrix}. \tag{6.3}$$

THEOREM 5.6.1. *If $\{f_j, j=0,1,\ldots\}$ are moments of a Hamburger series satisfying (6.2), the determinants $D(2m, n)>0$ for all $m, n>0$.*

Proof. The inequality

$$\int_{-\infty}^{\infty} u^{2m}(x_0 + x_1 u + \cdots + x_n u^n)^2 \, d\phi(u) > 0$$

is used, following precisely the method of Theorem 5.1.2.

Notice that unless $D(2m+1, n)>0$ for all $m, n \geq 0$, the series is not a Stieltjes series. The next theorem shows that Hamburger series have Padé approximants with poles on the real axis (not at the origin), positive residues on the negative real axis, and negative residues on the positive real axis.

THEOREM 5.6.2. *If the coefficients $\{f_j\}$ satisfy the inequalities $D(0, n)>0$ for all $n>0$, then $[M-1/M]$ Padé approximants of the formal power series $\sum_{j=0}^{\infty} f_j(-z)^j$ may be written as*

$$[M-1/M] = \sum_{j=1}^{M} \frac{\beta_i}{1+\gamma_i z} \tag{6.4}$$

with $\beta_i > 0$, γ_i real, for $i=1,2,\ldots, M$.

Proof. An inductive proof based on $D(0, n)>0$ and Sylvester's identity establishes that $D(2m, n)>0$ for all $m, n>0$. Then the method of Theorem 5.2.1 is used, but with the change of variable $w=-z^{-1}$. The Padé denominator is

$$Q^{[M-1/M]}(z) = \begin{vmatrix} f_0 + zf_1 & f_1 + zf_2 & \cdots & f_{M-1} + zf_M \\ f_1 + zf_2 & f_2 + zf_3 & \cdots & f_M + zf_{M+1} \\ \vdots & \vdots & & \vdots \\ f_{M-1} + zf_M & f_M + zf_{M+1} & \cdots & f_{2M-2} + zf_{2M-1} \end{vmatrix}$$

$$= w^{-M} \begin{vmatrix} wf_0 - f_1 & wf_1 - f_2 & \cdots & wf_{M-1} - f_M \\ wf_1 - f_2 & wf_2 - f_3 & \cdots & wf_M - f_{M+1} \\ \vdots & \vdots & & \vdots \\ wf_{M-1} - f_M & wf_M - f_{M+1} & \cdots & wf_{2M-2} - f_{2M-1} \end{vmatrix}.$$

$$\tag{6.5}$$

Following the ideas of method 2 of proof of Theorem 5.4.1, we consider the polynomial

$$\pi_M(w) = w^M Q^{[M-1/M]}(-1/w).$$

Its leading coefficient is $D(0, M-1)$, which is positive, and so each $\pi_M(w)$ is positive for w sufficiently large and positive. Sylvester's identity implies that

$$\pi_{M-1}(w)\pi_{M+1}(w) \leqslant 0 \qquad \text{if} \quad \pi_M(w)=0.$$

Since $\pi_0(w)=1$ and $\pi_1(w)=wf_0-f_1$, the interlacing property of the zeros of $\pi_M(w)$ follows by induction. Note the possibility of a zero at $w=0$. This situation corresponds to $Q^{[M-1/M]}(z)$ having true degree $M-1$ instead of M, a situation not prohibited by the Hamburger conditions. The signs of the residues follow using the method of Theorem 5.2.1, and the theorem is proved.

The motive for using $w=(-z)^{-1}$ in this proof is to borrow from the theory of orthogonal polynomials. Using the methods of Section 5.3, an identical proof shows that the polynomials $\pi_M(w)$ are orthogonal over a positive measure. The methods show that the zeros of $\pi_M(w)$ occur in a real interval including the origin, and that zeros of successive polynomials interlace.

Following the development of Section 5.2, we omit any analysis designed to prove convergence of a sequence of Padé approximants for $x>0$, which is irrelevant, and are led to Theorem 5.2.3 concerning uniform boundedness in $\mathcal{D}(\Delta)$ of paradiagonal sequences. The analogue is

THEOREM 5.6.3. *The sequence of* $[M-1/M]$ *Padé approximants to a Hamburger series is uniformly bounded as* $M \to \infty$ *in the domain* $\mathcal{D}'(\Delta)$, *a bounded, disconnected, two-component domain of the z-plane which is at least a distance* Δ *from the real axis.*

Remark. The two components of $\mathcal{D}'(\Delta)$ are shown in Figure 1.

Proof. The method is identical to that of Theorem 5.2.3.

The analogue of Theorem 5.2.4 is

THEOREM 5.6.4. *The sequence of* $[M-1/M]$ *Padé approximants to a Hamburger series is equicontinuous in* $\mathcal{D}'(\Delta)$.

Proof. The method is identical to that of Theorem 5.2.4.

Arzela's theorem is now applicable to the $[M-1/M]$ sequence, and we prove, using Theorems 5.6.3 and 5.6.4, the following result:

THEOREM 5.6.5. *At least an infinite subsequence of* $[M-1/M]$ *Padé approximants to a Hamburger series is uniformly convergent in* $\mathcal{D}'(\Delta)$ *to a function* $\tilde{f}(z)$ *analytic in that region.*

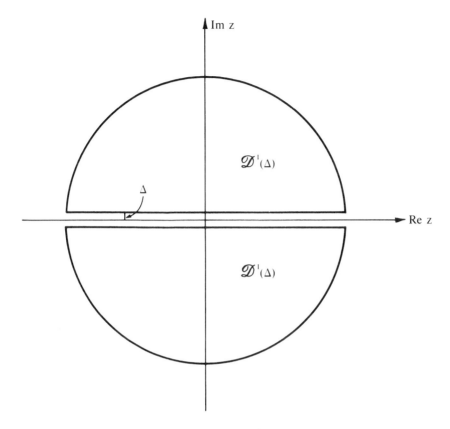

Figure 1.   The domain $D'(\Delta)$.

Theorem 5.6.5 leaves several questions unanswered. However, if the coefficients $\{f_j\}$ are such that $f(z)$ is analytic in $|z|<R$, the analysis is quite straightforward.

THEOREM 5.6.6. *If the moments $\{f_j\}$ are such that $D(0,n)>0$ for all $n>0$ and $\sum_{j=0}^{\infty} f_j(-z)^j$ is convergent in $|z|<R$, then*
(i) *a positive measure $d\phi(u)$ exists such that*

$$f_j = \int_{-R^{-1}}^{R^{-1}} u^j \, d\phi(u), \qquad j=0,1,2,\ldots,$$

(ii) *$d\phi(u)$ is unique,*

(iii)
$$\tilde{f}(z)=f(z)=\int_{-R^{-1}}^{R^{-1}} \frac{d\phi(u)}{1+zu}.$$

Proof. The construction of theorem 5.3.1 allows the representation

$$\tilde{f}(z) = \int_{-\infty}^{\infty} \frac{d\phi(u)}{1+zu} \tag{6.6}$$

since the determinantal conditions $D(0, n) > 0$ are satisfied. The methods of theorem 5.4.1 imply that the poles of $[M-1/M]$ Padé approximants of $\sum_{j=0}^{\infty} f_j(-z)^j$ are located on the cut, and method 2 of theorem 5.4.2 implies that $f(z)$ and $\tilde{f}(z)$ have an identical Maclaurin expansion. By hypothesis, this is convergent in $|z| < R$, and so $f(z) = \tilde{f}(z)$.

By a construction similar to (3.12), we establish that $\phi(u)$ is essentially unique, and the results (i), (ii) and (iii) of the theorem are proved.

Next we consider the case where $\sum_{j=0}^{\infty} f_j(-z)^j$ is not convergent in $|z| < R$ for any strictly positive value of R, and we seek conditions under which the result of Theorem 5.6.5 may be strengthened. This theorem only guarantees the existence of a function $\tilde{f}(z)$ which is the limit of a convergent subsequence of the $[M-1/M]$ Padé approximants of the given Hamburger series. Theorem 5.6.5 does not imply that there are no other convergent subsequences with different limits, nor does it assert that $\tilde{f}(z)$ has any expansion at all, let alone the given power series. We fill this gap by showing in the next theorem that the given power series is an asymptotic expansion of $\tilde{f}(z)$.

THEOREM 5.6.7. *Let the coefficients* $\{f_j\}$ *satisfy the determinantal conditions* $D(0, n) > 0$ *for all* $n = 0, 1, 2, \ldots$. *Let* $\tilde{f}(z)$ *be the limit of a convergent subsequence of* $[M-1/M]$ *Padé approximants of the formal series* $\sum_{j=0}^{\infty} f_j(-z)^j$. *Then* $\tilde{f}(z)$ *is a real symmetric function,* $\tilde{f}(z)$ *is analytic in the two half planes defined by* Im $z > 0$ *and* Im $z < 0$, *and* $\tilde{f}(z)$ *has the asymptotic expansion*

$$\tilde{f}(z) \simeq \sum_{j=0}^{\infty} f_j(-z)^j \qquad as \quad z \to 0, \quad z \in \mathcal{C}. \tag{6.7}$$

$\mathcal{C}$ *is any disk of radius* r, *center* $z = ir$, *as shown in Figure 2, or else* $\mathcal{C}$ *may be centered at* $z = -ir$.

Remark. The existence of at least one convergent subsequence of $[M-1/M]$ Padé approximants is guaranteed by Theorem 5.6.5.

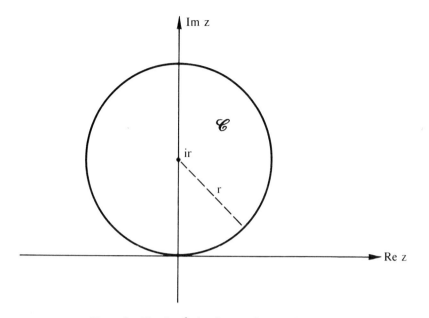

Figure 2.   The disc $\mathcal{C}$ of radius r and centered at $z = ir$.

Proof. We first prove that, for any positive integer k,

$$z^{-2k}\left\{[M-1/M] - \sum_{j=0}^{2k} f_j(-z)^j\right\} \to 0 \qquad \text{as} \quad z \to 0 \qquad (6.8)$$

uniformly for all $M > k$ and $z \in \mathcal{C}$.

Each Padé approximant has the representation (6.4),

$$[M-1/M] = \sum_{i=1}^{M} \frac{\beta_i}{1+\gamma_i z} = \sum_{i=1}^{M} \sum_{j=0}^{\infty} \beta_i(-\gamma_i z)^j, \qquad (6.9)$$

where the expansion is convergent in $|z| < R_M$, with

$$R_M = \min_{1 \leqslant i \leqslant M} |\gamma_i|^{-1}.$$

This expansion (6.9) agrees with the given series up to order z^{2M-1} inclusive, so that

$$[M-1/M] - \sum_{j=0}^{2k} f_j(-z)^j = \sum_{i=1}^{M} \beta_i \sum_{j=2k+1}^{\infty} (-z\gamma_i)^j$$

for $k<M$ and $|z|<R_M$. Therefore (6.8) holds for the particular case of $k=M-1$, for $k<M-1$, $|z|<R_M$, and $z\in\mathcal{C}$,

$$\left|[M-1/M]-\sum_{j=0}^{2k}f_j(-z)^j\right|-\left|\sum_{i=1}^{M}\beta_i\sum_{j=2k+1}^{\infty}(-z\gamma_i)^j\right|$$

$$\leq\left|\sum_{i-1}^{M}\beta_i(-z\gamma_i)^{2k+1}\right|+\left|\sum_{i=1}^{M}\sum_{j=2k+2}^{\infty}\beta_i(-z\gamma_i)^j\right|$$

$$\leq|z|^{2k+1}|f_{2k+1}|+\left|\sum_{i=1}^{M}\beta_i\sum_{j=2k+2}^{\infty}(-z\gamma_i)^j\right|$$

$$<|z|^{2k+1}|f_{2k+1}|+|z|^{2k+2}\left(\sum_{i=1}^{M}\beta_i\gamma_i^{2k+2}\right)\max_{1\leq i\leq m}\left|\frac{1}{1+z\gamma_i}\right|.$$

$$(6.10)$$

If $z\in\mathcal{C}$, then $\operatorname{Im}z\geq|\operatorname{Re}z|^2/(2r)$. Using analysis similar to that of Theorem 5.2.3, we find that

$$\sup_{\gamma}\left|\frac{z}{1+\gamma z}\right|=\frac{|z|^2}{|\operatorname{Im}z|}<4r\qquad\text{for}\quad z\in\mathcal{C}.$$

Hence we have shown that $|z(1+\gamma_i z)^{-1}|$ is uniformly bounded for all $z\in\mathcal{C}$ and independently of M.

By hypothesis, $k<M-1$, so that

$$\sum_{i=1}^{M}\beta_i\gamma_i^{2k+2}=f_{2k+2}.$$

Thus we deduce that, for $z\in\mathcal{C}$ and all $M>k$,

$$z^{-2k}\left|[M-1/M]-\sum_{j=0}^{2k}f_j(-z)^j\right|\to0\qquad\text{as}\quad z\to0,$$

uniformly in M. Taking the limit as $M\to\infty$, though values for which $[M-1/M]$ converges to $\tilde{f}(z)$ in the domain $\mathcal{D}'(\Delta)$, we discover that $\tilde{f}(z)$ has the property that

$$z^{-2k}\left|\tilde{f}(z)-\sum_{j=0}^{2k}f_j(-z)^j\right|\to0\qquad\text{as}\quad z\to0\qquad(6.11)$$

for $z \in \mathcal{C}$. Hence

$$z^{-k}\left|\tilde{f}(z) - \sum_{j=0}^{k} f_j(-z)^j\right| \to 0 \qquad \text{as} \quad z \to 0$$

for $z \in \mathcal{C}$ and each positive integer k. This establishes the asymptotic expansion

$$\tilde{f}(z) \simeq \sum_{j=0}^{\infty} f_j(-z)^j, \qquad z \to 0, \quad z \in \mathcal{C}.$$

$\tilde{f}(z)$ is, by Theorem 5.6.5, the uniform limit of a sequence of real symmetric Padé approximants each analytic in $\text{Im } z \neq 0$. Therefore $\tilde{f}(z)$ is real symmetric, and by Weierstrass's theorem (quoted in Section 5.2), $\tilde{f}(z)$ is analytic in $\text{Im } z \neq 0$.

In order to obtain an integral representation of $\tilde{f}(z)$, it is more convenient to use the variable $w = -z^{-1}$. We define

$$\tilde{F}(w) = \frac{1}{w}\tilde{f}\left(\frac{-1}{w}\right). \tag{6.12}$$

The sequence of approximants (6.9),

$$[M-1/M](z) = \sum_{i=1}^{M} \frac{\beta_i}{1 + \gamma_i z}, \qquad M = 1, 2, 3, \ldots,$$

defined by the formal series $\sum_{j=0}^{\infty} f_j(-z)^j$, now becomes

$$J_M(w) = \frac{1}{w}[M-1/M]\left(\frac{-1}{w}\right) = \sum_{i=1}^{M} \frac{\beta_i}{w - \gamma_i}, \qquad M = 1, 2, 3, \ldots. \tag{6.13}$$

Theorem 5.6.5 now implies that a subsequence of (6.13) converges in a domain in the w-plane corresponding to $z \in \mathcal{D}'(\Delta)$. We can prove

THEOREM 5.6.8. *Let the coefficients $\{f_j\}$ satisfy the determinantal conditions $D(0, n) > 0$ for all $n = 0, 1, 2, \ldots$. Let $\mathcal{D}^w(\delta)$ be a two-component domain in the w-plane,*

$$\mathcal{D}^w(\delta) = \{w : |\text{Im } w| > \delta\}$$

Then, for $w \in \mathcal{D}^w(\delta)$, there exists a function $\tilde{F}(w)$ which is the limit of a convergent subsequence of $\{\tilde{J}_M(w), M = 0, 1, 2, \ldots\}$, where $\tilde{J}_M(w)$ are Padé *approximants associated with the formal series* $\sum_{j=0}^{\infty} f_j w^{-j-1}$. $\tilde{F}(w)$ *is a real analytic function,* $\tilde{F}(w)$ *is analytic in* $\operatorname{Im} w \neq 0$, *and it has the asymptotic expansion*

$$\tilde{F}(w) \simeq \sum_{j=0}^{\infty} f_j w^{-j-1}, \qquad w \to \infty, \quad |\operatorname{Im} w| > \delta, \tag{6.14}$$

where ε is arbitrarily small but positive. Also,

$$\left\{ \operatorname{Im} \tilde{F}(w) \right\} \operatorname{Im} w < 0 \qquad \textit{for} \quad \operatorname{Im} w \neq 0. \tag{6.15}$$

Proof. The condition $\operatorname{Im} w > \delta$ corresponds to the disk $\mathcal{C}$ of Figure 2 of radius $(2\delta)^{-1}$. Thus we see that Theorem 5.6.8 is the expression of Theorem 5.6.7 in terms of the variable $z = -w^{-1}$.

We use $\mathfrak{M}$ to denote the values of M for which the subsequence of $J_M(w)$ converges in some specified domain $\mathcal{D}^w(\delta)$.

THEOREM 5.6.9. *The approximants* $\tilde{J}_M(w)$, *defined by* (6.13), *are uniformly bounded by the inequality*

$$\left| \tilde{J}_M(w) - \frac{f_0}{w} \right| < \frac{\sqrt{f_0 f_2}}{w |\operatorname{Im} w|}. \tag{6.16}$$

The limit function $\tilde{F}(w)$, *defined by Theorem 5.6.8, is uniformly bounded by*

$$\left| \tilde{F}(w) - \frac{f_0}{w} \right| \leqslant \frac{\sqrt{f_0 f_2}}{w |\operatorname{Im} w|}. \tag{6.17}$$

Proof. Using (6.13), we find that

$$\left| \tilde{J}_M(w) - \frac{f_0}{w} \right| = \left| \sum_{i=1}^{M} \frac{\beta_i \gamma_i}{w(w - \gamma_i)} \right|$$

$$< \frac{1}{w |\operatorname{Im} w|} \sum_{i=1}^{M} \beta_i |\gamma_i|.$$

Equation (6.16) follows from the Cauchy–Schwartz inequality

$$\left(\sum_{i=1}^{M} a_i b_i \right)^2 \le \left(\sum_{i=1}^{M} a_i^2 \right) \left(\sum_{i=1}^{M} b_i^2 \right),$$

with $a_i = \sqrt{\beta_i}$, $b_i = \sqrt{\beta_i |\gamma_i|}$. Since (6.16) is a uniform bound, (6.17) follows by letting $M \to \infty$, $M \in \mathfrak{M}$.

With Theorems 5.6.8 and 5.6.9 established, we proceed to our goal of constructing a representation of $\tilde{F}(w)$ as a Stieltjes integral. We need some preliminary results first.

LEMMA 1. *For any* $\delta > 0$,

$$\int_{-\infty}^{\infty} \operatorname{Im} \tilde{F}(u + i\delta)\, du = -\pi f_0. \tag{6.18}$$

Proof. Because $\tilde{F}(w)$ is analytic in the upper half plane, we are led to consider the contour integral

$$\int_C \tilde{F}(w)\, dw = 0$$

where C is the contour $ABCD$ shown in Figure 3. Point A is the point

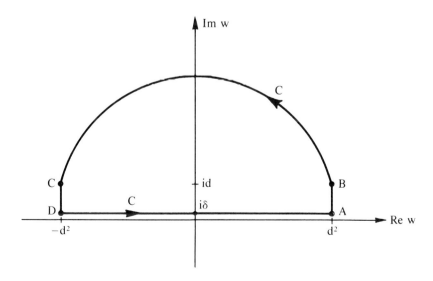

Figure 3.   The contour C in the w-plane.

$w = d^2 + i\delta$, and we define

$$w_A = d^2 + i\delta, \qquad w_B = d^2 + id,$$
$$w_C = -d^2 + id, \qquad w_D = -d^2 + i\delta.$$

The arc BC is the arc of a circle, centered at $w = 0$ and of radius $d\sqrt{d^2 + 1}$. Using (6.17), we estimate the integral over AB:

$$\left| \int_A^B \tilde{F}(w)\,dw \right| \leqslant f_0 \int_\delta^d \frac{dy}{|d^2 + iy|} + \sqrt{f_0 f_2} \int_\delta^d \frac{dy}{y|d^2 + iy|}$$

$$\leqslant \frac{f_0}{d} + \frac{\sqrt{f_0 f_2}}{d^2} \ln\left(\frac{\delta}{d}\right).$$

Likewise, we estimate the integral over the arc BC:

$$\left| \int_B^C \left\{ \tilde{F}(w) - \frac{f_0}{w} \right\} dw \right| \leqslant \sqrt{f_0 f_2} \int_B^C \frac{|dw|}{|w|d} \leqslant \frac{\pi\sqrt{f_0 f_2}}{d}.$$

We evaluate the integral

$$\int_B^C \frac{f_0}{w}\,dw = if_0 \left\{ \pi - 2\tan^{-1}\left(\frac{1}{d}\right) \right\}.$$

Using these results, it is straightforward to deduce that

$$\lim_{d \to \infty} \int_D^A \tilde{F}(w)\,dw = -i\pi f_0.$$

Equation (6.18) is the imaginary part of this formula.

As a consequence of Lemma 1, we are led to define

$$\phi(\delta, u) = -\frac{1}{\pi} \int_{-\infty}^u \tilde{F}(t + i\delta)\,dt. \tag{6.19}$$

Theorem 5.6.8 and Lemma 1 imply that

$$\frac{d\phi(\delta, u)}{du} = -\frac{1}{\pi} \tilde{F}(u + i\delta) > 0 \qquad \text{for} \quad \delta > 0,$$

and that

$$\int_{-\infty}^\infty \phi(\delta, u) = f_0.$$

Consequently, $d\phi(\delta, u)$ is a bounded Hamburger measure.

LEMMA 2. *For* $\operatorname{Im} w > \delta > 0$,

$$\tilde{F}(w) = \frac{1}{\pi} \int_{-\infty}^{\infty} \frac{\operatorname{Im} \tilde{F}(u+i\delta)\,du}{u+i\delta-w} = \int_{-\infty}^{\infty} \frac{d\phi(\delta, u)}{w-u-i\delta}. \tag{6.20}$$

Proof. We use the contour C of Figure 3, and let w be an interior point of C. Then

$$\tilde{F}(w) = \frac{1}{2\pi i} \int_C \frac{\tilde{F}(w')\,dw'}{w'-w}.$$

Using the previous analysis, we let $d \to \infty$ and deduce that

$$\tilde{F}(w) = \frac{1}{2\pi i} \int_{-\infty}^{\infty} \frac{\tilde{F}(u+i\delta)\,du}{u+i\delta-w}. \tag{6.21}$$

If w lies within C, the point $w' - w^* + 2i\delta$ lies outside C; Cauchy's theorem shows that

$$0 = \frac{1}{2\pi i} \int_{-\infty}^{\infty} \frac{\tilde{F}(u+i\delta)\,du}{u+i\delta-(w^*+2i\delta)}.$$

Taking the complex conjugate,

$$0 = \frac{1}{2\pi i} \int_{-\infty}^{\infty} \frac{\{\tilde{F}(u+i\delta)\}^*\,du}{u+i\delta-w} \tag{6.22}$$

Subtracting (6.22) from (6.21) yields

$$\tilde{F}(w) = \frac{1}{\pi} \int_{-\infty}^{\infty} \frac{\operatorname{Im} \tilde{F}(u+i\delta)\,du}{u+i\delta-w}.$$

Equation (6.20) is established by the definition (6.19).

LEMMA 3. *Let* $\phi(\delta, u)$ *be a family of bounded nondecreasing functions of* u *defined on* $-\infty < u < \infty$ *for* $\delta > 0$. *Then there exists a sequence* $\{\delta_i, i = 1, 2, 3, \ldots,\}$ *such that* $\delta_i \to 0$ *and* $\phi(\delta_i, u) \to \phi(u)$ *as* $i \to \infty$, *where* $\phi(u)$ *is bounded and nondecreasing on* $-\infty < u < \infty$.

Remark. $\phi(\delta, u)$, defined by (6.19), satisfies the conditions of Lemma 3.

Proof. Take a set of points $U = \{u_k\}$ which is dense on $(-\infty, \infty)$. Let D_0 be the sequence $\{1/i, i = 1, 2, 3, \ldots\}$. The sequence of values $\{\phi(\delta, u_1),$

$\delta \in D_0$} is a bounded sequence, and we may choose D_1 to be a subsequence of D_0 for which

$$\{\phi(\delta, u_1), \delta \in D_1\} \text{ is convergent.}$$

Let $\lim_{\delta \to 0, \delta \in D_1} \phi(\delta, u_1) = \phi(u_1)$.
  Likewise, we choose a subsequence D_2 of D_1 for which

$$\{\phi(\delta, u_2), \delta \in D_2\} \to \phi(u_2).$$

Similarly, we choose D_k so that $D_k \subset D_{k-1} \subset \cdots \subset D_1$ and for which

$$\{\phi(\delta, u_k), \delta \in D_k\} \to \phi(u_k).$$

Thus we define iteratively a bounded nondecreasing function $\phi(u)$ on a dense set of points in $-\infty < u < \infty$.

  Although $\phi(\delta, u)$ is continuous, the limit function $\phi(u)$ need not be continuous. If $\phi(u)$ is discontinuous at $u = v$, v is called a point of discontinuity of ϕ. Because $\phi(u)$ is increasing, we may define

$$\phi(v_-) = \lim_{u \to v} \phi(u) \qquad \text{for} \quad u \in U, \quad u < v,$$

$$\phi(v_+) = \lim_{u \to v} \phi(u) \qquad \text{for} \quad u \in U, \quad u > v,$$

and

$$\Delta\phi(v) = \phi(v_+) - \phi(v_-).$$

There are only, at most, a countable set of points of discontinuity of ϕ: not more than two points with $\Delta\phi \geq f_0/2$, not more than four points with $f_0/4 \leq \Delta\phi < f_0/2$, not more than eight points with $f_0/8 \leq \Delta\phi < f_0/4$, etc. The points of discontinuity may be dense on the entire interval; nevertheless, Lemma 3 still defines $\phi(u)$ at these points. At all points in $-\infty < u < \infty$ other than its points of discontinuity, $\phi(u)$ is continuous; such points are naturally called points of continuity of $\phi(u)$. We have now set up the framework for a substantial theorem:

THEOREM 5.6.10. *Let the coefficients* $\{f_j\}$ *satisfy the determinantal conditions* $D(0, n) > 0$ *for all* $n = 0, 1, 2, \ldots$. *Let* $\bar{F}(w)$ *be the limit of a convergent subsequence of* $\{\bar{J}_M(w), M = 0, 1, 2, \ldots\}$ *of Padé approximants defined for the formal series*

$$\sum_{j=0}^{\infty} f_j w^{-j-1}.$$

Then $\tilde{F}(w)$ has the Hamburger integral representation

$$\tilde{F}(w) = \int_{-\infty}^{\infty} \frac{d\phi(u)}{w-u}, \tag{6.23}$$

where $\phi(u)$ is bounded and nondecreasing on $-\infty < u < \infty$.

Proof. The fact that the definition (6.19) fits the conditions of Lemma 3 allows $\phi(u)$ to be constructed. We obtain (6.23) from (6.20) in the limit $\delta \to 0$. Because $\tilde{F}(w)$ is real symmetric, it is valid for all $\mathrm{Im}\, w \neq 0$.

We take the opportunity to quote the Stieltjes inversion formula, a simple consequence of Lemma 2 and Lemma 3, which expresses $\phi(u)$ in terms of $\tilde{F}(w)$ for Hamburger functions. The formula is

$$\frac{\phi(t_+) + \phi(t_-)}{2} = \frac{1}{\pi} \lim_{\delta \to 0} \int_{-\infty}^{t} \mathrm{Im}\, \tilde{F}(u+i\delta)\, du.$$

With the integral representation (6.23) established, it is convenient to revert to our standard variable $z = -1/w$. Under the conditions of Theorem 5.6.7, we have proved that

$$\lim_{\substack{M \to \infty \\ M \in \mathfrak{M}}} [M-1/M] = \tilde{f}(z) = \int_{-\infty}^{\infty} \frac{d\phi(u)}{1+zu} \simeq \sum_{j=0}^{\infty} f_j(-z)^j. \tag{6.24}$$

Next, we prove the standard integral formula for each coefficient f_j, which follows from Hamburger's theorem:

THEOREM 5.6.11. *Let $f(z)$ be defined by the Hamburger integral representation*

$$f(z) = \int_{-\infty}^{\infty} \frac{d\phi(u)}{1+zu}. \tag{6.25}$$

Then a necessary and sufficient condition for the moments

$$f_j = \int_{-\infty}^{\infty} u^j d\phi(u) \tag{6.26}$$

to be finite is that $f(z)$ has the asymptotic expansion

$$f(z) \simeq \sum_{j=0}^{\infty} f_j(-z)^j \tag{6.27}$$

defined in the sectors $\varepsilon \leqslant |\arg(z)| \leqslant \pi - \varepsilon$ for arbitrarily small $\varepsilon > 0$.

Remarks. In the context of this theorem, we understand the purpose in making the distinction between a Stieltjes function and a Stieltjes series, and

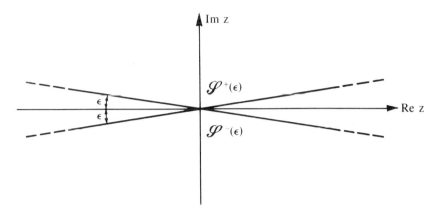

Figure 4. The sectors $S^+(\varepsilon)$ and $S^-(\varepsilon)$. $S^+(\varepsilon)$ is the sector $\varepsilon \leqslant \arg(z) \leqslant \pi - \varepsilon$, and $S^-(\varepsilon)$ is the sector $-\pi + \varepsilon \leqslant \arg(z) \leqslant \varepsilon$.

between a Hamburger function and a Hamburger series. A Hamburger function is defined by the representation (6.25), provided the moments (6.26) are finite. A Hamburger series is defined by the right hand side of (6.27), provided the moments (6.26) are finite. Hamburger's theorem asserts that a Hamburger function $f(z)$ defined by (6.25) has a Hamburger series defined uniquely by the right-hand side of (6.27) using the definition (6.26). The theorem also asserts that at least one Hamburger function, defined by (6.25), may be associated with a given Hamburger series defined by (6.26) and (6.27).

Proof. Assuming the representation (6.25) and the definition (6.26), it follows that

$$R_k(z) \equiv z^{-k} \left[f(z) - \sum_{j=0}^{k} f_j(-z)^j \right]$$

$$= (-)^{k+1} z \int_{-\infty}^{\infty} \frac{u^{k+1} d\phi(u)}{1 + zu}$$

$$= (-)^{k+1} \left[z f_{k+1} - z^2 \int_{-\infty}^{\infty} \frac{u^{k+2} d\phi(u)}{1 + zu} \right]. \qquad (6.28)$$

Hence, if k is odd,

$$|R_k(z)| \leqslant |z| f_{k+1} + |z|^2 \int_{-\infty}^{\infty} \frac{|u|^{k+2} d\phi(u)}{|u| |\mathrm{Im}\, z|}$$

$$\leqslant |z| f_{k+1} + \frac{|z|^2 f_{k+1}}{|\mathrm{Im}\, z|}$$

$$\leqslant |z| f_{k+1} (1 + |\csc \varepsilon|).$$

We deduce that for k odd, $R_k(z) \to 0$ uniformly as $|z| \to 0$ in the sectors $S^+(\varepsilon)$ or $S^-(\varepsilon)$. If k is even, an extra term must be included in (6.28), with the consequence that

$$|R_k(z)| \leqslant |z| |f_{k+1}| + |z| f_{k+2}(1 + |\csc \varepsilon|),$$

and (6.27) is proved.

For the converse, we assume that an asymptotic Hamburger series is given, with moments defined by (6.26). The representation (6.26) allows the construction of $f(z)$ for $\operatorname{Im} z \neq 0$ according to (6.25). The estimates of $|R_k(z)|$ establish that (6.27) is the asymptotic series of $f(z)$.

We now state our principal result:

THEOREM 5.6.12. *Let the coefficients $\{f_j\}$ satisfy the determinantal conditions $D(0, n) > 0$ for all $n = 0, 1, 2 \ldots$. Let $\mathcal{D}^1(\Delta)$ be a bounded domain, distant at least Δ from the real axis, as shown in Figure 1. Then the sequence of $[M-1/M]$ Padé approximants, $M = 1, 2, 3, \ldots$, of the formal series $\sum_{j=0}^{\infty} f_j(-z)^j$ contains a subsequence, denoted by $M \in \mathfrak{M}$, which converges to $\tilde{f}(z)$ in $\mathcal{D}^1(\Delta)$. The function $\tilde{f}(z)$ is real symmetric, is analytic in $\mathcal{D}^1(\Delta)$, and has the representations*

$$\lim_{\substack{M \to \infty \\ M \in \mathfrak{M}}} [M-1/M] = \tilde{f}(z) = \int_{-\infty}^{\infty} \frac{d\phi(u)}{1 + zu} \simeq \sum_{j=0}^{\infty} f_j(-z)^j, \qquad (6.29)$$

where Figure 4 shows the sectors for the asymptotic expansion and

$$f_j = \int_{-\infty}^{\infty} u^j \, d\phi(u), \qquad j = 0, 1, 2, \ldots, \qquad (6.30)$$

and $\phi(u)$ is bounded and nondecreasing on $-\infty < u < \infty$.

Proof. Given $\Delta > 0$, notice that a positive δ may always be found such that $D^1(\Delta) \subset \{z: -z^{-1} \equiv w \in D^W(\delta)\}$; see also exercise 5. Theorem 5.6.12 then follows from Theorems 5.6.5, 5.6.7, 5.6.11 and (6.24).

What has not been established in Theorem 5.6.12 is that $\tilde{f}(z)$ is uniquely represented by (6.29) and (6.30). In the previous section we gave examples to show that, even in the case of Stieltjes series, the conditions of the theorem are insufficient to define $\phi(u)$ and $\tilde{f}(z)$ uniquely. A sufficient condition for uniqueness is Carleman's criterion for Hamburger series, which we quote:

CARLEMAN'S CRITERION.   If the coefficients $\{f_j\}$ are real and satisfy the determinantal inequalities $D(0, n) > 0$ for all $n \geqslant 0$ and also the condition that

$$\sum_{j=1}^{\infty} f_{2j}^{-1/(2j)} \text{ diverges,} \qquad (6.31)$$

then there exists a unique Hamburger function $f(z)$ with the asymptotic expansion

$$f(z) \simeq \sum_{j=0}^{\infty} f_j(-z)^j, \qquad \text{Im } z \neq 0.$$

If Carleman's criterion (6.31) is satisfied, the representations (6.29) and (6.30) are uniquely determined and the moment problem is said to be determinate. If Carleman's criterion is not satisfied, we refer the reader to the books of Akhiezer [1965] and Wall [1948] for further details.

The connection between the sequence of $[M-1/M]$ Padé approximants which converge to $\tilde{f}(z)$ and the J-fractions of (4.4.9) is established using the variable $\omega = z^{-1}$. Following (6.29), we define

$$\bar{J}(\omega) = z^{-1}\tilde{f}(z) = \int_{-\infty}^{\infty} \frac{d\phi(u)}{\omega + u}. \tag{6.32}$$

Consequently $\bar{J}(\omega)$ is the limit of a sequence of convergents given by

$$J_M(\omega) = z^{-1}[M-1/M]_{\tilde{f}}(z), \qquad M \in \mathfrak{M}, \quad \text{Im } \omega \neq 0.$$

The algebraic derivation of (4.5.9) and (4.5.10) is valid (provided the fractions are well defined) for Hamburger series as well as for Stieltjes series. Using (4.4.23)–(4.4.25), (4.5.1)–(4.5.4) and (1.12), we deduce that

$$J_M(\omega) = \frac{k_1}{l_1 + \omega} - \frac{k_2}{l_2 + \omega} - \frac{k_3}{l_3 + \omega} - \cdots - \frac{k_M}{l_M + \omega},$$

where

$$
\begin{aligned}
k_1 &= a_1 = f_0, \\
l_1 &= a_2 = f_1/f_0, \\
k_2 &= a_2 a_3 = \frac{D(0,1)}{D(0,0)^2},
\end{aligned}
\tag{6.33a}
$$

$$l_2 = a_3 + a_4 = \frac{D(0,1)}{D(0,0)D(1,0)} + \frac{D(1,1)D(0,0)}{D(1,0)D(0,1)}, \tag{6.33b}$$

and for $j = 3, 4, 5, \ldots,$

$$k_j = \frac{D(0, j-3)D(0, j-1)}{D(0, j-2)^2},$$

$$l_j = \frac{D(0, j-1)D(1, j-3)}{D(0, j-2)D(1, j-2)} + \frac{D(1, j-1)D(0, j-2)}{D(1, j-2)D(0, j-1)}. \tag{6.33c}$$

We use (6.33) to prove

THEOREM 5.6.13. *Given the real J-fraction*

$$J(\omega) = \frac{k_1}{l_1 + \omega} - \frac{k_2}{l_2 + \omega} - \frac{k_3}{l_3 + \omega} - \cdots \qquad (6.34)$$

with $k_i > 0$ and l_i real for all $i \geq 1$, then the sequence of convergents of $J(\omega)$ contains a convergent subsequence, with the property

$$\lim_{\substack{M \to \infty \\ M \in \mathfrak{M}}} J_M(\omega) = \bar{J}(\omega) = \int_{-\infty}^{\infty} \frac{d\phi(u)}{\omega + u}, \qquad (6.35)$$

where $\phi(u)$ is bounded and non-decreasing on $-\infty < u < \infty$.

If $J(\omega)$ given by (34) is a convergent J-fraction, then it has the Hamburger representation

$$J(\omega) = \int_{-\infty}^{\infty} \frac{d\phi(u)}{\omega + u}. \qquad (6.36)$$

Proof. By expansion of (6.34), we find that $J(\omega)$ has the formal expansion

$$J(\omega) = \frac{1}{\omega} \sum_{j=0}^{\infty} f_j(-\omega)^{-j}.$$

An inductive proof based on (6.33) shows that the coefficients f_j satisfy the determinantal conditions that $D(0, n) > 0$ and $D(1, n)$ is real. Therefore the coefficients f_j are real and satisfy the Hamburger series conditions. By Theorem 5.6.12, a subsequence of the convergents of $J(\omega)$ is convergent. If the fraction $J(\omega)$ is convergent, then all subsequences converge to a common limit.

It is interesting to notice in the latter proof that the hypothesis of convergence of the fraction (6.34) replaces Carleman's criterion as the condition for uniqueness.

Herglotz functions are a class of functions which are intimately connected with Hamburger series. We define these functions, and conclude this section by quoting the main theorem.

DEFINITION. $h(z)$ is defined to be a Herglotz function if

  (i) $h(z)$ is analytic for $\mathrm{Im}(z) \neq 0$
  (ii) $h(z)$ is real symmetric
  (iii) $\mathrm{Im}\, h(z)$ and $\mathrm{Im}\, z$ have the same sign for all z.

We quote Herglotz's theorem in the form [Stone, 1932, p. 573]

THEOREM 5.6.14. *A Herglotz function has the representation*

$$h(z)=\alpha z+\beta+\int_{-\infty}^{\infty}\frac{zu+1}{u-z}\,d\sigma(u),\qquad(6.37)$$

where $\sigma(u)$ is bounded and nondecreasing on $-\infty<u<\infty$, β is real, and $\alpha\geqslant0$. When this representation (6.37) exists, it is unique.

COROLLARY. Provided that the first moment

$$\mu_1=\int_{-\infty}^{\infty}u\,d\sigma(u)$$

is finite, $h(z)$ has the representation

$$h(z)=\alpha z+\beta'\int_{\infty}^{\infty}\frac{d\phi(u)}{u-z},$$

where $\beta'=\beta-\mu_1$ and $d\phi(u)=(1+u^2)\,d\sigma(u)$.

For further details of the applications of Herglotz functions, we refer to Narcowich and Allen [1975] and Allen and Narcowich [1975].

For further details of the construction of the density function $\phi(u)$ in the context of this section, we refer to the books of Perron [1957] and Riesz and Nagy [1955].

Techniques and results are known for treating a series which is the difference of two Stieltjes series [Baker and Gammel, 1971; Barnsley, 1976].

Exercises 1. Let $f(z)$ have the formal power series

$$f(z)=\sum_{j=0}^{\infty}f_j(-z)^j,$$

in which the coefficients f_j satisfy the conditions $D(0,n)>0$ for all $n\geqslant0$. Define $g(z)$ by the identity

$$f(z)=\frac{f_0}{1+f_1f_0^{-1}z-z^2g(z)}$$

so that

$$g(z)=\sum_{j=0}^{\infty}g_j(-z)^j,\qquad\text{formally.}$$

Use Hadamard's formula (1.6.9) to show that the coefficients g_j satisfy the conditions

$$D_g(0, n) = \begin{vmatrix} g_0 & g_1 & \cdots & g_m \\ g_1 & g_2 & \cdots & g_{m+1} \\ \vdots & \vdots & & \vdots \\ g_m & g_{m+1} & \cdots & g_{2m} \end{vmatrix} > 0.$$

Exercise 2. Let $f(z)$ be a Hamburger function, according to the definition (6.25). Let $w = -z^{-1}$ and $F(w) = -zf(z)$. Deduce that $-f(z)$ and $-F(w)$ are Herglotz functions.

Exercise 3. Prove that $\phi(u)$, defined in Lemma 3, is continuous at its points of continuity.

Exercise 4. Construct a Stieltjes density function $\phi(u)$ with the properties that

(i) $\phi(u)$ is a constant except on the interval $0 \leqslant u \leqslant 1$.

(ii) the points of discontinuity of $\phi(u)$ are dense on $0 \leqslant u \leqslant 1$.

Exercise 5. Show that Hamburger's theorem is valid if the condition that z lies in the sector $S^+(\varepsilon)$ or $S^-(\varepsilon)$ is replaced by the condition that z lies in the disk $\mathcal{C}$ of Figure 2.

Exercise 6. Let $\sum_{j=0}^\infty f_j(-z)^j$ be a Hamburger series, let $\omega = z^{-1}$, and define, according to (6.4),

$$J_M(\omega) = z^{-1}[M-1/M](z) = \sum_{i=1}^M \frac{\beta_i}{\omega + \gamma_i}.$$

Prove that, for $M = 1, 2, 3, \ldots$,

(i) the poles of $J_M(\omega)$ have positive residues,

(ii) the zeros of $J_M(\omega)$ are real,

(iii) the poles and zeros of $J_M(\omega)$ interlace.

5.7 Pólya Frequency Series

We have already seen that explicit forms for the numerator $A^{[L/M]}(z)$ and the denominator $B^{[L/M]}(z)$ of the Padé approximant of $\exp(z)$ are known. The results may be obtained from the explicit forms (1.2.12) or from collocation (II.3.3.21); continued fractions (4.6.1) provide yet another equivalent representation. The formula (1.2.12) for the Padé approximant of $\exp(z)$ may be derived from the determinantal representations (1.1.8) and (1.1.9), using (1.2.6):

$$C(L/M) = (-1)^{M(M-1)/2} \prod_{k=1}^M \frac{1}{k(k+1)\cdots(k+L-1)}$$

with $L, M \geqslant 1$, using the methods of Section 5.2. Notice that the sign of $C(L/M)$ is $(-1)^{M(M-1)/2}$, which does not depend on L; for $M = 1, 2, 3, \ldots$, the signs are $+, -, -, +, +, -, -, \ldots$, and this pattern of signs of $C(L/M)$ characterizes a class of functions known as Pólya frequency series. (They are also known as totally positive series: see Exercise 1.) Every function of the class has the representation

$$f(z) - a_0 e^{\gamma z} \prod_{j=1}^{\infty} (1 + \alpha_j z)(1 - \beta_j z)^{-1} \qquad (7.1)$$

with $a_0 > 0$, $\gamma \geqslant 0$, $\alpha_j \geqslant 0$, $\beta_j \geqslant 0$, and

$$\sum_{j=1}^{\infty} (\alpha_j + \beta_j) \text{ convergent.}$$

The present point of interest is that these functions have not only convergent Padé approximants, but also the numerators and denominators of the Padé approximants converge along any ray of the Padé table.

THEOREM 5.7.1 [Arms and Edrei, 1970]. *If $f(z)$ is a Pólya frequency series defined by* (7.1), *and* $L_k \to \infty$, $M_k \to \infty$ *with* $(M_k/L_k) \to \omega$ *as* $k \to \infty$, *then*

$$A^{[L_k/M_k]}(z) \to a_0 \exp\left(\frac{\gamma}{1+\omega} z \right) \prod_{j=1}^{\infty} (1 + \alpha_j z),$$

$$B^{[L_k/M_k]}(z) \to \exp\left(\frac{-\gamma\omega}{1+\omega} z \right) \prod_{j=1}^{\infty} (1 - \beta_j z)$$

uniformly on any compact region of the z-plane.

COROLLARY 1. *If $f(z) = e^{\gamma z}$,*

$$A^{[L_k/M_k]} \to \exp\left(\frac{\gamma z}{1+\omega} \right) \quad and \quad B^{[L_k/M_k]} \to \exp\left(\frac{-\gamma\omega z}{1+\omega} \right).$$

COROLLARY 2. *Provided $\alpha_i \neq -\beta_j$ for all i, j, a finite number of α_i and β_j may be arbitrary and complex.*

Corollary 1 is a special case of the theorem, and Corollary 2 is a generalization. The proofs are in EPA and in Arms and Edrei [1970]. Excellent accounts of the basic theory are given by Edrei [1953] and Karlin [1968].

Partly because the exponential function has known explicit forms for its Padé approximants and partly because of its role as a solution of the most elementary ordinary differential equation, $dy/dx=y$, considerable attention has been paid to the location of poles and zeros of the Padé approximants of $\exp(z)$. If the $[L/M]$ Padé approximant of $\exp(z)$ has a pole in the left half plane, then this Padé approximant most definitely does not define an A-stable method for the solution of ordinary differential equations (see Part II, Section 3.4). Simply because $\exp(-z)=[\exp(z)]^{-1}$, the duality theorem (1.5.1) shows that a zero of the $[L/M]$ Padé approximant of $\exp(z)$ is a pole of the $[M/L]$ Padé approximant of $\exp(-z)$. Thus attention is usually directed to the location of only the zeros of the Padé approximant to $\exp(z)$.

Because $\exp(+\infty)=+\infty$ and $\exp(-\infty)=0$, we expect the poles of the Padé approximants of $\exp(x)$ to accumulate at $x=+\infty$ and the zeros at $x=-\infty$, with increasing order of Padé approximation. If $L_1>L_2$, there are more zeros of $[L_1/M]$ than $[L_2/M]$ to be accommodated, and so we expect a smaller zero free region for $[L_1/M]$. These simple ideas are borne out by the following theorems.

THEOREM 5.7.2 (Sectorial theorem) [Saff and Varga, 1975]. *Let* $z=re^{i\theta}$ *denote a generic point in the complex plane. The* $[L/M]$ *Padé approximant to* $\exp(z)$ *is zero free in the sector given by*

$$\cos\theta \geqslant \frac{L-M-2}{L+M}, \qquad provided \quad M\geqslant 2.$$

Example 1. If $L=3M+2$, the $[3M+2/M]$ Padé approximants are zero free in the region $\cos\theta\geqslant\frac{1}{2}$, i.e. $|\theta|\leqslant60°$. Clearly, all Padé approximants with $L\leqslant3M+2$ are also zero free in this region, as shown in Figure 1.

THEOREM 5.7.3 (Parabola theorem) [Saff and Varga, 1976b]. *Let* $z=x+iy$ *denote a generic point in the complex plane. The* $[L/M]$ *Padé approximant to* $\exp(z)$ *is zero free in the parabolic region*

$$y^2 \leqslant 4(M+1)(x+M+1). \tag{7.2}$$

Remark. This theorem asserts that no $[L/M]$ Padé approximant in the $(M+1)$th row of the Padé table has a zero in the region given by (7.2). For the first row, it is obvious that the truncated Maclaurin series of $\exp(x)$ is positive for $x\geqslant0$, and so the zero free region includes the positive real axis. The theorem asserts that there are no zeros in $y^2\leqslant4(x+1)$. For $[L/3]$ Padé approximants, the parabolic zero free region is larger: the region is specified by the equation $y^2\leqslant16(x+4)$ and it is shown in Figure 2.

Theorem 5.7.3 can be reviewed by rescaling the axes by the factor $M+1$, leading to the following restatement:

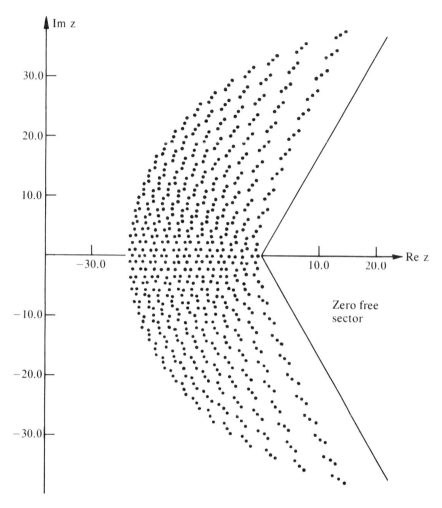

Figure 1. Zeros of $[L/M]$ Padé approximants of $\exp(z)$ with $M \leqslant [\frac{1}{3} L]$, for $L = 1, 2, 3, \ldots, 36$.

COROLLARY. The $[L/M]$ Padé approximant of $\exp\{(M+1)z\}$ is zero free in the region

$$\left(\frac{y}{2}\right)^2 \leqslant x + 1.$$

The locations of the zeros of $[L/M]$ Padé approximants with $L \leqslant 25$, $M \leqslant 25$ are shown in Figure 3, demonstrating visually the optimal nature of Theorem 5.7.3. The remarkable trajectories of the zeros are not, as yet, fully explained.

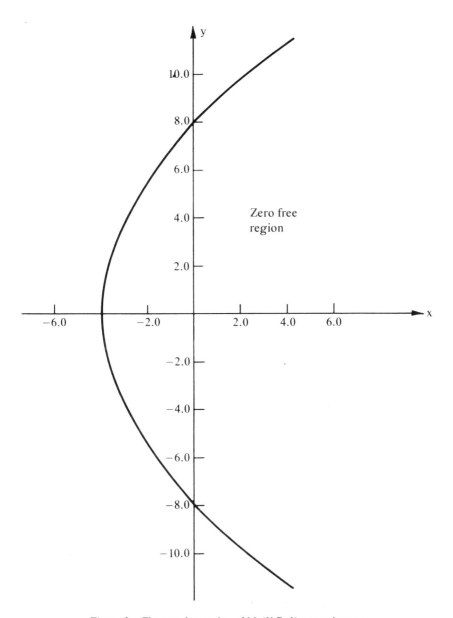

Figure 2. The zero free region of $[L/3]$ Padé approximants.

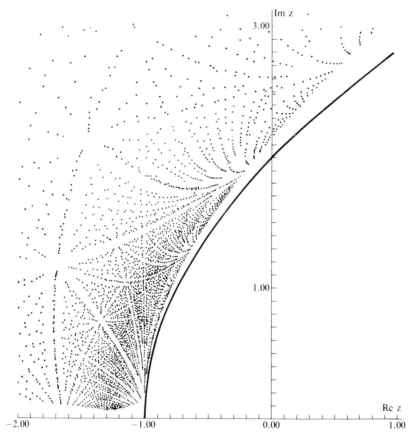

Figure 3. The zero free region of $[L/M]$ Padé approximants of $\exp\{(M+1)z\}$.

THEOREM 5.7.4 (Annulus theorem of Saff and Varga, [1977]). *For any $L \geq 1$ and any $M \geq 0$, all the zeros of the $[L/M]$ Padé approximants of $\exp(z)$ lie in the annulus $(L+M)\mu < |z| < L+M+\frac{4}{3}$, where μ is the (unique) positive root of $te^{1+t} = 1$, which occurs at $t = \mu = 0.278465$.*

THEOREM 5.7.5 [Saff and Varga, 1975, 1977]. *An $[L/M]$ Padé approximant of $\exp(z)$ with $L \leq M+4$ has all its zeros in the left half plane. The $[5/0]$ Padé approximant has two complex conjugate zeros in the right half plane. Let z_0 be any zero of an $[L/M]$ approximant. Then $\mathrm{Re}\, z_0 < L-M$ for all $L \geq 1$ and $M \geq 0$.*

The content of this theorem is rather obvious. Our last theorem for the exponential function concerns the behavior of Padé approximants lying on rays in the Padé table and with z confined to the negative real axis. We define an approximation error on this axis by

$$\eta^{[L/M]} = \sup_{-\infty \leqslant x \leqslant 0} |\exp(x) - [L/M](x)|. \tag{7.3}$$

THEOREM 5.7.6 [Saff, Varga and Ni, 1976]. *If a ray sequence is defined by* $(L_k/M_k) \to \omega$ *as* $k \to \infty$, *with* $0 \leqslant \omega \leqslant 1$, *then*

$$\left\{ \eta^{[L_k/M_k]} \right\}^{1/M_k} \to \omega^\omega \left(\frac{1-\omega}{2} \right)^{1-\omega} \equiv g(\omega).$$

COROLLARY. Observe that $\min\{g(\omega)\} = \frac{1}{3}$ for $0 \leqslant \omega \leqslant 1$, with minimization achieved at $\omega = \frac{1}{3}$, $g(\omega) = \frac{1}{3}$. The ray sequence of $[L/3L]$ Padé approximants to $\exp(z)$ has the best rate of convergence on the negative real axis, with convergence characterized by

$$\eta^{[L/3L]} \simeq \left(\tfrac{1}{3} \right)^L.$$

The reader is referred to the original papers for complete statements and proofs of the foregoing theorems. The proofs are based on the explicit forms and the interrelations which exist for the Padé numerator and denominator. This concludes our discussion on the location of poles on zeros of Padé approximants to $\exp(z)$.

We turn our attention to trigonometric polynomials having real roots. The discussion is based on the formulas

$$\sin z = z \prod_{n=1}^{\infty} \left(1 - \frac{z^2}{n^2 \pi^2} \right), \tag{7.4}$$

$$\cos z - 1 = -2\sin^2\left(\frac{z}{2} \right) = \frac{-z^2}{2} \prod_{n=1}^{\infty} \left(1 - \frac{z^2}{4n^2\pi^2} \right), \tag{7.5}$$

$$\cos z - \cos \alpha = (1 - \cos \alpha) \prod_{n=-\infty}^{\infty} \left(1 - \frac{z^2}{(2n\pi + \alpha)^2} \right), \qquad 0 < \alpha \leqslant \pi. \tag{7.6}$$

Any real cosine polynomial

$$C(z) = c_0 + c_1 \cos z + c_2 \cos 2z + \cdots + c_N \cos Nz \tag{7.7}$$

or any real sine polynomial

$$S(z) = s_1 \sin z + s_2 \sin 2z + \cdots + s_N \sin Nz, \tag{7.8}$$

if it has only real zeros, may be reexpressed using (7.6). We may write

$$C(z) = K(\cos z - 1)^p \prod_{k=1}^{k_{max}} (\cos z - \cos \alpha_k)^{p_k}, \tag{7.9}$$

where p, p_k are nonnegative integer powers. $S(z)$ may be expressed quite generally as

$$S(z) = C(z) \sin z, \tag{7.10}$$

and so we consider (7.6), (7.7), and (7.9). We see that

$$C(z) = K\left(\frac{-z^2}{2}\right)^p \left\{ \prod_{n=1}^{\infty} \left(1 - \frac{z^2}{4n^2\pi^2}\right)^{2p} \right\}$$

$$\times \prod_{k=1}^{k_{max}} \left\{ (1 - \cos \alpha_k)^{p_k} \prod_{n=-\infty}^{\infty} \left(1 - \frac{z^2}{(2n\pi + \alpha_k)}\right)^{p_k} \right\}, \tag{7.11}$$

and (7.11) fills the bill for the Arms–Edrei theorem, leading to the following result:

THEOREM 5.7.7 [Edrei, 1975a]. *The denominators of the* $[L_k/M_k]$ *Padé approximants of the normalized real cosine polynomial having real zeros,* $C(z)z^{-2p}$, *defined by* (7.7), (7.9), *and* (7.11) *tend to unity uniformly on any compact domain of the z-plane provided* $L_k \to \infty$ *and* $M_k \to \infty$ *as* $k \to \infty$. *An identical result under identical conditions holds for* $S(z)z^{-2p-1}$ *defined by* (7.8), (7.10), *and* (7.11).
For the function $T(z) = z^{-2} \tan^2 z$,

$$A^{[L_k/M_k]} \to z^{-2} \sin^2 z \quad and \quad B^{[L_k/M_k]} \to \cos^2 z$$

under the same conditions.

In this section, we have considered several theorems about what appear to be special cases. Primarily for this reason, no proofs have been given. However, it is tempting to speculate that a theory as complete as that for Stieltjes series will eventually be forthcoming, and that methods of general applicability will emerge in the process.

Exercise Show that the Toeplitz determinants of the coefficients of Polyá frequency series, defined by

$$T(L/M) \equiv \begin{vmatrix} c_L & c_{L+1} & \cdots & c_{L+M-1} \\ c_{L-1} & c_L & \cdots & c_{L+M-2} \\ \vdots & \vdots & & \vdots \\ c_{L-M+1} & c_{L-M+2} & \cdots & c_L \end{vmatrix}$$

satisfy $T(L/M) \geq 0$ for all $L, M \geq 0$.

CHAPTER 6

Convergence Theory

6.1 Introduction to Convergence Theory: Rows

This chapter is concerned with what is known about convergence of sequences of Padé approximants to complex functions.

For row sequences, de Montessus's theorem proves convergence for functions meromorphic in a disk, as explained in Section 6.2.

Diagonal sequences are the natural choice for meromorphic functions in the absence of further information. Simply because meromorphy of a function implies meromorphy of its reciprocal, the symmetric choice of diagonal approximants is natural, especially in view of the duality theorem. Paradiagonal sequences, $[M+J/M]$, with J fixed and $M \to \infty$ are an obvious generalization, usually motivated by the requirement of an asymptotic approximation of z^J as $z \to \infty$.

Ray sequences of $[L/M]$ Padé approximants with $L = \lambda M$, λ fixed, are useful in special circumstances, and parabolic sequences, $[M^2/M]$, are worth considering sometimes. Hence interest settles on general sequences $[L_k/M_k]$ which may be particularized to suit special needs. The most natural convergence theorems for general sequences of Padé approximants involve convergence in capacity, a difficult concept. Instead of attempting to prove pointwise convergence for a class of functions, the theorems prove that the region of bad approximation becomes arbitrarily small. In no way do the these theorems imply pointwise convergence, but they do prove that the Padé method converges in a real sense in very general circumstances.

Recognizing that Padé approximants to meromorphic functions are simply rational approximations, another natural development is the notion of convergence of the function values on the Riemann sphere, which is treated

ENCYCLOPEDIA OF MATHEMATICS and Its Applications, Gian-Carlo Rota (ed.). Vol. 13: George A. Baker, Jr., and Peter R. Graves-Morris, Padé Approximants: Basic Theory, Part I   ISBN 0-201-13512-4

in Section 6.4. This allows the value ∞ to be treated on an equal footing with any other function value.

The Perron and Gammel counterexamples are clear warnings about what may not be proved convergence of Padé approximants to entire functions, and an understanding of these surprising results saves us from ill-founded and wasted optimism about mythical theorems. In fact, the Padé conjecture, due to Baker, Gammel, and Wills, is widely accepted, and is the foundation of many calculations and applications. It asserts the convergence of a subsequence of diagonal approximants, and is more fully discussed in Section 6.7.

It is interesting to consider a few theorems about rows of the Padé table; with the advantage of having explicit expressions for the Padé approximants, one or two optimal results and several significant theorems have been proved which seem to act as signposts to the complete theory. A discussion of the convergence of row sequences indicates clearly the type of results expected in general.

The first row of the Padé table consists of $[L/0]$ Padé approximants, which are truncated Maclaurin series. They converge within (but not necessarily on) a circle of convergence of which the radius may be zero, finite, or infinite.

The second row of $[L/1]$ approximants is governed by a theorem of Beardon [1968b].

THEOREM 6.1.1. *Let $f(z)$ be analytic in $|z| \leq R$. Then an infinite subsequence of $[L/1]$ Padé approximants converges to $f(z)$ uniformly in $|z| \leq R$.*

Proof. By hypothesis, $f(z)$ is analytic in $|z| \leq R$, and consequently within a larger circle, $|z| < \rho$, with $\rho > \rho' > R$. Let

$$f(z) = \sum_{i=0}^{\infty} c_i z^i \quad \text{with} \quad c_i = O((\rho')^{-i}). \tag{1.1}$$

The second row of Padé approximants are given by

$$[L/1] = c_0 + c_1 z + \cdots + c_{L-1} z^{L-1} + \frac{c_L z^L}{1 - c_{L+1} z / c_L}. \tag{1.2}$$

If a subsequence of coefficients $\{c_{L_j}, j = 1, 2, \ldots\}$ are zero, then $[L_j - 1/1]$ are truncated Maclaurin expansions which converge to $f(z)$ uniformly in $|z| \leq R < \rho'$. So we assume that no infinite subsequence of $\{c_L\}$ vanishes, and consider $r_L = c_{L+1}/c_L$ which is well defined for all sufficiently large L. Because

$$c_L z^L = O\left(\frac{R}{\rho'}\right)^L,$$

the sequence of $[L/1]$ Padé approximants given by (1.2) converges uni-
formly in $|z| \leqslant R$ *unless* there exists a sequence of values of L for which
$1 - c_{L+1} z c_L^{-1} = 0$ within $|z| < \rho'$. Thus either a subsequence of the second
row converges uniformly, or else for some L_0 and all $L > L_0, |c_L/c_{L+1}| > \rho'$.
In the latter case

$$\left| \frac{c_{L_0}}{c_L} \right| = \prod_{i=L_0}^{L-1} \left| \frac{c_i}{c_{i+1}} \right| > (\rho')^{L-L_0},$$

contradicting (1.1), so the theorem is proved.

PERRON'S COUNTEREXAMPLE [Perron, 1957, Chapter 4]. A function $f(z)$
$= \sum_{i=0}^{\infty} c_i z^i$ is defined by selecting a sequence of points $\{z_n, n = 1, 2, \ldots\}$ in
the complex plane, and the coefficients c_i are defined in triples by

$$\left. \begin{array}{l} c_{3n} = z_n/(3n+2)! \\ c_{3n+1} = 1/(3n+2)! \\ c_{3n+2} = 1/(3n+2)! \end{array} \right\} \quad \text{if} \quad |z_n| \leqslant 1$$

or

$$\left. \begin{array}{l} c_{3n} = 1/(3n+2)! \\ c_{3n+1} = 1/(3n+2)! \\ c_{3n+2} = z_n^{-1}/(3n+2)! \end{array} \right\} \quad \text{if} \quad |z_n| > 1.$$

Since $|c_i| \leqslant (i!)^{-1}$, the comparison test shows that $f(z)$ is entire. Since

$$\frac{c_{3n}}{c_{3n+1}} = z_n \text{ if } |z_n| < 1 \quad \text{or} \quad \frac{c_{3n+1}}{c_{3n+2}} = z_n \text{ if } |z_n| > 1,$$

(1.2) shows that either $[3n/1]$ or $[3_{n+1}/1]$ has a pole at the selected point z_n.
By choosing $\{z_n\}$ dense in the plane, repeating its values if necessary, a
holomorphic function is constructed for which the $[L/1]$ Padé approximants converge in no open set, however small, in the z-plane.

Perron's counterexample shows that Beardon's theorem is an optimal
result. The following conjecture of Baker and Graves-Morris [1977] summarizes what is expected to hold good in general.

CONJECTURE.   Let $f(z)$ be a meromorphic function which is analytic at
the origin, and let $|z| \leqslant R$ be a disk which contains not more than M poles
of $f(z)$. Then at least an infinite subsequence of the row of $[L/M]$ Padé

approximants converges to $f(z)$ uniformly in $|z| \leqslant R$ except on arbitrarily small open sets enclosing the poles of $f(z)$.

No proof of this conjecture yet exists, except for special cases such as Beardon's theorem. It is clear that Perron's example, which may easily be generalized to apply to any row, shows that any attempt to prove convergence of an entire row sequence is futile. This is precisely why the conjecture only asserts convergence of a subsequence. The approximants of the conjecture have sufficiently many poles to represent the poles of $f(z)$ within $|z| \leqslant R$; the poles of the Padé approximants either congregate near the poles of $f(z)$ or lie outside the circle $|z| = R$.

It is interesting to compare this conjecture with de Montessus's theorem in the next section, and in particular the comparison between Beardon's theorem and de Montessus's theorem for the second row is important. For general reviews, we refer to EPA, Chapter 11, Graves-Morris [1975], and Wallin [1972].

Exercise 1. Construct the sequence of $[L/1]$ Padé approximants explicitly for $f(z) = 1/((z-a)(z-b))$ with a, b complex. Prove that $[L/1] \to f(z)$ in $|z| < |b|$ provided $|a| < |b|$, and state precisely what is the domain of uniform convergence. Explain what happens if $|a| = |b|$.

Exercise 2. Generalize Perron's function to provide a counterexample for an arbitrary row of the Padé table.

6.2   de Montessus's Theorem

Before embarking on de Montessus's theorem, we will state the Cauchy–Binet formula. This is a formula for calculating the determinant of the product of two matrices. It combines a number of familiar results, which we give as examples, in a unified scheme [Gragg, 1972].

We define a multiindex $\boldsymbol{\alpha} = (\alpha_1 \alpha_2 \ldots \alpha_k)$ which belongs to the class $\mathcal{I}\binom{m}{k}$ of multiindices with k elements chosen from $(1, 2, \ldots, m)$, and which obeys the extra condition

$$1 \leqslant \alpha_1 < \alpha_2 < \cdots < \alpha_k \leqslant m.$$

We define $A(\boldsymbol{\alpha}, \boldsymbol{\beta})$ to be the submatrix of A formed from the rows $\{\alpha_i\}$ and columns $\{\beta_j\}$ of the original matrix A.

We define $\mathfrak{M}(m \times n)$ to be the class of matrices with m rows and n columns. If

$$\boldsymbol{\alpha} \in \mathcal{I}\binom{m}{l}, \qquad \boldsymbol{\beta} \in \mathcal{I}\binom{n}{l}, \quad \text{and} \quad A \in \mathfrak{M}(m \times n),$$

then $A(\boldsymbol{\alpha}, \boldsymbol{\beta})$ is an $l \times l$ matrix belonging to $\mathfrak{M}(l \times l)$.

THE CAUCHY–BINET THEOREM.   Let $A \in \mathfrak{M}(m \times n)$, $B \in \mathfrak{M}(m \times k)$, $C \in \mathfrak{M}(k \times n)$, $A = BC$,

$$\alpha \in \mathscr{I}\binom{m}{l}, \quad \beta \in \mathscr{I}\binom{n}{l} \quad \text{and} \quad \gamma \in \mathscr{I}\binom{k}{l}.$$

These conditions ensure that $A(\alpha, \beta)$, $B(\alpha, \gamma)$, and $C(\gamma, \beta)$ are $l \times l$ matrices, and

$$\det A(\alpha, \beta) = \sum_{\gamma \in \mathscr{I}\binom{k}{l}} \det B(\alpha, \gamma) \det C(\gamma, \beta).$$

Example 1.   Let A, B, C be $m \times m$ matrices, and let $\alpha, \beta, \gamma \in \mathscr{I}\binom{m}{m}$ be the full m-dimensional multiindex. Then the Cauchy–Binet formula reduces to

$$\det A = \det B \det C.$$

Example 2.   Let A, B, C be square $m \times m$ matrices, and let $l = 1$, so that α, β, γ reduce to ordinary indices. Then the formula becomes

$$A_{\alpha\beta} = \sum_{\gamma} B_{\alpha\gamma} C_{\gamma\alpha}.$$

which is the matrix multiplication rule.

Example 3.   If $l > k$, the set $\mathscr{I}\binom{k}{l}$ is empty and

$$\det A(\alpha, \beta) = 0.$$

This corresponds to the property that if the rows of a matrix are linearly dependent, then its determinant is zero.

Example 4.   Let $A, B, C \in \mathfrak{M}(3, 3)$, $A = BC$, $\alpha = (1, 2)$, and $\beta = (1, 3)$. Then

$$\det \begin{pmatrix} a_{11} & a_{12} & a_{13} \\ a_{21} & a_{22} & a_{23} \\ a_{31} & a_{32} & a_{33} \end{pmatrix} = \det \left[\begin{pmatrix} b_{11} & b_{12} & b_{13} \\ b_{21} & b_{22} & b_{23} \\ b_{31} & b_{32} & b_{33} \end{pmatrix} \begin{pmatrix} c_{11} & c_{12} & c_{13} \\ c_{21} & c_{22} & c_{23} \\ c_{31} & c_{32} & c_{33} \end{pmatrix} \right]$$

$$= \sum_{j=1}^{3} \det \begin{pmatrix} b_{11} & b_{12} & b_{13} \\ b_{21} & b_{22} & b_{23} \\ b_{31} & b_{32} & b_{33} \end{pmatrix} \det \begin{pmatrix} c_{11} & c_{12} & c_{13} \\ c_{21} & c_{22} & c_{23} \\ c_{31} & c_{32} & c_{33} \end{pmatrix} \leftarrow [\text{row } j]$$

$$[\text{col } j]$$

The Cauchy–Binet theorem is useful in the proof we give of de Montessus's theorem.

de Montessus's theorem applies to a function $f(z)$ which is meromorphic in a circle with M poles within a circle. In practical terms, provided M is known, de Montessus's theorem asserts convergence of the row of $[L/M]$ approximants within the circle, which is all that can possibly be established. If the number of poles of $f(z)$ within the circle is not known *a priori*, the theorem is usually unhelpful. The complete statement of de Montessus's theorem [1902] for simple poles is:

THEOREM 6.2.1. *Let $f(z)$ be a function which is meromorphic in the disk $|z| \leqslant R$, with precisely M simple poles at distinct points $z_1, z_2, \dots z_M$, where*

$$0 < |z_1| \leqslant |z_2| \leqslant \cdots \leqslant |z_M| < R.$$

Then

$$\lim_{L \to \infty} [L/M] = f(z) \tag{2.1}$$

uniformly on any compact subset of

$$\mathcal{D}_M = \{ z, |z| \leqslant R, z \neq z_i, i = 1, 2, \dots, M \}. \tag{2.2}$$

Proof. Since $|z_1| > 0$ and $f(z)$ is analytic in $|z| < |z_1|$, $f(z)$ has a power-series expansion

$$f(z) = \sum_{i=0}^{\infty} c_i z^i. \tag{2.3}$$

Define the polynomial

$$B(z) = \sum_{i=0}^{M} B_i z^i = \prod_{i=1}^{M} \left(1 - \frac{z}{z_i} \right). \tag{2.4}$$

Then the series for $A(z)$ defined by

$$\left(\sum_{i=0}^{M} B_i z^i \right) \left(\sum_{i=0}^{\infty} c_i z^i \right) = \sum_{i=0}^{\infty} A_i z^i \equiv A(z) \tag{2.5}$$

is convergent in $|z| \leqslant R$, and we have thereby reexpressed the function $f(z)$ as

$$f(z) = A(z)/B(z) \qquad \text{with} \quad B_0 = 1. \tag{2.6}$$

Equating coefficients of z^i in (2.5) for $i = L+1, L+2, \ldots, L+M$ with $L \geqslant M$ gives

$$\sum_{j=0}^{M} B_j c_{i-j} = A_i, \qquad i = L+1, L+2, \ldots, L+M. \tag{2.7}$$

Now the $[L/M]$ Padé approximant is

$$[L/M] = \frac{\displaystyle\sum_{i=0}^{L} a_i^{[L/M]} z^i}{\displaystyle\sum_{j=0}^{M} b_j^{[L/M]} z^j} = \frac{A^{[L/M]}(z)}{B^{[L/M]}(z)}$$

and is defined by the Padé equations

$$\sum_{j=0}^{M} b_j^{[L/M]} c_{i-j} = 0, \qquad i = L+1, L+2, \ldots, L+M, \tag{2.8}$$

with $b_0^{[L/M]} = 1$. A major part of the proof of de Montessus's theorem consists of proving that $b_j^{[L/M]} \to B_j$ as $L \to \infty$. Define

$$\Delta_j = B_j - b_j^{[L/M]}, \qquad j = 0, 1, \ldots, M, \tag{2.9}$$

and note that $\Delta_0 = 0$. From (2.7) and (2.8),

$$\sum_{j=0}^{M} \Delta_j c_{i-j} = A_i, \, i = L+1, L+2, \ldots L+M, \tag{2.10}$$

which are M equations for $\Delta_1, \Delta_2, \ldots, \Delta_M$. We define

$$C = \begin{pmatrix} c_L & \cdots & c_{L-M+1} \\ \vdots & & \vdots \\ c_{L+M-1} & \cdots & c_L \end{pmatrix}, \qquad \Delta = \begin{pmatrix} \Delta_1 \\ \vdots \\ \Delta_M \end{pmatrix}, \qquad A = \begin{pmatrix} A_{L+1} \\ \vdots \\ A_{L+M} \end{pmatrix}, \tag{2.11}$$

so that (2.10) becomes $C\Delta = A$, and $\Delta = C^{-1}A$ if C is nonsingular. Then

$$\Delta_i = \sum_{j=1}^{M} (C^{-1})_{ij} A_j = \frac{\displaystyle\sum_{j=1}^{M} C(j,i) A_j}{\det C} = \frac{\det C(i \to A)}{\det C}, \tag{2.12}$$

where $C(j, i)$ is the cofactor of the element of C in row j and column i, and $C(i \to \mathbf{A})$ is the matrix with column i replaced by $\mathbf{A}$. To establish that $\Delta_i \to 0$, we must consider the behavior of $C(j, i)$ and $\det C$ as $L \to \infty$ [Hadamard, 1892].

Because $f(z)$ has precisely M simple poles in $|z| < R$, we consider explicitly the contribution of these poles to the Maclaurin coefficients by writing

$$\sum_{i=0}^{\infty} c_i z^i = \sum_{k=1}^{M} r_k \left(1 - \frac{z}{z_k} \right)^{-1} + \sum_{i=0}^{\infty} c_i' z^i, \tag{2.13}$$

where $\sum_{i=0}^{\infty} c_i' z^i$ is convergent in $|z| \leq R$, and $c_i' = O(R^{-i})$. By expanding $(1 - z/z_k)^{-1}$, it follows from (2.13) that

$$c_i = d_i + O(R^{-i}) \tag{2.14}$$

where

$$d_i = \sum_{k=1}^{M} r_k z_k^{-i} \tag{2.15}$$

is the dominant part of the coefficients c_i. We define a matrix D by

$$D_{ij} = d_{L+i-j} = \sum_{k=1}^{M} r_k z_k^{-L-i+j}, \qquad i, j = 1, 2, \ldots, M, \tag{2.16}$$

which is the dominant part of C.

Example. For $M = 2$

$$D = \begin{pmatrix} r_1 z_1^{-L} + r_2 z_2^{-L} & r_1 z_1^{-L+1} + r_2 z_2^{-L+1} \\ r_1 z_1^{-L-1} + r_2 z_2^{-L-1} & r_1 z_1^{-L} + r_2 z_2^{-L} \end{pmatrix}$$

$$= \begin{pmatrix} 1 & 1 \\ z_1^{-1} & z_2^{-1} \end{pmatrix} \begin{pmatrix} r_1 z_1^{-L} & r_1 z_1^{-L+1} \\ r_2 z_2^{-L} & r_2 z_2^{-L+1} \end{pmatrix}$$

$$= \begin{pmatrix} 1 & 1 \\ z_1^{-1} & z_2^{-1} \end{pmatrix} \begin{pmatrix} r_1 z_1^{-L} & 0 \\ 0 & r_2 z_2^{-L} \end{pmatrix} \begin{pmatrix} 1 & z_1 \\ 1 & z_2 \end{pmatrix}$$

We are led to define Vandermonde matrices V, V' and a diagonal matrix D' by

$$V_{ij} = (z_j)^{1-i}, \quad V_{ij}' = (z_i)^{j-1}, \quad D_{ij}' = r_i z_i^{-L} \delta_{ij} \qquad (i, j = 1, 2, \ldots, M),$$

$$\tag{2.17}$$

so that D may be written as

$$D - VD'V' \tag{2.18}$$

and

$$\det D = \det V \det D' \det V'. \tag{2.19}$$

We need the formulas

$$\det V = \prod_{i=2}^{M} \prod_{j=1}^{i} \left(z_i^{-1} - z_j^{-1} \right) \tag{2.20}$$

and

$$\det V' = \prod_{i=2}^{M} \prod_{j=1}^{i} \left(z_i - z_j \right), \tag{2.21}$$

which are two well-known expansions of Vandermonde determinants. Their important characteristic is that they are nonzero constants, independent of L. Hence

$$\det D = K \prod_{i=1}^{M} \left(z_i \right)^{-L}, \tag{2.22}$$

where K is a nonzero constant.

Now consider the evaluation of $\det C$. Its dominant contribution is given by (2.24) below, but (2.14) shows that there are also a large but finite number of terms given by replacing a d_i by an $O(R^{-i})$ term. Hence we define $p = |z_N|$ to be the modulus of the pole furthest from the origin in $|z| < R$, and

$$\det C = K \prod_{i=1}^{M} \left(z_i \right)^{-L} \left(1 + O\left(\frac{p}{R} \right)^m \right) \tag{2.23}$$

In a similar way, using the factorization (2.18) and the Cauchy–Binet theorem, we may show that

$$C(j, i) = O\left(\prod_{i=1}^{M-1} |z_i|^{-L} \right) = O\left(p^L \prod_{i=1}^{M} |z_i|^{-L} \right). \tag{2.24}$$

Since (2.5) converges in $|z| \leq R$, $A_i = O(R^{-i})$. Hence, from (2.12), (2.23), and (2.24),

$$\Delta_i = O\left(\frac{p}{R} \right)^i, \qquad i = 1, 2, \dots, M, \tag{2.25}$$

which establishes that $\Delta_i \to 0$ uniformly as $L \to \infty$.

It now remains to consider

$$f(z) - [L/M] = \frac{A(z)B^{[L/M]}(z) - B(z)A^{[L/M]}(z)}{B(z)B^{[L/M]}(z)}. \tag{2.26}$$

Now

$$A(z)B^{[L/M]}(z) = \left(\sum_{i=0}^{\infty} A_i z^i \right) \left(\sum_{j=0}^{M} b_j^{[L/M]} z^j \right)$$

$$= \sum_{j=0}^{M} b_j^{[L/M]} \sum_{i=j}^{\infty} A_{i-j} z^i.$$

By using the formal identity defining the approximants and noting that $B(z)A^{[L/M]}$ is a polynomial of degree $L+M$, it follows that

$$A(z)B^{[L/M]}(z) = \sum_{j=0}^{M} b_j^{[L/M]} \sum_{i=L+M+1}^{\infty} A_{i-j} z^j.$$

Hence

$$|A(z)B^{[L/M]}(z)| \leqslant \sum_{j=0}^{M} |b_j^{[L/M]}| \cdot \left| \sum_{i=L+M+1}^{\infty} A_{i-j} z^j \right|.$$

Since $\Delta_i \to 0$, $b_j^{[L/M]}$ are bounded as $L \to \infty$. Since $A(z)$ is analytic in $|z| \leqslant R$, for some $R' > R$

$$|A(z)B^{[L/M]}(z) - B(z)A^{[L/M]}(z)| = O\big((z/R')^L\big), \tag{2.27}$$

and the numerator of (2.26) tends to zero as $L \to \infty$ for $|z| \leqslant R$. To treat the denominator, we consider

$$|B(z) - B^{[L/M]}(z)| \leqslant \sum_{j=1}^{M} |B_j - b_j^{[L/M]}| |z^j|. \tag{2.28}$$

The right-hand side of (2.28) tends to zero as $L \to \infty$ for $|z| \leqslant R$. Let z belong to a compact subset of this disk for which $|B(z)| > 2\varepsilon$ for any given positive ε. Hence $L(\varepsilon)$ exists such that $|B^{[L/M]}(z)| > \varepsilon$ for all $L \to L(\varepsilon)$. Then the modulus of the denominator of (2.26) is bounded below by $2\varepsilon^2$ independently of L for $L > L(\varepsilon)$. Together with (2.27), it follows from (2.26) that $f(z) - [L/M] \to 0$ as $L \to \infty$ uniformly an any compact subset of $\mathcal{D}_M$, and the theorem is proved.

Now we proceed with the full theorem of R. de Montessus de Ballore. This theorem allows the possibility of multiple poles instead of simple poles, and this inclusion makes the proof technically complicated.

THEOREM 6.2.2 [de Montessus, 1902]. *Let $f(z)$ be a function which is meromorphic in the disk $|z| \leqslant R$, with m poles at distinct points $z_1, z_2, \ldots, z_m$ with*

$$0 < |z_1| \leqslant |z_2| \leqslant \cdots \leqslant |z_m| < R.$$

Let the pole at z_k have multiplicity μ_k, and let the total multiplicity $\sum_{k=1}^{m} \mu_k = M$ precisely. Then

$$f(z) = \lim_{L \to \infty} [L/M]$$

uniformly on any compact subset of

$$\mathcal{D}_m = \{ z, |z| \leqslant R, z \neq z_k, k = 1, 2, \ldots, m \}.$$

Proof. The beginning and the end of the proof are substantially the same as for Theorem 6.2.1.

We define

$$B(z) = \sum_{i=0}^{M} B_i z^i = \prod_{k=1}^{m} \left(1 - \frac{z}{z_k} \right)^{\mu_k} \tag{2.29}$$

so that $f(z) = A(z)/B(z)$ with $A(z)$ analytic in $|z| \leqslant R$. We replace (2.13) by

$$\sum_{i=0}^{\infty} c_i z^i = \sum_{k=1}^{m} \sum_{\tau=1}^{\mu_k} r_\tau^{(k)} \left(1 - \frac{z}{z_k} \right)^{-\tau} + \sum_{i=0}^{\infty} c_i' z^i, \tag{2.30}$$

so that $c_i = d_i + O(R^{-i})$, with

$$d_i = \sum_{k=1}^{m} \sum_{\tau=1}^{\mu_k} r_\tau^{(k)} \binom{-\tau}{i} (-z_k)^{-i}.$$

The binomial coefficient is

$$\binom{-\tau}{i} = \frac{(-\tau)(-\tau-1)\cdots(-\tau-i+1)}{1 \times 2 \times \cdots \times i} = \frac{(i+1)(i+2)\cdots(i+\tau-1)}{1 \times 2 \times \cdots \times (\tau-1)}(-1)^i$$

and obeys the identity [Riordan, 1966, p. 9], see also exercise 2.

$$\binom{-\tau}{L+i-j} = \sum_{\sigma=0}^{\tau-1} \binom{-j}{\sigma}\binom{L+i+\tau-1}{i+\sigma}. \tag{2.31}$$

Hence the elements d_{L+i-j} $(i, j=1,2,\ldots, M)$ of the matrix D are

$$d_{L+i-j} = \sum_{k=1}^{m} \sum_{\tau=1}^{\mu_k} r_\tau^{(k)}(z_k)^{-L-i+j} \sum_{\sigma=0}^{\tau-1} \binom{-j}{\sigma}\binom{L+i+\tau-1}{\tau-\sigma-1},$$

which may be cast as the element of a matrix product·

$$d_{L+i-j} = \sum_{k=1}^{m} \sum_{\sigma=0}^{\mu_k-1} \left[\underbrace{\sum_{\tau=\sigma+1}^{\mu_k} r_\tau^{(k)}(z_k)^{-L-i}\binom{L+i+\tau-1}{\tau-\sigma-1}}_{\substack{\text{row suffix } i \\ \text{column label } (k,\sigma)}} \right] \left[\underbrace{(z_k)^j\binom{-j}{\sigma}}_{\substack{\text{row label } (k,\sigma) \\ \text{column suffix } j}} \right].$$

Furthermore,

$$\sum_{\tau=\sigma+1}^{\mu_k} r_\tau^{(k)}(z_k)^{-L-i}\binom{L+i+\tau-1}{\tau-\sigma-1}$$

$$= \sum_{\tau=\sigma+1}^{\mu_k} r_\tau^{(k)}(z_k)^{-L-i} \sum_{\sigma'=0}^{\tau-\sigma-1} \binom{L+i}{\sigma'}\binom{\tau-1}{\tau-1-\sigma-\sigma'}$$

$$= \sum_{\sigma'=0}^{\mu_k-\sigma-1} (z_k)^{-i}\binom{L+i}{\sigma'} \sum_{\tau=\sigma'+\sigma+1}^{\mu_k} r_\tau^{(k)}(z_k)^{-L}\binom{\tau-1}{\sigma+\sigma'}$$

$$= \underbrace{\sum_{\sigma'=0}^{\mu_k-1} (z_k)^{-i}\binom{L+i}{\sigma'}}_{\substack{\text{row label } i \\ \text{column label } \sigma'}}$$

$$\times \underbrace{\sum_{\tau=\sigma+\sigma'+1}^{\mu_k} \theta(\mu_k-\sigma-\sigma'-1)r_\tau^{(k)}(z_k)^{-L}\binom{\tau-1}{\sigma+\sigma'}}_{\substack{\text{row label } \sigma' \\ \text{column label } (k,\sigma)}}.$$

This provides a complete breakdown of D as

$$D_{ij} = \sum_{(k'\sigma')} \sum_{(k\sigma)} V_{i,(k'\sigma')} D'_{(k'\sigma')(k\sigma)} V'_{(k\sigma)j}$$

where

$$D'_{(k'\sigma')(k\sigma)} = \delta_{kk'} \sum_{\tau=\sigma'+\sigma+1}^{\mu_k} \theta(\mu_k - \sigma - \sigma' - 1) r_\tau^{(k)} (z_k)^{-L} \binom{\tau-1}{\sigma+\sigma'},$$

$$V_{i,(k'\sigma')} = (z_{k'})^{-i} \binom{L+i}{\sigma'},$$ (2.32a)

and

$$V'_{(k\sigma)j} = (z_k)^{j} \binom{-j}{\sigma}.$$ (2.32b)

V and V' are basically Vandermonde matrices, and may be verified to have nonzero determinants, see exercise 1. D' takes block diagonal forms given by the Kronecker factor of $\delta_{kk'}$. Each block of D' has triangular form, expressed by the Heaviside factor $\theta(\mu_k - \sigma - \sigma' - 1)$. Hence it follows that

$$\det D = K \prod_{k=1}^{m} \left[(z_k)^{\mu_k} \right]^{-L},$$ (2.33)

where K is a nonzero constant. Notice that only the residues $r_{\mu_k}^{(k)}$ are required to be nonzero. Equation (2.31) is the equivalent of (2.22) in the theorem for simple poles, and the remainder of the proof follows that of the main theorem very closely from that point onward.

Example. Let $f(z)$ have a triple pole at $z=z_1$ and a double pole at $z=z_2$. Then $\mu_1 = 3$, $\mu_2 = 2$, and the $(k'\sigma')$ indices are $(1,0)$, $(1,1)$, $(1,2)$, $(2,0)$, $(2,1)$. We have

$$V_{i,(k'\sigma')}$$

$$= \begin{vmatrix} z_1^{-1} & (L+1)z_1^{-2} & \frac{1}{2}(L+1)(L+2)z_1^{-3} & z_2^{-4} & (L+1)z_2^{-5} \\ z_1^{-1} & (L+2)z_1^{-2} & \frac{1}{2}(L+2)(L+3)z_1^{-3} & z_2^{-4} & (L+2)z_2^{-5} \\ z_1^{-1} & (L+3)z_1^{-2} & \frac{1}{2}(L+3)(L+4)z_1^{-3} & z_2^{-4} & (L+3)z_2^{-5} \\ z_1^{-1} & (L+4)z_1^{-2} & \frac{1}{2}(L+4)(L+5)z_1^{-3} & z_2^{-4} & (L+4)z_2^{-5} \\ z_1^{-1} & (L+5)z_1^{-2} & \frac{1}{2}(L+5)(L+6)z_1^{-3} & z_2^{-4} & (L+5)z_2^{-5} \end{vmatrix}.$$

(2.34)

The $(1, \sigma') \times (1, \sigma)$ block of $D_{(k'\sigma')(k\sigma)}$ is

$$
\begin{array}{c}
 \\
\sigma'=0 \\
\sigma'=1 \\
\sigma'=2
\end{array}
\left(
\begin{array}{ccc}
\sigma=0 & \sigma=1 & \sigma=2 \\
\left[r_1^{(1)}+r_2^{(1)}+r_3^{(1)}\right]z_1^{-L} & -\left[r_2^{(1)}+2r_3^{(1)}\right]z_1^{-L} & r_3^{(1)}z_1^{-L} \\
-\left[r_2^{(1)}+2r_3^{(1)}\right]z_1^{-L} & r_3^{(1)}z_1^{-L} & 0 \\
r_3^{(1)}z_1^{-L} & 0 & 0
\end{array}
\right),
$$

and the triangular structure is self-evident.

There are several features of de Montessus's theorem which are worth emphasis. Firstly, we assumed meromorphy in $|z| \leqslant R$, and that all the poles lie in $|z| < R$. Although it is unusual to assume meromorphy on closed sets, and indeed our felony is compounded by using the implication of analyticity in a slightly larger annulus, it is quite convenient, and we prove the usual result about uniform convergence in $|z| \leqslant R$ except in neighborhoods of the poles. The alternative assumption (analyticity on an open set) leads to uniform convergence on a compact subset.

Secondly, there are many generalizations of the theorem. Much work has been done on the problem of one mild singularity on the boundary $|z| = R$, for which the dominant part of the coefficients c_i is a "smooth" function of i, and controllable in the spirit of the proof given [Wilson, 1928a, b, 1930].

Thirdly, we repeat that the number of poles M in the circle $|z| < R$ is fixed precisely. The necessity of this condition is seen by the counterexample $f(z) = (1 - z^2)^{-1}$. Inside $|z| < 1$, $f(z)$ is analytic and the $[L/0]$ Padé approximants converge. In $|z| \leqslant R$ with $R > 1$, the $[L/2]$ Padé approximants are exact. But the $[L/1]$ Padé approximants are Maclaurin polynomials if L is odd and nonexistent if L is even. In fact, the $[L/1]$ are less use than any other row, and in general an ill-chosen row sequence for a function having poles equidistant from the origin is best avoided.

Lastly, we observe that we have given a kind of constructive proof of de Montessus's theorem. Indeed, an error formula follows by this method Gragg [1972]. There is also an intimate connection with symmetric polynomials [EPA, p. 135]. The proof generalizes, in a way, to more than one dimension [Chisholm and Graves-Morris, 1975; Graves-Morris, 1977]. For these reasons, the preceeding proof is important, but in the next section we given a simpler, implicit proof.

Exercise 1. Prove that the determinant of the generalized Vandermonde matrix (2.34) is nonzero, and find its value. Such determinants are sometimes called confluent alternants. An account of them is given by Aitken [1964, p. 120], see also Chisholm and Graves-Morris [1977].

Exercise 2. Prove (2.31) by considering the identity

$$(1+x)^{L+i+\tau-1-j}=(1+x)^{L+i+\tau-1}(1+x)^{-j}.$$

Notice the conventions, implicit in (2.31), that $j, \sigma, \tau - 1$ and $L+i-j$ are non-negative integers.

6.3   Hermite's Formula and de Montessus's Theorem

In this section we prove Hermite's formula for Maclaurin's expansion and show that it leads to a useful formula for the error incurred in Padé approximation. Then these results are used in Saff's elegant proof of de Montessus's theorem.

HERMITE'S FORMULA.   If $f(z)$ is analytic inside a contour Γ enclosing the origin and $f(z)$ is continuous on Γ, then its $[L/0]$ Padé approximant is given by

$$[L/0]=\frac{1}{2\pi i}\int_\Gamma \frac{v^{L+1}-z^{L+1}}{v-z}\frac{f(v)}{v^{L+1}}\,dv. \tag{3.1}$$

COMMENT This formula is the special case of Hermite's more general formula, Part II Theorem 1.1.1, which is needed in this section.

Proof.

$$\frac{v^{L+1}-z^{L+1}}{v-z}=\sum_{j=0}^{L} v^{L-j}z^{j},$$

and so $[L/0]$ defined by (3.1) is a polynomial in z of degree L. In fact,

$$[L/0]=\sum_{j=0}^{L} z^{j}\frac{1}{2\pi i}\int_\Gamma v^{-j-1}f(v)\,dv=\sum_{j=0}^{L} z^{j}\frac{f^{(j)}(0)}{j!},$$

which proves the formula for the truncated Maclaurin expansion of $f(z)$.

COROLLARY. An error formula for Padé approximation is given by

$$f(z)-[L/M]=\frac{z^{L+M+1}}{2\pi iQ^{[L/M]}(z)R_M(z)}\int_\Gamma \frac{f(v)Q^{[L/M]}(v)R_M(v)}{v^{L+M+1}(v-z)}\,dv \tag{3.2}$$

with the same hypothesis that $f(z)$ is analytic in and continuous on a contour Γ enclosing the origin. $R_M(z)$ is an arbitrary polynomial of degree at most M, but not identically zero. [$R_M(z)$ may be taken to be $R_M(z)=1$, or whatever is most convenient in context.]

Method 1. Interpolate $f(z)Q^{[L/M]}(z)R_M(z)$ to order $L+M$ using Hermite's formula

$$\pi_{L+M}(z)=\frac{1}{2\pi i}\int_\Gamma \frac{v^{L+M+1}-z^{L+M+1}}{v-z}\frac{f(v)Q^{[L/M]}(v)R_M(v)}{v^{L+M+1}}\,dv$$

But, by Cauchy's theorem,

$$f(z)Q^{[L/M]}(z)R_M(z)=\frac{1}{2\pi i}\int_\Gamma \frac{1}{v-z}f(v)Q^{[L/M]}(v)R_M(v)\,dv,$$

and subtraction yields

$$f(z)Q^{[L/M]}(z)R_M(z)-\pi_{L+M}(z)$$
$$=\frac{1}{2\pi i}\int\left(\frac{z}{v}\right)^{L+M+1}f(v)Q^{[L/M]}(v)R_M(v)\,dv$$

$$(3.3)$$

The Padé equations require

$$f(z)Q^{[L/M]}(z)R_M(z)=\left\{P^{[L/M]}(z)+O(z^{L+M+1})\right\}R_M(z)$$
$$=P^{[L/M]}(z)R_M(z)+O(z^{L+M+1}),$$

and it follows that

$$\pi_{L+M}(z)=P^{[L/M]}(z)R_M(z)+O(z^{L+M+1})$$
$$=P^{[L/M]}(z)R_M(z),$$

since $R_M(z)$ is a polynomial of order not greater than M. The result (3.2) now follows from (3.3) by division.

Method 2. We present a second method of proof to explain a paradox occurring in the use of this equation. Apply Cauchy's theorem to the function:

$$\phi(z)=R_M(z)\left\{f(z)Q^{[L/M]}(z)-P^{[L/M]}(z)\right\}z^{-L-M-1}.$$

The Padé equations ensure that $\phi(z)$ is analytic in a neighborhood of the origin, so

$$\phi(z)=\frac{1}{2\pi i}\int_\Gamma \frac{\phi(v)}{v-z}\,dv,$$

where Γ is any simple closed contour enclosing the origin. Hence

$$f(z)-[L/M]=\frac{z^{L+M+1}}{2\pi iQ^{[L/M]}(z)R_M(z)}$$

$$\times\left\{\int_\Gamma\frac{R_M(v)f(v)Q^{[L/M]}(v)\,dv}{(v-z)v^{L+M+1}}+E(z)\right\} \qquad (3.4)$$

where

$$E(z)=\int_\Gamma\frac{R_M(v)P^{[L/M]}(v)\,dv}{v^{L+M+1}(v-z)}.$$

The integrand of $E(z)$ is analytic in $|v|>|z|$, and so the contour Γ may be expanded to infinity. The integrand is $O(v^{-2})$, since $R_M(v)$ has order no greater than M, and therefore $E(z)=0$, explaining the paradox of (3.4) and proving the error formula (3.2).

We now apply these results to a second proof of de Montessus's theorem in (Section 6.2).

Proof. (Staff's method).   The hypothesis of de Montessus's theorem is that $f(z)$ is meromorphic in $|z|\leqslant R$ with poles of total multiplicity M, and it is convenient to choose the polynomial $R_M(z)$ to annihilate these poles.

We consider first the case when the poles of $f(z)$ are simple and are located at $\alpha_1,\alpha_2,\ldots,\alpha_M$. Form the polynomials

$$R_0(z)=1,$$

$$R_k(z)=(z-\alpha_1)(z-\alpha_2)\cdots(z-\alpha_k) \qquad \text{for}\quad k=1,2,\ldots,M.$$

Then it follows that $R_M(z)f(z)$ is analytic in $|z|<R$. We will also need the polynomial $t^{(L,M)}(z)=R_M(z)+\sum_{k=1}^M a_k^{(L,M)}R_{k-1}(z)$ of degree M defined in terms of the arbitrary coefficients $a_k^{(L,M)}$, which will be fixed presently.

Hermite's formula is used to construct the Maclaurin polynomial of order $L+M$ which interpolates $t^{(L,M)}(z)R_M(z)f(z)$. Since this function is analytic in $|z|<R$,

$$\pi_{L+M}(z)=\frac{1}{2\pi i}\int_{|v|=R}\frac{v^{L+M+1}-z^{L+M+1}}{v-z}\frac{t^{(L,M)}(v)R_M(v)f(v)}{v^{L+M+1}}\,dv.$$

$$-\frac{1}{2\pi i}\int_{|v|=R}\frac{1-(z/v)^{L+M+1}}{v-z}\left\{R_M(v)+\sum_{k=1}^M a_k^{(L,M)}R_{k-1}(v)\right\}$$

$$\times R_M(v)f(v)\,dv. \qquad (3.5)$$

The idea is to choose the coefficients $a_k^{(L,M)}$ so that $R_M(z)$ is a factor of $\pi_{L+M}(z)$. This factorization is accomplished by setting $\pi_{L+M}(\alpha_i)=0$ for

$i = 1, 2, \ldots, M$, and leads to the set of linear equations for $a_k^{(L, M)}$

$$\sum_{k=0}^{M-1} c_{jk}^{(L, M)} a_k^{(L, M)} = d_j^{(L, M)}, \tag{3.6}$$

where

$$c_{jk}^{(L, M)} = \frac{1}{2\pi i} \int_{|v|=R} \frac{1 - (\alpha_j/v)^{L+M+1}}{v - \alpha_j} R_{k-1}(v) R_M(v) f(v)\, dv \tag{3.7}$$

and

$$d_j^{(L, M)} = \frac{-1}{2\pi i} \int_{|v|=R} \frac{1 - (\alpha_j/v)^{L+M+1}}{v - \alpha_j} [R_M(v)]^2 f(v)\, dv. \tag{3.8}$$

The principal feature of (3.7) and (3.8) is their well-defined limit as $L \to \infty$. From (3.8),

$$d_j^{(L, M)} \to d_j = \frac{-1}{2\pi i} \int_{|v|=R} \frac{[R_M(v)]^2 f(v)\, dv}{v - \alpha_j}$$

$$= 0, \qquad \text{using Cauchy's theorem,}$$

and from (3.7),

$$c_{jk}^{(L, M)} \to c_{jk} = \frac{1}{2\pi i} \int_{|v|=R} \frac{R_k(v) R_M(v) f(v)\, dv}{v - \alpha_j}. \tag{3.9}$$

If $j < k$, $c_{jk} = 0$. If $j = k$, $c_{jj} \neq 0$, and this value depends only on the residues and positions of the M poles of $f(z)$. We discover, first, that for L sufficiently large, the system of equations (3.6) for $a_k^{(L, M)}$ is nonsingular, and second, that $a_k^{(L, M)} \to 0$ as $L \to \infty$. Consequently, $\{a_k^{(L, M)}\}$ exists such that $R_M(z)$ is a factor of $\pi_{L+M}(z)$, and from the specification of (3.5),

$$\frac{\pi_{L+M}(z)/R_M(z)}{t^{(L, M)}(z)} - \frac{P^{[L/M]}(z)}{Q^{[L/M]}(z)} = f(z) + O(z^{L+M+1}),$$

and furthermore $t^{(L, M)}(z) \to R_M(z)$ as $L \to \infty$. To complete the proof, the error formula gives

$$f(z) - [L/M] = \frac{z^{L+M+1}}{2\pi i Q^{[L/M]}(z) R_M(z)} \int_{|v|=R} \frac{f(v) Q^{[L/M]}(v) R_M(v)}{v^{L+M+1}(v - z)}\, dv$$

$$= O\left(\left(\frac{z}{R}\right)^{L+M+1}\right),$$

which tends to zero for $|z| < R$.

Finally we consider the case in which at least one of the poles is a multipole. Suppose that α_1 is a pole of order p. Then $(z - \alpha_1)^p$ is a factor of

$\pi_{L+M}(z)$ provided

$$\left(\frac{d}{d\alpha_1}\right)^n \pi_{L+M}(\alpha_1)=0, \qquad n=0,1,\ldots,p-1$$

in (3.5). The polynomials $\{R_k(z), k=0,1,\ldots,M\}$ are defined by confluence; for example, $R_k(z)=(z-\alpha_1)^k, k=0,1,\ldots,p$. The equivalent of (3.9) is

$$c_{jk}^{(L,M)} \to c_{jk} - \frac{j!}{2\pi i}\int_{|v|=R}\frac{R_k(v)R_M(v)f(v)\,dv}{(v-\alpha_1)^{j+1}}, j,k=0,1,\ldots,p-1.$$

if $j<k$, $c_{jk}=0$ and if $j=k$, $c_{jj}\neq 0$. Any other multipoles are treated similarly, and the proof for the case of simple poles is otherwise unchanged.

The special significance of this method is that it enables de Montessus's theorem to be generalized to Saff's theorem. Saff's theorem proves convergence of row sequences of multipoint Padé approximants (see Part II, Section 1.1) on a simply connected point set $\mathcal{E}$, except at the M poles of the given function and subject to an appropriate selection of the interpolation points within the region of analyticity of the given function.

THEOREM 6.3.1 [Saff, 1972]. *Let $\mathcal{E}$ be a closed bounded set in the complex z-plane whose complement $\mathcal{K}$, (with $\infty \in \mathcal{K}$), is connected. Further, we suppose that a sequence of interpolation points*

$$\beta_1^{(0)}$$
$$\beta_1^{(1)}\beta_2^{(1)}$$
$$\vdots \tag{3.10}$$
$$\beta_1^{(n)}\beta_2^{(n)}\cdots\beta_{n+1}^{(n)}$$

are defined, with no limit point in $\mathcal{K}$, such that

$$\lim_{n\to\infty}\left|\prod_{i=1}^{n+1}\left(z-\beta_i^{(n)}\right)\right|^{1/n}=\mathrm{cap}(\mathcal{E})\exp\{G(z)\} \tag{3.11}$$

uniformly in z on each compact subset of $\mathcal{K}$, $\mathrm{cap}(\mathcal{E})$ is defined in Section 6.6. For each $\sigma>1$, let Γ_σ denote the locus $G(z)=\ln\sigma$, and let $\mathcal{E}_\sigma$ denote the interior of Γ_σ. Let $f(z)$ be analytic in $\mathcal{E}$ and meromorphic with precisely M poles, counting multiplicity, in $\mathcal{E}_\rho$ for some $\rho>1$. Then, for all L sufficiently large, there exists a unique rational function $r^{[L/M]}(z)$ which interpolates $f(z)$ at $\beta_1^{(L+M)}, \beta_2^{(L+M)},\ldots,\beta_{L+M+1}^{(L+M)}$. Each $r^{[L/M]}(z)$ has M finite poles which converge to the poles of $f(z)$ in $\mathcal{E}_\rho$. Furthermore, $r^{[L/M]}(z) \to f(z)$ uniformly on any compact subset of $\mathcal{E}_\rho$ not containing any pole of $f(z)$, and

$$\lim_{n\to\infty}\left(\sup_{z\in\mathcal{E}_\rho}|f(z)-r^{[L/M]}(z)|\right)^{1/n}\leq\frac{1}{\rho}.$$

Remarks. de Montessus's Theorem, as expressed by Theorem 6.2.2 is a corollary of Saff's Theorem, in which we take $\mathcal{E}=\{z:|z|\leqslant R/\rho\}$, and the limit $\rho\rightarrow\infty$ and $\mathrm{cap}(\mathcal{E})\rightarrow0$. Saff's theorem is to be understood in terms of the concepts and definitions of Section 6.6 and Part II, Section 1. We refer to Warner [1976] for a discussion of how the interpolation points $\beta_i^{(n)}$ in (3.10) may be consistently chosen within $\mathcal{E}$, for a set theoretic generalization of Saff's Theorem, and for the interesting connection with Runge's phenomenon in polynomial interpolation. We simply note that $G(z)$ is well defined by (3.11) if we take $\{\beta_i^{(n-1)}, i=1,2,...,n\}$ to be the zeros of the Chebychev polynomial of order n for the region $\mathcal{E}$. The proof of Theorem 6.3.1 follows the previous method almost exactly.

The contour integral methods of this section provide an "instant" proof of the orthogonality results for the polynomials (3.12) associated with Stieltjes and Hamburger series having non-zero radius of convergence, without the paraphernalia of the complete theory. We start with the Hamburger function

$$f(z)=\int_{-R}^{R}\frac{d\phi(u)}{1+zu},$$

and let $Q^{[m-1/m]}(z)$ be the denominator of its $[m-1/m]$ Padé approximant. The mth-order polynomial, $\pi_m(u)$, is defined by

$$\pi_m(u)=u^mQ^{[m-1/m]}(-1/u),\qquad m=1,2,3,...,\qquad(3.12)$$

and $\pi_0(u)=1$. The set $\{\pi_m(u),\ m=0,1,2,...\}$ are orthogonal in the sense that

$$I_{mk}=\int_{-R}^{R}\pi_m(u)\pi_k(u)\,d\phi(u)=0\qquad\text{for }m\neq k.\qquad(3.13)$$

Quick proof [Zinn-Justin, 1970]. As in Section 5.3, let $w=-1/z$, and define

$$F(w)=\int_{-R}^{R}\frac{d\phi(u)}{w-u}=-z\int_{-R}^{R}\frac{d\phi(u)}{1+uz}=-zf(z).\qquad(3.14)$$

Using Cauchy's theorem,

$$\frac{1}{2\pi i}\int_{|w|=2R}\pi_m(w)\pi_k(w)F(w)\,dw=\int_{-R}^{R}\pi_m(u)\pi_k(u)\,d\phi(u)=I_{mk}.$$

Substituting $z=-1/w$, we find

$$I_{mk}=\frac{1}{2\pi i}\int_{|z|=\frac{1}{2R}}(-z)^{-m-k-1}Q^{[m-1/m]}(z)Q^{[k-1/k]}(z)f(z)\,dz.$$

$$(3.15)$$

Without loss of generality, we may choose $m > k$. Since

$$Q^{[m-1/m]}(z)f(z) - P^{[m-1/m]}(z) + O(z^{2m}),$$

we may collapse the contour of (3.15) to the origin and find that $I_{mk} = 0$, establishing (3.13).

6.4   Uniqueness of Convergence

A striking feature of any Padé approximant is that it is a rational function. Consequently, if the limit of a sequence of Padé approximants is to be at all useful, this limit should be meromorphic (or even holomorphic) in some substantial region of the complex plane. The expectation is that the pole positions and the residues of the Padé approximants converge to those of the limit function. But because the values of the Padé approximants near the poles of the limit function are arbitrarily large, ordinary convergence at the pole is not possible. The easiest technical device which circumvents this difficulty is the use of the chordal metric. The notion of convergence is replaced by that of convergence on the sphere. Many theorems about convergence of Padé approximants to meromorphic functions are more elegantly reexpressed in terms of convergence on the sphere, and then the poles are not classed as belonging to an exceptional set [Baker, 1974]. For example, de Montessus's theorem takes the form

If $f(z)$ is meromorphic in $|z| \leq R$, has poles of total multiplicity M in $|z| < R$, and is analytic at the origin, then the row sequence $[L/M] \rightarrow f(z)$ uniformly on the sphere in $|z| \leq R$.

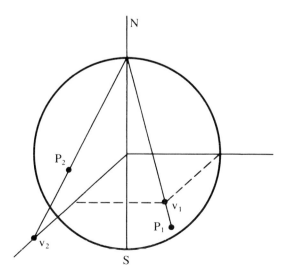

Figure 1. The Riemann sphere.

To define the chordal metric [Ostrowki, 1925; Hille, 1959, p. 42], let v_1 and v_2 be two complex numbers. These might be the values $v_1 = f(z_1)$, $v_2 = f(z_2)$ of a meromorphic function. Then $\chi(v_1, v_2)$, which defines the distance between v_1 and v_2 in the chordal metric, is the chord length $P_1 P_2$ on the Riemann sphere in Figure 1. P_1, P_2 are the points where the unit sphere centered at the origin intersects the line from ts north pole N to v_1, v_2 in the equatorial plane. The origin, $z = 0$, is mapped to the south pole S, and $z = \infty$ mapped to the unique point N. A geometrical calculation shows that

$$\chi(v_1, v_2) = P_1 P_2 = \frac{2|v_1 - v_2|}{\sqrt{|1 + v_1^* v_2|^2 + |v_1 - v_2|^2}}$$

$$= \frac{2|v_1 - v_2|}{\sqrt{1 + |v_1|^2}\sqrt{1 + |v_2|^2}}. \tag{4.1}$$

Proof. Figure 2 shows a vertical plane containing the north and south poles, and the complex point v_1. By simple trigonometry $NV_1 = \sec \alpha_1$ and $NP_1 = 2 \cos \alpha_1$, and therefore $NV_1 \cdot NP_1 = 2$.

Figure 3 shows the plane of the lines $NV_1 P_1$ and $NV_2 P_2$ as well as the chord $P_1 P_2$ required.

Because $NV_1 \cdot NP_1 = NV_2 \cdot NP_2 = 2$, the triangles $NV_1 V_2$ and $NP_2 P_1$ are similar and so

$$\frac{NV_1}{NP_2} = \frac{NV_2}{NP_1} = \frac{V_1 V_2}{P_1 P_2}.$$

Thus $P_1 P_2 = V_1 V_2 \cos \alpha_1 (2 \cos \alpha_2)$, and Equation (4.1) follows immediately.

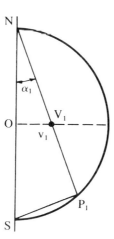

Figure 2. A section of the Riemann sphere. The point V_1 corresponds to the value v_1 in the complex plane.

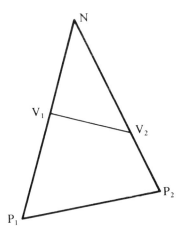

Figure 3. The triangle NP_1P_2. The points V_1, V_2 correspond to values, v_1, v_2 in the complex plane.

The mapping of the Riemann sphere onto the complex plane is especially useful when the limit $z \to \infty$ is uniquely defined, because this limit corresponds to the unique limit $\mathbf{r} \to ON$.

The chordal metric obeys an inequality which follows directly from (4.1):

$$\chi(v_1, v_2) < 2|v_1 - v_2| \tag{4.2}$$

This proves mathematically that nearby points in the complex plane are mapped to nearby points on the sphere. Elementary geometry can also be used to prove that

$$\chi(\infty, v) = \frac{2}{\sqrt{1+|v|^2}} \tag{4.3}$$

and that

$$\chi(v_1, v_2) \leqslant 2 \tag{4.4}$$

for all v_1 and v_2. The geometrical proof of (4.4) is quite obvious from Figure 1.

From the definition of the chordal metric, we will prove

THEOREM 6.4.1. *Any meromorphic function is continuous on the sphere.*

Proof. Let $f(z)$ be the meromorphic function. First, let z_0 be any regular point of $f(z)$, so that $f(z)$ is analytic in $|z - z_0| < \rho$ for some $\rho > 0$. Then $f(z)$

is continuous at $z=z_0$, and for any $\varepsilon>0$, a δ may be found such that

$$|f(z)-f(z_0)|<\tfrac{1}{2}\varepsilon \qquad \text{for} \quad |z-z_0|<\delta,$$

and from (4.2),

$$\chi(f(z),f(z_0))<\varepsilon \qquad \text{for} \quad |z-z_0|<\delta,$$

proving continuity in the chordal metric at regular points.

If z_0 is not a regular point, it is a pole of $f(z)$. In this case, m exists such that $\phi(z)=(z-z_0)^m f(z)$ is regular at $z=z_0$, $\phi(z_0)\neq0$, and m is an integer with $m\geqslant1$. Given any $\varepsilon>0$,

$$|f(z)|>2/\varepsilon \quad \text{and} \quad \chi(f(z),\infty)<\varepsilon$$

for all z such that both

$$|z-z_0|<\left|\frac{\varepsilon}{4}\phi(z_0)\right|^{1/m}$$

and

$$|\phi(z)-\phi(z_0)|<\tfrac{1}{2}|\phi(z_0)|.$$

Since $\phi(z)$ is continuous at $z=z_0$, we have found implicitly a neighborhood in which $\chi(f(z),\infty)<\varepsilon$, proving continuity at any pole. In other words, the mapping $z\to f(z)$ is continuous in the chordal metric.

COROLLARY. Any meromorphic function defined on a compact region is uniformly continuous on the sphere.

Proof. The mapping $z\to P(f(z))$, where $P(f(z))$ is the point P on the Riemann sphere representing the complex number $f(z)$, is continuous on a compact region and so is uniformly continuous.

We now turn to three theorems which apply when z is confined to a compact region and uniform continuity may be exploited.

THEOREM 6.4.2. *If* $\{P_n(z)\}$ *is any sequence of meromorphic functions which converge uniformly on the sphere for all z in a compact region $\mathfrak{R}$, then*

(i) *the limit function is uniformly continuous on the sphere, in the region $\mathfrak{R}$,*
(ii) $P_n(z)$ *are equicontinuous on the sphere and in $\mathfrak{R}$,*
(iii) *the limit function is meromorphic in the interior of $\mathfrak{R}$.*

Proof. The chordal metric satisfies the triangle inequality. Let the limit of $\{P_n(z)\}$ be $f(z)$. To prove continuity on the sphere of $f(z)$ at $z-z_0$, consider

$$\chi(f(z), f(z_0)) \leqslant \chi(f(z), P_N(z)) + \chi(P_N(z), P_N(z_0)) + \chi(P_N(z_0), f(z_0)).$$

Given any $\varepsilon > 0$, uniform convergence on the sphere of $P_n(z)$ to $f(z)$ gives an N for which

$$\left. \begin{array}{l} \chi(f(z), P_n(z)) < \varepsilon/3, \\ \chi(f(z_0), P_n(z_0)) < \varepsilon/3 \end{array} \right\} \quad \text{for all } n \geqslant N \text{ and for all } z, z_0 \in \mathfrak{R}.$$

Uniform continuity on the sphere of $P_N(z)$ gives a δ for which $\chi(P_N(z), P_N(z_0)) < \varepsilon/3$ for all $z, z_0 \in \mathfrak{R}$ such that $|z - z_0| < \delta$. Hence $\chi(f(z), f(z_0)) < \varepsilon$, proving (i), which is that $f(z)$ is uniformly continuous on the sphere for $z \in \mathfrak{R}$.

To prove (ii), the equicontinuity property, consider the inequality

$$\chi(P_n(z), P_n(z_0)) < \varepsilon. \tag{4.5}$$

The triangle inequality shows that (4.5) is true, provided

$$\chi(P_n(z), f(z)) < \varepsilon/3, \tag{4.6}$$

$$\chi(f(z), f(z_0)) < \varepsilon/3, \tag{4.7}$$

and

$$\chi(f(z_0), P_n(z_0)) < \varepsilon/3. \tag{4.8}$$

Now (4.6) and (4.8) are true for all $n > N(\varepsilon)$ independently of z, z_0, and (4.7) is true for all $z, z_0 \in \mathfrak{R}$ such that $|z - z_0| < \delta(\varepsilon)$. Hence (4.5) is true for all $n > N(\varepsilon)$, provided $z, z_0 \in \mathfrak{R}$ and $|z - z_0| < \delta(\varepsilon)$.

Since $P_k(z)$ is uniformly continuous on the sphere for $z \in \mathfrak{R}$, k fixed,

$$\chi(P_k(z), P_k(z_0)) < \varepsilon \text{ for } |z - z_0| < \delta_k(\varepsilon).$$

By choosing

$$\delta_{\min}(\varepsilon) = \min\{\delta_1, \delta_2, \ldots \delta_N, \delta(\varepsilon)\},$$

it follows that $\chi(P_n(z), P_n(z_0)) < \varepsilon$ for all $z, z_0 \subset \mathfrak{R}$ with $|z - z_0| < \delta_{\min}$, and (ii) is proved.

To show that $f(z)$ is meromorphic in the interior of $\mathfrak{R}$, take any point z_0 in the interior. If $f(z_0)$ is finite, there is a closed neighborhood, $|z - z_0| \leqslant \delta$, in which $f(z)$ is the uniform limit of a sequence of analytic functions, and so $f(z)$ is analytic in the interior of this region by Weierstrass's theorem. If $f(z_0)$ is infinite, we consider the reciprocal functions, $(P_n(z))^{-1}$, which are analytic at $z=z_0$ for sufficiently large n; hence $(f(z))^{-1}$ is analytic at $z=z_0$ for an identical reason based on Weierstrass's theorem, and hence $f(z)$ is meromorphic in the interior of $\Re$.

Next we use Arzela's theorem (applied to compact regions rather than bounded domains) to establish a sequence to which Theorem 6.4.2 is applied.

THEOREM 6.4.3 [Hille, 1962, p. 241]. *If $\{P_n(z)\}$ is an infinite sequence of meromorphic functions which is equicontinuous on the sphere for z in a compact region $\Re$, then an least a subsequence of $\{P_n(z)\}$ converges uniformly to a limit $f(z)$ which is uniformly continuous on the sphere for $z \in \Re$ and is meromorphic in the interior of $\Re$.*

Proof. Using the construction of the proof of Arzela's theorem (Theorem 5.2.5), we obtain a subsequence of $\{P_n(z)\}$ which converges uniformly in $\Re$ and satisfies the conditions of Theorem 6.4.2, which in turn proves the result.

This theorem completes the buildup necessary for the convergence uniqueness theorem. It follows the style of the theorems for Stieltjes series to the extent that Arzela's theorem leads to a convergent subsequence. This result is combined with the proof that the whole sequence converges at the origin to yield convergence in a larger domain of the whole sequence.

THEOREM 6.4.4. *Let $P_k(z)=[L_k/M_k]$ be a sequence of Padé approximants of a function $f(z)$ which is regular at the origin, such that $L_k + M_k \to \infty$ as $k \to \infty$. If $\{P_k(z)\}$ is equicontinuous on the sphere in a simply connected and compact region $\Re$, of which the origin is an interior point, then the domain of definition of $f(z)$ may be extended so that $f(z)$ is meromorphic in the interior of $\Re$ and $[L_k/M_k]$ converges to $f(z)$ on the sphere for all $z \in \Re$.*

Proof. Theorem 6.4.3 asserts the existence of at least one limit function $\tilde{f}(z)$, which is meromorphic in the interior of $\Re$ and with $f(0)=\tilde{f}(0)$. Hence ρ exists such that $f(z)$ and $\tilde{f}(z)$ are analytic in $|z|<\rho$; $\tilde{f}(z)$ is the uniform limit of an equicontinuous subsequence and so is meromorphic in the interior of $\Re$. Consider now the entire sequence $\{P_k(z), k=1,2,3,\dots\}$. Given any $\varepsilon>0$, equicontinuity of the sequence on the sphere requires that $\delta=\delta(\varepsilon)$ exists such that

$$|P_k(z)-f(0)|<\varepsilon \qquad \text{for all} \quad |z|\leqslant\delta<\rho.$$

Thus $|P_k(y)|$ is uniformly bounded on $|y|=\delta$. Using Cauchy's theorem,

$$\frac{P_k(z)}{z^{L_k+M_k}} - f(z) = \frac{1}{2\pi i}\int_C \frac{P_k(y)-f(y)}{y^{L_k+M_k}}\frac{dy}{y-z},$$

where C is the circle $|y|=\delta$ and z is restricted to $|z|\leqslant\frac{1}{2}\delta$. Hence

$$|P_k(z)-f(z)| < \left(\tfrac{1}{2}\right)^{L_k+M_k}\cdot\text{constant}$$

for $|z|\leqslant\frac{1}{2}\delta$ and all k. By the uniqueness of analytic continuation, this proves that the entire sequence $\{P_k(z)\}$ converges on the sphere to $f(z)=\tilde{f}(z)$ as $k\to\infty$ and that $f(z)$ is meromorphic in the interior of $\mathcal{R}$.

Omitting the detailed conditions, we see that establishing convergence of a sequence of Padé approximants to a meromorphic function depends mainly on proving that the approximants form an equicontinuous sequence. Such results are most naturally expressed in terms of convergence on the Riemann sphere. Our final theorem is a similar result, but involving ordinary convergence.

THEOREM 6.4.5 [Baker, 1965]. *Let* $P_k(z)=[L_k/M_k]$ *be a member of a sequence of Padé approximants of a function* $f(z)$ *which is analytic at the origin, and such that* $|P_k(z)|$ *is uniformly bounded on a simply connected bounded domain* $\mathcal{D}$ *containing the origin. Provided* $L_k+M_k\to\infty$ *as* $k\to\infty$, *then* $P_k(z)\to f(z)$ *uniformly on any compact region* $\mathcal{R}\subset\mathcal{D}$, *thereby extending the domain of definition of* $f(z)$, *and* $f(z)$ *is analytic in* $\mathcal{R}$.

Proof. We first prove that the sequence $\{P_k(z)\}$ is equicontinuous on $\mathcal{D}$. If $|P_k(z)|$ is bounded on $\mathcal{D}$, then $P_k(z)$ is analytic on $\mathcal{D}$ because it is rational. We use the formula

$$\frac{dP_k(z)}{dz} = \frac{1}{2\pi i}\int_C \frac{P_k(w)\,dw}{(z-w)^2}$$

where C is a circle with center z, radius δ. We have arranged that we may choose $\delta>0$ independently of z for $z\in\mathcal{R}$. Hence $|P_k'(z)|$ is uniformly bounded in $\mathcal{R}$. Consequently $\{P_k(z)\}$ is an equicontinuous sequence. Theorem 6.4.4 shows that $\{P_k(z)\}\to f(z)$ uniformly on the sphere. As the limit of a uniformly bounded sequence, $f(z)$ is not merely meromorphic but analytic in $\mathcal{R}$, and the theorem is proved.

Theorem 6.4.4 asserts that the limit function is meromorphic in $\mathcal{R}$. If the given function, which is proved to be the limit function, is not meromorphic in $\mathcal{R}$, but the other requirements are met, it follows from the theorem that the Padé approximants are not equicontinuous on the sphere. Likewise, if $f(z)$ is not analytic, but the other conditions of the theorem are met, the Padé approximants of $f(z)$ cannot be uniformly bounded. A similar result is given by Jones and Thron [1975], and the generalization to a class of finite Laurent series is given in Jones and Thron [1979]. Other important theorems governing the convergence of Padé approximants of functions analytic at the origin and with various regions of meromorphy have been given by Chisholm [1966], Beardon [1968a], and Baker, [1970].

Exercise 1. Find three points v, v_1, and v_2 in the complex plane such that $|v_1-v|<|v_2-v|$ but $\chi(v_1,v)>\chi(v_2,v)$.

Exercise 2. Prove that the chordal metric satisfies the triangle inequality.

Exercise 3. Express the results described in the final paragraph of this section carefully as two self-contained theorems.

6.5 Convergence in Measure

We will make repeated use of the formula (3.2):

$$f(z)-[L/M]=\frac{z^{L+M+1}}{2\pi iQ^{[L/M]}(z)R_M(z)}\int_\Gamma\frac{f(t)Q^{[L/M]}(t)R_M(t)\,dt}{t^{L+M+1}(t-z)} \quad (5.1)$$

for the error of Padé approximation. The idea is that we should bound $|Q^{[L/M]}(z)|^{-1}$ in (5.1) using a known bound for an arbitrary polynomial of degree M. This bound is given by

$$|q_m(z)|\geqslant\eta^m, \quad (5.2)$$

where $q_m(z)$ is a polynomial with leading coefficient unity, and is true *except* for z belonging to a set $\mathcal{E}$ of measure at most $\pi\eta^2$. Usually we know nothing about the location of this exceptional set, but we have the given bound on its area.

The theorems we are about to prove next are valid except on sets of arbitrarily small measure. This statement means that this section contains no assertion about convergence at any particular point in the z-plane. The results are all to the effect that the "area of disruption" caused by unwanted pole-and-zero combinations of the approximants is arbitrarily small.

As an example of (5.2), we consider $q_m(z)=(z-a)^m$, which is a polynomial of degree m with leading coefficient unity. Then $|q_m(z)|\geqslant\eta^m$ except for $|z-a|<\eta$, which is a disk of area $\pi\eta^2$. This is an extreme example, in which all the roots of $q_m(z)$ are at $z=a$, and also shows that the theorem is the best result of its kind. If the roots of $q_m(z)$ are well separated, the inequality measure$(\mathcal{E})<\pi\eta^2$ gives a substantial overestimate of the area of the multi component region in which $|q_m(z)| \prec \eta^m$. Equation (5.1) leads directly to a bound on $|f(z) - [L/M]|$ if the inequality (5.2) is made use of. This approach [Zinn-Justin, 1971] leads to a variety of interesting theorems about convergence of various sequences of Padé approximants. They seem, at the time to of writing, to be the most natural general results for convergence of Padé approximants to analytic functions.

Throughout this section, we assume that the Padé approximants needed exist, and the theorems are to be understood in this sense. There is nothing in the right-hand side of (5.1) which implies the existence of $[L/M]$. Because, for example, a subsequence of Padé approximants of any row exists, the theorem about convergence in measure of a row may be re-phrased as a theorem about convergence in measure of an infinite subsequence of extant Padé approximants in the row. To avoid this cumbersome phrasing, we state clearly that the results of this section apply to Padé approximants which exist, and no theorem is to be understood as asserting the existence of a particular Padé approximant.

The first theorem is an analogue of de Montessus's theorem in the context of convergence in measure, and then we turn to analogues of diagonal sequences. An interesting feature of the proofs is the possibility of bounding the area of disruption, δ_k, of the $[L_k/M_k]$ Padé approximant of a sequence so that $\sum_{k=1}^{\infty} \delta_k < \infty$ and $\sum_{k=K}^{\infty} \delta_k < \delta$, where δ is a preassigned arbitrarily small positive number. Then one has the convergence in measure of the sequence $[L_k/M_k]$, $k = 1, 2, \ldots$, provided the Padé approximants exist. It is important to notice which theorems assert convergence in measure of an entire sequence of extant Padé approximants and which theorems refer to subsequences only.

The first theorem asserts convergence in measure of a row sequence of Padé approximants of a meromorphic function. It is a weaker form of de Montessus's theorem applicable when the degree of the denominator is known to be greater than or equal to (instead of precisely equal to) the number of poles within the circle of convergence of the row of Padé approximants.

THEOREM 6.5.1. *Let $f(z)$ be analytic at the origin and also in a given disk $|z| \leq R$, except for m poles, counting multiplicity. Consider a row of $[L/M]$ Padé approximants of $f(z)$ with M fixed, $M \geq m$, and $L \to \infty$. Suppose that arbitrarily small, positive ε and δ are given. Then L_0 exists such that $|f(z) - [L/M]| < \varepsilon$ for any $L > L_0$ and for all $|z| < R$ except for $z \in \mathcal{E}_L$, where $\mathcal{E}_L$ is a set of points in the z-plane of measure less than δ.*

Proof. Let $f(z)$ have poles at $z = \alpha_1$, $z = \alpha_2, \ldots$, $z = \alpha_m$ within $|z| < R$, and define

$$R_m(z) = (z - \alpha_1)(z - \alpha_2) \cdots (z - \alpha_m),$$

so that $R_m(z)f(z)$ is analytic in $|z| \leq R$. Then Hermite's formula is directly applicable as

$$f(z) - [L/M] = \frac{z^{L+M+1}}{2\pi i Q_M(z) R_m(z)} \int_{|t|=R} \frac{f(t)Q_M(t)R_m(t)}{t^{L+M+1}(t-z)} dt, \quad (5.3)$$

where $Q_M(z) \equiv Q^{[L/M]}(z)$. Since $|z| < R$, define

$$K = \frac{1}{1-|z/R|} \sup_{|t|=R} |f(t)|. \quad (5.4)$$

and then (5.3) and (5.4) yield

$$|f(z) - [L/M]| \leq \frac{K}{|Q_M(z)R_m(z)|} \left| \frac{z}{R} \right|^{L+M+1} \sup_{|t|=R} |Q_M(t)R_m(t)| \quad (5.5)$$

The next step is to bound some factors on the right-hand side of (5.5). Separate the roots of $Q_M(z)$ into those with

$$|z_i| < 2R, \quad i = 1, 2, \ldots, M',$$

and those with

$$|z_i| \geq 2R, \quad i = M'+1, M'+2, \ldots, M.$$

Then

$$\sup_{|t|=R} \prod_{i=M'+1}^{M} \left| \frac{t-z_i}{z-z_i} \right| \leq \prod_{i=M'+1}^{M} \frac{1+R/|z_i|}{1-R/|z_i|} \leq 3^{M-M'}$$

and

$$\sup_{|t|=R} \left| \frac{Q_M(t)}{Q_M(z)} \right| = \sup_{|t|=R} \prod_{i=1}^{M'} \frac{|t-z_i|}{|z-z_i|} \prod_{i=M'+1}^{M} \frac{|t-z_i|}{|z-z_i|}$$

$$\leq \left\{ \prod_{i=1}^{M'} \frac{3R}{|z-z_i|} \right\} 3^{M-M'} = \frac{3^M R^{M'}}{\prod\limits_{i=1}^{M'} |z-z_i|}.$$

Because the zeros of $R_m(t)$ lie in $|z| < R$,

$$\frac{\sup\limits_{|t|=R} |Q_M(t)R_m(t)|}{|Q_M(z)R_m(z)|} < \frac{R^{M'}3^M \times (2R)^m}{|R_m(z)| \prod\limits_{i=1}^{M'} |z-z_i|} \quad (5.6)$$

with $M' \leqslant M$. The denominator of the right-handside of (5.6) contains a polynomial with leading coefficient unity and is bounded by

$$|R_m(z)| \prod_{i=1}^{M'} |z - z_i| > \eta^{M'+m} \tag{5.7}$$

except for $z \in \mathcal{E}_l$ where measure$(\mathcal{E}_l) \leqslant \pi \eta^2 = \delta$. Clearly, this set $\mathcal{E}_l$ depends on the values of z_i, which in turn depend on L. Assembling (5.5), (5.6), and (5.7),

$$|f(z) - [L/M]| \leqslant \left|\frac{z}{R}\right|^{L+M+1} \left[\frac{KR^{M'} 3^M \times (2R)^m}{\eta^{M'+m}}\right], \tag{5.8}$$

except for $z \in \mathcal{E}_l$. The factor in square brackets on the right-hand side of (5.8) is bounded independently of L, and as $L \to \infty$ for $|z| < R$,

$$|f(z) - [L/M]| \to 0$$

except for $z \in \mathcal{E}_l$ where measure$(\mathcal{E}_l) \leqslant \delta$. The theorem is now proved.

We will now extend the theorem with two corollaries.

COROLLARY 1. *With the hypotheses of the theorem, the more general sequence* $[L_k/M_k]$ *satisfies*

$$|f(z) - [L_k/M_k]| < \varepsilon$$

for any $k > k_0$ *and for all* $|z| < R$ *excepting* $z \in \mathcal{E}_k$, *where* $\mathcal{E}_k$ *is a set of measure less than* δ, *provided*

(i) $L_k/M_k \to \infty$ *as* $k \to \infty$ $(M_k \neq 0)$, *and*
(ii) $M_k \geqslant M$ *for all* $k > k_0$.

Proof. We assume, without loss of generality, that $R/\eta > 1$, because a smaller value of η gives a stronger result. Since $M_k \geqslant M$, (5.8) is valid, and since $M' < M$,

$$|f(z) - [L_k/M_k]| \leqslant \left|\frac{z}{R}\right|^{L_k+M_k+1} \left(\frac{3R}{\eta}\right)^{M_k} K\left(\frac{2R}{\eta}\right)^m.$$

Because

(i) $L_k/M_k \to \infty$,
(ii) $|z/R| < 1$,

given any $\varepsilon'>0$, we may choose k_0 such that

$$\frac{L_k}{M_k}\ln\left|\frac{z}{R}\right|+\ln\left(\frac{3R}{\eta}\right)<\ln\varepsilon'$$

for all $k>k_0$. Hence

$$\left|\frac{z}{R}\right|^{L_k}\left(\frac{3R}{\eta}\right)^{M_k}<(\varepsilon')^{M_k}.$$

By choosing $\varepsilon'=\min\{1,(\varepsilon/K)(\eta/2R)^m\}$, which is positive, it follows that

$$|f(z)-[L_k/M_k]|<\varepsilon$$

for any $k>k_0$.

COROLLARY 2. *With the hypotheses of the theorem for $f(z)$, ε, δ, and M, an L_0 exists such that*

$$|f(z)-[L/M]|<\varepsilon$$

for all $L>L_0$ and for all $|z|<R$ except for $z\in\mathcal{E}$, where $\mathcal{E}$ is a set of measure less than δ.

Remark. This result is substantially stronger than that of the main theorem, being a result about convergence of all the extant approximants of the $(M+1)$th row beyond $L=L_0$.

Proof. From (5.8), the $[l+L_1/M]$ Padé approximant satisfies

$$|f(z)-[l+L_1/M]|<\left|\frac{z}{R}\right|^{l+L_1+M+1}\left[\frac{KR^{M'}3^M(2R)^m}{\eta_l^{M'+m}}\right] \qquad (5.9)$$

except on a set $\mathcal{E}_l$ of measure at most $\pi\eta_l^2$. We assume, without loss of generality, that $R/\eta_l>1$. Recalling that $M'\leqslant M$, we choose L_1 such that

$$\left|\frac{z}{R}\right|^{L_1}(3R)^M K(2R)^m<\varepsilon,$$

where ε is a preassigned small positive number.

Then, the requirement on η_l that

$$\left|\frac{z}{R}\right|^l\left(\frac{1}{\eta_l}\right)^{M+m}=1 \qquad (5.10)$$

is sufficient for (5.9) to reduce to

$$|f(z) - [l + L_1/M]| < \varepsilon$$

except on a set $\mathcal{E}_l$ of measure at most $\pi \eta_l^2$. Equation (5.10) may be written as

$$2\pi \eta_l^2 = 2\pi \left| \frac{z}{R} \right|^{2l/(M+m)}$$

and

$$\sum_{l=l'}^{\infty} 2\pi \eta_l^2 = 2\pi \frac{|z/R|^{2l'/(M+m)}}{1 - |z/R|^{2/(M+m)}}. \qquad (5.11)$$

We may certainly choose l' so that

$$\sum_{l=l'}^{\infty} 2\pi \eta_l^2 < \delta,$$

and this asserts that the total measure of all the individual exceptional sets of all the $[l + L/M]$ Padé approximants with $l > l'$ is arbitrarily small. Thus convergence in measure of a row is proved.

THEOREM 6.5.2. *Let* $f(z)$ *be analytic at the origin, and also in the circle* $|z| \leqslant R$, *except for* m *poles, counting multiplicity. Consider a sequence* $[L_k/M_k]$ *of Padé approximants of* $f(z)$ *with* $M_k \geqslant m$ *and* $L_k/M_k \to \infty$ *as* $k \to \infty$ *(* $M_k \neq 0$ *). Let* ε, δ *be arbitrarily small positive given numbers. Then* k_0 *exists such that*

$$|f(z) - [L_k/M_k]| < \varepsilon$$

for all $k > k_0$ *and all* $|z| < R$ *except for* $z \in \mathcal{E}$, *where* $\mathcal{E}$ *is a set of points of the* z-*plane of measure less than* δ.

Theorem 6.5.2 embodies the previous theorem and its two corollaries. The method of proof is a conflation of the proofs given.

Corollary 2 of Theorem 6.5.1 about convergence of rows is a powerful result. Recall that the exceptional sets have the poles of the approximants in $|z| < R$ as interior points. Convergence of a row is expected in $|z| < R$ except in neighborhoods of limit points of poles of the approximants and in neighborhoods of the poles of $f(z)$. In this light, the theorem may be regarded as a statement that the exceptional set of an entire row is a set of arbitrarily small measure which is in turn the union of sets enclosing poles of $f(z)$ and other limit points of the poles of the approximants. In this language, it is plain that Corollary 2 embodies many earlier results expressed in terms of ordinary convergence.

Having treated row sequences and their relatives, we turn to diagonal sequences, ray sequences, and their relatives. As an introduction, we prove Theorem 6.5.3, which is a simple theorem, and subsequently we generalize it substantially. This theorem was originally proved by Nuttall [1970b], using Szego's theorem, but the proof given by Zinn-Justin [1971], based on Hermite's formula, is somewhat simpler.

THEOREM 6.5.3. *Let $f(z)$ be a meromorphic function. Suppose that ε, δ are given positive numbers. Then M_0 exists such that any $[M/M]$ Padé approximant satisfies*

$$|f(z) - [M/M]| < \varepsilon$$

for all $M > M_0$ on any compact set of the z-plane except for a set $\mathcal{E}_M$ of measure less than δ.

Proof. We will set the scale of the z-plane by proving convergence for $|z| < 1$ except on a set $\mathcal{E}_M$ of measure less than δ. Define $\eta - \frac{1}{2}\sqrt{\delta}/\pi$. We may assume $0 < \eta < 1$ without loss of generality. Define

$$R_{\min} \equiv (3/\eta)^3.$$

For some Δ, no matter how small, in the range $0 < \Delta < 1$, R exists satisfying

(i) $f(z)$ is analytic in $R - \Delta < |z| < R + \Delta$,
(ii) $R > R_{\min}$,
(iii) $f(z)$ has $m = m(R)$ poles located at $z = u_i$, $i = 1, 2, \ldots, m$, in $|z| < R$.

Next we need Hermite's interpolation formula using the polynomial

$$R_m(z) = \prod_{i=1}^{m} (z - u_i),$$

and consequently for $M \geqslant m$, $Q_M(z) \equiv Q^{[M/M]}(z)$,

$$f(z) - [M/M] = \frac{z^{2M+1}}{2\pi i Q_M(z) R_m(z)} \oint_{|t|=R} \frac{Q_M(t) R_m(t) f(t)}{(t-z) t^{2M+1}} dt.$$

We have chosen to consider $|z| < 1$, $R > R_{\min} > 2$, and so

$$|f(z) - [M/M]| \leqslant K R^{-(2M+1)} \frac{\sup_{|t|=R} |Q_M(t) R_m(t)|}{|Q_M(z) R_m(z)|}, \qquad (5.12)$$

where $K = \sup_{|t|=R} |f(t)|$.

From (5.6), with some $M' \leqslant M$, we proved that

$$\frac{\sup_{|t|=R} |Q_M(t)R_m(t)|}{|Q_M(z)R_m(z)|} < \frac{R^{M'} 3^M \times (2R)^m}{|R_m(z)| \prod\limits_{i=1}^{M'} |z-u_i|}. \tag{5.13}$$

The denominator of the right-hand side of (5.13) is a polynomial with leading coefficient unity, and is bounded by

$$|R_m(z)| \prod_{i=1}^{M'} |z-z_i| > \eta^{M'+m} \tag{5.14}$$

except for z in a set $\mathscr{E}_M$ of measure $\pi\eta^2 = \delta$. Assembling (5.12), (5.13), and (5.14),

$$|f(z) - [M/M]| < \left(\frac{1}{R}\right)^{2M+1} \frac{(3R)^M (2R)^m}{\eta^{M+n}}$$

$$< \left(\frac{3}{\eta}\right)^{M+m} \frac{1}{R^{M-m}}.$$

Provided $M > 2m$, and recalling that $R > (3/\eta)^3$,

$$|f(z) - [M/M]| < \left(\frac{3}{\eta}\right)^{3M/2} \frac{1}{R^{M/2}} < \varepsilon$$

for any $M > $(some M_0), except on the set $\mathscr{E}_M$ of measure less than δ.

As already stated, Theorem 6.5.3 is a weak form of both what is known to be true and what is expected to be true about convergence in measure of ray sequences. Nonetheless, it provides a basis for further development.

First, the diagonal sequence may be replaced by the sequence $[L_k/M_k]$, $k = 1, 2, \ldots$, provided that, for any λ in the range $0 < \lambda < 1$, however small,

$$\lambda < \frac{L_k}{M_k} < \lambda^{-1}. \tag{5.15}$$

Equation (5.15) confines the Padé approximants to a fan-shaped region of the Padé table as shown in Figure 1. Provided (5.15) holds and $L_k + M_k \to \infty$, this weaker constraint is sufficient to allow convergence in measure.

Second, $f(z)$ need not be meromorphic, but may also have a countable number of isolated essential singularities. This means that $\exp[-(1-z)^{-1}]$ and $\exp[z\Gamma(z)]$ are allowable functions, but not functions whose singularities have a limit point in the finite z-plane.

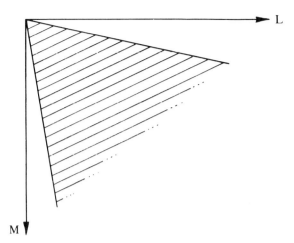

Figure 1. The allowed domain of the Padé table for the approximants in Theorem 6.5.4.

THEOREM 6.5.4 [Pommerenke, 1973]. *Let $f(z)$ be a function which is analytic at the origin and analytic in the entire z-plane except for a countable number of isolated poles and essential singularities. Suppose $\varepsilon > 0$ and $\delta > 0$ are given. Then M_0 exists such that any $[L/M]$ Padé approximant of the ray sequence with $L/M = \lambda$ ($\lambda \neq 0$, $\lambda \neq \infty$) satisfies*

$$|f(z) - [L/M]| < \varepsilon$$

for any $M \geqslant M_0$, on any compact set of the z-plane except for a set $\mathcal{E}_M$ of measure less than δ.

Proof. As in Theorem 6.5.3, we take $|z| < 1$, define $\eta = \frac{1}{2}\sqrt{\delta/\pi}$, and assume $0 < \eta < 1$. We choose

$$R_{\min} = \left(\frac{3}{\eta}\right)^{1+2/\lambda} \tag{5.16}$$

and some $\Delta > 0$, $R > R_{\min}$, such that

(i) $f(z)$ is analytic in $R - \Delta < |z| < R + \Delta$,
(ii) $f(z)$ has $m = m(R)$ poles located at $z = u_i$, $i = 1, 2, \ldots, m$, in $|z| < R$,
(iii) $f(z)$ has $\mu = \mu(R)$ essential singularities located at $z = w_i$, $i = 1, 2, \ldots, \mu$, in $|z| < R$.

Define the polynomial

$$R_M(z) = \prod_{i=1}^{m} (z - u_i) \prod_{i=1}^{\mu} (z - w_i)^p, \tag{5.17}$$

where $p=p(M)$ is defined so that the ratio $\rho=p/M$ satisfies

$$0<\frac{p}{M}<\min\left\{-\frac{\ln 2}{\ln \Delta},\frac{\frac{1}{2}\lambda}{\mu}\right\}. \tag{5.18}$$

The purpose of this is to be able subsequently to let $M\to\infty$ and $p\to\infty$ simultaneously but keep $R_M(z)$ of sufficiently low degree. At any rate, $R_M(z)$ has leading coefficient unity and degree less than M. We use Hermite's formula, with $Q_M(z)\equiv Q^{[L/M]}(z)$,

$$f(z)-[L/M]=\frac{z^{L+M+1}}{2\pi i Q_M(z)R_M(z)}\oint_C \frac{Q_M(t)R_M(t)f(t)}{(t-z)t^{L+M+1}}dt,$$

where C is a closed contour containing the origin and no singularities of $Q_M(z)f(z)$. By enlarging the contour so as to enclose the essential singularities, we find

$$f(z)-[L/M]=\frac{z^{L+M+1}}{2\pi i Q_M(z)R_M(z)}$$

$$\times\left\{\int_{|t|=R}\frac{Q_M(t)R_M(t)f(t)\,dt}{(t-z)t^{L+M+1}}-\sum_{k=1}^{\mu}I_k(z)\right\} \tag{5.19}$$

where

$$I_k(z)=\int_{|t-w_k|=\delta_k}\frac{R_M(t)Q_M(t)f(t)}{(t-z)t^{L+M+1}}dt. \tag{5.20}$$

Equation (5.20) is a contour integral round a small circle of radius δ_k enclosing the essential singularity at $z=w_k$. To bound $I_k(z)$, we require that $|z-w_k|>2\delta_k$. Using the maximum-modulus theorem for the polynomial $Q_M(t)R_M(t)(t-w_k)^{-p}$,

$$\left|\frac{Q_M(t)R_M(t)}{(t-w_k)^p}\right|<\sup_{|t|=R}\left\{\frac{|Q_M(t)R_M(t)|}{|t-w_k|^p}\right\}<\frac{\sup_{|t|=R}|Q_M(t)R_M(t)|}{\Delta^p}.$$

Hence, for $|z-w_k|>2\delta_k$,

$$|I_k(z)|\leqslant 2\pi\frac{\sup_{|t|=R}|R_M(t)Q_M(t)|}{\Delta^p}\frac{\sup_{|t-w_k|=\delta_k}|f(t)|\delta_k^p}{(|w_k|-\delta_k)^{L+M+1}}. \tag{5.21}$$

We now specify the radii δ_k of the small circles by defining

$$\delta_k = \min\left\{\left|\frac{w_k}{2R}\right|^{(\lambda+1)/\rho}, \frac{1}{2}|w_k|\right\}. \tag{5.22}$$

Equation (5.16) ensures that the regions $|z - w_k| < 2\delta_k$ surrounding the essential singularities in $|z| < 1$ have arbitrarily small total measure.

From (5.22),

$$\delta_k^p < \left(\frac{|w_k| - \delta_k}{R}\right)^{L+M}.$$

We define

$$\sup_{|t-w_k|=\delta_k} |f(t)| = K_k,$$

where K_k is independent of M, and then

$$|I_k(z)| < K' \frac{\sup_{|t|=R} |R_M(t)Q_M(t)|}{\Delta^p R^{L+M+1}}, \tag{5.23}$$

where K' is independent of M and k. Assembling (5.19) and (5.23), we find

$$|f(z) - [L/M]| < \left|\frac{z}{R}\right|^{L+M+1} K'' \frac{\sup_{|t|=R} |Q_M(t)R_M(t)|}{|Q_M(z)R_M(z)|} \left\{1 + \sum_{k=1}^{\mu} \frac{1}{\Delta^p}\right\}, \tag{5.24}$$

where K'' is also independent of M.

Again, when $Q_M(t)$ has M' zeros within $|t| \leqslant 2R$, we note that $M' \leqslant M$ and $R_M(t)$ is a polynomial of degree $m + p\mu$, so that (5.6) and (5.7) yield

$$\frac{\sup_{|t|=R} |Q_M(t)R_M(t)|}{|Q_M(z)R_M(z)|} < \frac{(3R)^M (2R)^{m+p\mu}}{\eta^{m+p\mu+M}} \tag{5.25}$$

provided $z \notin \mathcal{E}_M$, a set of measure less than $\pi\eta^2$. Since $L = \lambda M$ and $|z| < 1$, (5.18), (5.24), and (5.25) yield

$$|f(z) - [L/M]| \leqslant \left(\frac{1}{R}\right)^{\lambda+1 M} K''' \left(\frac{3R}{\eta}\right)^{M+m+p\mu}, \tag{5.26}$$

where K''' is also independent of M. From (5.18), $p\mu < \frac{1}{2}\lambda M$, and then

(5.26) gives

$$| f(z) - [L/M] | \leqslant K''' R^{-\frac{1}{2}\lambda M + m} \left(\frac{3}{\eta} \right)^{M(1 + \frac{1}{2}\lambda) + m}$$

Hence, provided $R > (3/\eta)^{1 + 2/\lambda}$ and M is sufficiently large, $| f(z) - [L/M] | < \varepsilon$ except on the set $\mathscr{E}_M$ and the small circles enclosing the essential singularities of $f(z)$ in $|z| < 1$.

COROLLARY 1. *This theorem can also be generalized to treat arbitrary sequences in the region of the Padé table shown Figure 1.*

COROLLARY 2 [Zinn-Justin, 1971]. *Let $f(z)$ be meromorphic in $|z| \leqslant R$. Let the number of zeros of $Q_M(z)$ in $|z| < R$ be q_M. Then if*

$$\frac{q_M \ln M}{M} \to 0 \qquad as \quad M \to \infty,$$

the $[M/M]$ Padé approximants of $f(z)$ converge in measure in $|z| < R/\sqrt{3}$.

COROLLARY 3 [Zinn-Justin, 1971]. *If $f(z)$ is an analytic function of exponential order less than $2/\lambda$, then the sequence of $[\lambda M/M]$ approximants converges on any compact set of the z-plane except on a set of arbitrarily small measure.*

We present these corollaries without proof. The second and third are interesting because they show that further restrictions on the class of functions considered lead to stronger convergence results. However, no theorem yet proved gives convergence in measure of diagonal Padé approximants in $|z| < R$ for functions known only to be meromorphic in $|z| < R$. We refer to Edrei [1975b] for a result which allows the essential singularities to be limit points of pole sequences, rather than isolated essential singularities as in Pommerenke's theorem.

6.6 Lemniscates, Capacity, and Measure

The purpose of this section is not to give a detailed and rigorous account of the foundations of capacity and measure, but rather to indicate why the results of the previous section are much stronger if rephrased in terms of capacity or Hausdorff measure.

The basis of Section 6.5 is the result related to Cartan's lemma [Cartan, 1928; Nuttall, 1970h] that, for a polynomial $q_m(z)$ with leading coefficient unity, $|q_m(z)| > \eta^m$ except on a set $\mathscr{E}$ of measure at most $\pi\eta^2$. The boundary on which $|q_m(z)| = \eta^m$ is usually called a lemniscate, and so all the results of

Section 6.5 apply except on lemniscatic regions which are arbitrarily small. We will see that the natural measure of the size of a lemniscatic region is its capacity.

To begin with, let us recall Tchebycheff's result for the minimax polynomial on an interval $-1 \leqslant x \leqslant 1$. The problem is to find the polynomial $T_n(x)$ in the class P_n of all polynomials $p_n(x)$ of degree n and leading coefficient unity for which the limit

$$\inf_{p_n(x) \in P_n} \sup_{a \leqslant x \leqslant b} |p_n(x)|$$

is attained. The solution is well known. For $n \geqslant 1$,

$$T_n(x) = \frac{1}{2^{n-1}} \cos(n \cos^{-1} x)$$

is a polynomial of degree n of leading coefficient unity, and

$$\inf_{p_n(x) \in P_n} \sup_{-1 \leqslant x \leqslant 1} |p_n(x)| = \frac{1}{2^{n-1}}. \tag{6.1}$$

A linear change of variable

$$x' = -\frac{a(x-1)}{2} + \frac{b(x+1)}{2} \tag{6.2}$$

leads to the result that

$$\inf_{p_n(x) \in P_n} \sup_{a \leqslant x \leqslant b} |p_n(x)| = \left(\frac{b-a}{4}\right)^{n-1}\left(\frac{b-a}{2}\right). \tag{6.3}$$

To generalize these ideas to an arbitrary compact set in the z-plane, we have the following theorem.

THEOREM 6.6.1. *Let $\mathcal{E}$ be a compact set in the complex plane (containing infinitely many points). Then there is a unique Tchebycheff polynomial for which*

$$M_n \equiv \inf_{p_n(z) \in P_n} \sup_{z \in \mathcal{E}} |p_n(z)| \tag{6.4}$$

is attained. The minimax polynomial, $p_n(z) = T_n(z)$, may be written as

$$T_n(z) = \prod_{i=1}^{n} (z - z_i). \tag{6.5}$$

The zeros z_i of $T_n(z)$ lie in the convex hull of $\mathcal{E}$, and the maximum value M_n of $|T_n(z)|$ is achieved at least n times on the boundary of $\mathcal{E}$.

DISCUSSION. We shall not prove this important theorem, [Hille, 1962, p 265], but elaborate it with a few remarks.

It is easy to see that all the z_i lie within the convex hull of $\mathcal{E}$. For suppose not, and let z_1 lie outside the convex hull $\mathcal{H}$ of $\mathcal{E}$ and the points $z_2, z_3, \ldots, z_n$. By considering a point z_1' nearer to $\mathcal{H}$, and the point z' for which the

$$\sup_{z \in \mathcal{E}} \left\{ |z - z_1'| \prod_{i=2}^{n} |z - z_i| \right\} = m'$$

is attained at $z = z'$, we find that

$$M_n = \sup_{z \in \mathcal{E}} \prod_{i=1}^{n} |z - z_i| > |z' - z_1| \prod_{i=2}^{n} |z' - z_i|$$

$$> |z' - z_1'| \prod_{i=2}^{n} |z' - z_i| = m',$$

which contradicts the minimum property of M_n with respect to variation of the z_i. Hence the zeros of $T_n(z)$ lie in the convex hull of $\mathcal{E}$.

The proof of the existence of a minimax polynomial is based on "tracing" the roots to extremal positions in the closed convex hull. We omit the proofs of existence and uniqueness of $T_n(z)$ in the general case. Since $\mathcal{E}$ is compact, it is obvious that the maximum of $|T_n(z)|$ is attained in $\mathcal{E}$. Further, the maximum-modulus theorem shows that the maximum is attained on the boundary of $\mathcal{E}$. That $|T_n(z)|$ should equal M_n at n distinct points on the boundary is another significant result we state without proof.

If $\mathcal{E}$ is a finite point set, a possibility excluded by the hypothesis of the theorem, then the Tchebycheff polynomial of order n is zero on $\mathcal{E}$ for $n \geqslant N$. Furthermore, it is not unique for $n > N$, and so this degenerate case is naturally excluded by the hypothesis.

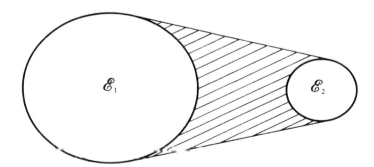

Figure 1. A set $\mathcal{E} = \mathcal{E}_1 \cup \mathcal{E}_2$. The union of the shaded region and the set $\mathcal{E}$ comprises the convex hull of $\mathcal{E}$.

COROLLARY. The Tchebycheff polynomial $T_n(z)$ for a set $\mathcal{E}$ defines a lemniscate by $|T_n(z)| = M_n$ and a lemniscatic region $\mathcal{L}_n$ by

$$|T_n(z)| \leqslant M_n \qquad \text{for all} \quad z \in \mathcal{L}_n.$$

Then $\mathcal{E} \subset \mathcal{L}_n$, and the boundary of $\mathcal{L}_n$ has at least n points in common with $\mathcal{E}$.

DISCUSSION. The important idea is that $\mathcal{E}$ becomes a subset of the lemniscatic region $\mathcal{L}_n$ defined by $|T_n(z)| \leqslant M_n$, and the proof follows immediately from the maximum-modulus theorem and the main theorem.

We now proceed to three theorems which we prove, because the proofs illustrate the structure of lemniscates and the idea of capacity. Repeated use is made of the *maximum-modulus theorem*, which states that the maximum modulus of an analytic function in a compact region is achieved on the boundary.

THEOREM 6.6.2. *Let $\mathcal{E}$ be a compact set in the complex z-plane (containing infinitely many points). Let $T_n(z)$ be the Tchebycheff polynomials defined on $\mathcal{E}$, and let*

$$M_n(\mathcal{E}) = \sup_{z \in \mathcal{E}} |T_n(z)|. \tag{6.6}$$

Then the capacity of $\mathcal{E}$ is uniquely defined by

$$\lim_{n \to \infty} \left[M_n(\mathcal{E}) \right]^{1/n} = \text{cap}(\mathcal{E}) \tag{6.7}$$

Proof. The only proof required is that the limit (6.7) is well defined. To do this, let

$$\alpha = \liminf \left[M_n(\mathcal{E}) \right]^{1/n},$$
$$\beta = \limsup \left[M_n(\mathcal{E}) \right]^{1/n}.$$

If δ is the maximum diameter of the compact set $\mathcal{E}$ in the usual sense, then from (6.4),

$$M_n(\mathcal{E}) = \sup_{z \in \mathcal{E}} |T_n(z)| = \inf_{z_i} \sup_{z \in \mathcal{E}} \prod_{i=1}^{n} |z - z_i| \leqslant \delta^n.$$

Hence $0 \leqslant \alpha \leqslant \beta \leqslant \delta$.

Given $\varepsilon>0$, we may find N such that

$$\alpha+\varepsilon>M_N^{1/N}$$

and

$$|T_N(z)|<(\alpha+\varepsilon)^N \qquad \text{for all} \quad z\in\mathcal{E}.$$

For any positive integers m, k but with $m<N$,

$$|z^m[T_n(z)]^k|<K(\alpha+\varepsilon)^{Nk+m} \qquad \text{for all} \quad z\in\mathcal{E},$$

where

$$K=K(N,\mathcal{E})$$

is a constant independent of k. Recognising $z^m[T_n(z)]^k$ as a polynomial of order $m+nk$, we see that

$$(M_{m+nk})^{1/(m+NK)}\leqslant K^{1/(m+Nk)}(\alpha+\varepsilon),$$

and by taking the limit as $k\to\infty$,

$$\limsup_{k\to\infty}[M_{m+Nk}]^{1/(m+Nk)}\leqslant\alpha+\varepsilon. \tag{6.8}$$

Equation (6.8) holds for any positive $m<N$, and any $\varepsilon>0$. Hence $\beta=\alpha$, and $\mathrm{cap}(\mathcal{E})$ is well defined by (6.7).

The quantity $\mathrm{cap}(\mathcal{E})$ is a measure of the magnitude of the set $\mathcal{E}$. It is called the logarithmic capacity, colloquially abbreviated to capacity or transfinite diameter in different contexts.

The next theorem shows the key role of lemniscates in the theory of capacity.

THEOREM 6.6.3. *If $\mathcal{E}$ is a lemniscatic region given by*

$$|p_n(z)|\leqslant\eta^n \qquad \text{for} \quad z\in\mathcal{E},$$

then $\mathrm{cap}(\mathcal{E})=\eta$.

Proof. The maximum modulus theorem show that $\partial\mathcal{E}$ (the boundary of $\mathcal{E}$) is given by $|p_n(z)|=\eta^n$. Let $T_n(z)$ be the nth-order Tchebycheff polynomial defined on $\mathcal{E}$, so that

$$|T_n(z)|\leqslant(\eta')^n \qquad \text{for all} \quad z\in\mathcal{E}$$

and consequently $\text{cap}(\mathcal{E})=\eta'$. By definition of $T_n(z)$, $\eta'\leqslant\eta$. If $\eta'<\eta$, then

$$|T_n(z)|<|p_n(z)| \qquad \text{for all} \quad z\in\partial\mathcal{E}.$$

Recall

ROUCHÉ'S THEOREM.   If $f(z)$ and $g(z)$ are analytic inside and on a closed contour C and $|g(z)|<|f(z)|$ on C, then $f(z)$ and $f(z)+g(z)$ have the same number of zeros inside C.

In this case $f(z)=T_n(z)$, $g(z)=-p_n(z)$, and the theorem implies that a polynomial of degree $n-1$ has n zeros in $\mathcal{E}$. This is impossible, hence $\eta'=\eta$ and $\text{cap}(\mathcal{E})=\eta$, proving the theorem.

Before giving examples of the capacity of a set, we prove the major result which improves the theorems of the previous section.

THEOREM 6.6.4.   *Let $\mathcal{E}$ be a compact set. Then*

$$\text{meas}(\mathcal{E})\leqslant\pi\big[\text{cap}(\mathcal{E})\big]^2. \tag{6.9}$$

Proof. Let $T_n(z)$ be the nth Chebychev polynomial defined on $\mathcal{E}$. For any $\varepsilon>0$, (6.7) implies that a sufficiently large n exists such that

$$|T_n(z)|\leqslant\big[\text{cap}(\mathcal{E})+\varepsilon\big]^n \qquad \text{for all} \quad z\in\mathcal{E}.$$

This inequality defines a lemniscatic region $\mathcal{L}_n$, of capacity $\eta=\text{cap}(\mathcal{E})+\varepsilon$, which has $\mathcal{E}$ as a subset. If we prove that

$$\text{meas}(\mathcal{L}_n)\leqslant\pi\big[\text{cap}(\mathcal{L}_n)\big]^2, \tag{6.10}$$

it follows that

$$\text{meas}(\mathcal{E})\leqslant\text{meas}(\mathcal{L}_n)\leqslant\pi\big[\text{cap}(\mathcal{L}_n)\big]^2\leqslant\pi\big[\text{cap}(\mathcal{E})+\varepsilon\big]^2.$$

Hence the theorem is true provided we prove (6.10). For brevity, let the lemniscate be defined by

$$|p(z)|=\eta^n, \tag{6.11}$$

where $p(z)=T_n(z)$ and $\eta^n=M_n$. We will consider the family of lemniscates given by $|p(z)|=\rho$ for all positive ρ. This consists of at most n disjoint closed curves $K_1, K_2,\ldots, K_m$. The maximum-modulus theorem shows that

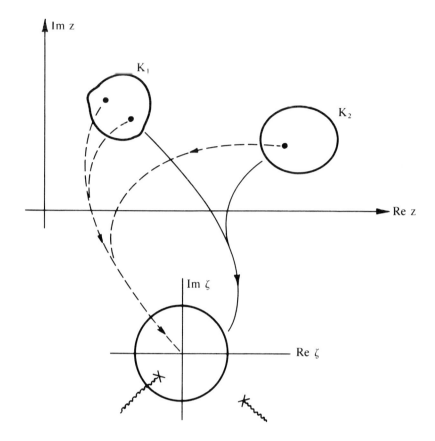

Figure 2. The mapping of K_1 and K_2 onto the circle $|\zeta|=\rho$, showing two branch points in the ζ-plane.

the curves lie outside each other. There is at least one root of $p(z)$ in each curve K_k: let there be precisely μ_k roots in K_k. As the point z moves around K_k, $p(z)=\Pi_{i=1}^{n}(z-z_i)=\zeta$ moves μ_k times around the circle $|\zeta|=\rho$ and $\zeta'=\zeta^{1/\mu_k}$ moves once around the circle $|\zeta'|=\rho^{1/\mu_k}$.

Now consider the mappings $z\to p(z)=\zeta\to\zeta^{1/\mu_k}=\zeta'$ and the inverse mapping $\zeta'\to z=f(\zeta')$. Here $p(z)$ is clearly a single-valued function of z, but the inverse map $z=z(\zeta)$ is not, and $z=z(\zeta)$ has $n-1$ branch points occurring at $p'(z)=0$. However, assuming that $|\zeta|=\rho$ avoids these branch points, one of the branches of $z=f(\zeta')$ is regular on $|\zeta'|=\rho^{1/\mu_k}$, and has the Laurent expansion

$$f(\zeta')=\sum_{j=-\infty}^{\infty}a_j^{(k)}\zeta'^j,$$

i.e.,

$$z = \sum_{j=-\infty}^{\infty} a_j^{(k)} \zeta^{j/\mu_k}. \tag{6.12}$$

To find the area of K_k,

$$\text{meas}(K_k) = \tfrac{1}{2} \oint_{K_k} (x\, dy - y\, dx)$$

$$= \tfrac{1}{2} \oint_{K_k} \text{Re} \frac{z^* \, dz}{i}$$

$$= \tfrac{1}{2} \text{Re} \oint_{K_k} \left(\sum_{j=-\infty}^{\infty} a_j^{(k)*} \rho^{j/\mu_k} e^{-ij\phi/\mu_k} \right)$$

$$\times \left(\sum_{j=-\infty}^{\infty} a_j^{(k)} \rho^{j/\mu_k} e^{ij\phi/\mu_k} \right) \frac{j \, d\phi}{\mu_k}$$

$$= \pi \sum_{j=-\infty}^{\infty} j |a_j^{(k)}|^2 \rho^{2j/\mu_k}.$$

Therefore

$$\sum_{k-1}^{m} \text{meas}(K_k) = \sum_{k=1}^{m} \pi \left[\sum_{j=1}^{\infty} j |a_j^{(k)}|^2 \rho^{2j/\mu_k} - \sum_{j=1}^{\infty} j |a_{-j}^{(k)}|^2 \rho^{-2j/\mu_k} \right]. \tag{6.13}$$

Equation (6.13) is an increasing and continuous function of ρ, and so is bounded by its value at $\rho = \infty$. For ρ sufficiently large, K_1 contains all n roots of $p(z)$, and therefore $\mu_1 = n$ and $m = 1$.
Since $\zeta = p(z) = z^n + c_1 z^{n-1} + \cdots + c_n$,

$$z = f(\zeta) - \zeta^{1/n} + a_0 + a_1 \zeta^{-1/n} + \cdots, \tag{6.14}$$

which makes (6.12) more explicit, and (6.13) becomes

$$\sum_{k=1}^{m} \text{meas}(K_k) = \pi \rho^{2/n} - \sum_{j=1}^{\infty} j |a_{-j}|^2 \rho^{-2j/n} \qquad \text{for large } \rho. \tag{6.15}$$

Hence $\text{meas}(\mathcal{L}_n) \leqslant \pi \eta^2$ if $\eta = \rho^{1/n}$.

The result is proved, but it is interesting to note from (6.15) that equality is attained when $a_0 = a_1 = \cdots = 0$ and (6.14) shows that this locus is the circle $z^n = \rho$.

We now turn to a few examples.

Example 1. The capacity of the interval $a \leqslant x \leqslant b$ is $(b-a)/4$.

DISCUSSION. We assume the result (6.1) [Cheney, 1966, p. 61; Rivlin, 1969, Chapter 1] for the proof. Then result (6.3) follows from the substitution (6.2), and from (6.16) it follows that

$$\text{cap}(-1 \leqslant x \leqslant 1) = \tfrac{1}{2}$$

and the theorem is proved.

Example 2. The capacity of the disk $|z| \leqslant R$ and the circle $|z| = R$ is each equal to R.

Proof. The circle $|z| - R$ is a lemniscate, given by $|z^m| = R^m$, and so the results follow from Theorem 6.6.3 (and also from an inspection of the proof of Theorem 6.6.4).

Example 3. If $\mathcal{E}$ is a countable compact set, $\text{cap}(\mathcal{E}) = 0$. We leave the proof as an exercise.

We next state two very important theorems about capacity, without proof. Each gives considerable insight into the magnitude of the capacity of an arbitrary point set.

THEOREM 6.6.5 [Hille, 1962, p. 268–273]. *Let $\mathcal{E}$ be a compact set. Then*

$$\text{cap}(\mathcal{E}) = \lim_{n \to \infty} \left[\max_{z_j \in \mathcal{E}} \prod_{1 \leqslant j \leqslant k \leqslant n} |z_j - z_k| \right]^{\frac{2}{n(n-1)}}$$

DISCUSSION. The points $z_1, z_2, \ldots, z_n$ in $\mathcal{E}$ are chosen so as to maximize

$$\prod_{1 \leqslant j \leqslant k \leqslant n} |z_j - z_k|,$$

which contains $n(n-1)/2$ terms in its expansion. Hence it is clear that $\text{cap}(\mathcal{E})$ is bounded by the maximum diameter of $\mathcal{E}$. Furthermore, $\text{cap}(\mathcal{E})$ is some sort of geometric mean of the distance apart of the points of $\mathcal{E}$, and so it is called the transfinite diameter.

Two corollaries of the theorem are self-evident:

COROLLARY 1 (Monotonicity). *If $\mathcal{D}$ and $\mathcal{E}$ are compact sets with $\mathcal{D} \subset \mathcal{E}$, then $\text{cap}(\mathcal{D}) \leqslant \text{cap}(\mathcal{E})$.*

COROLLARY 2 (Homogeneity). *If* $z' = az + b$ *maps* $\mathcal{E}$ *onto* $\mathcal{E}'$, *then* $\text{cap}(\mathcal{E}')$ $= |a| \text{cap}(\mathcal{E})$.

THEOREM 6.6.6 [Hille, 1962, p. 280–289]. *Let* $\mu(z)$ *be a normalized measure defined on* $\mathcal{E}$, *and define*

$$I[\mu] = \int_E \int_E \ln\left(|z_1 - z_2|^{-1}\right) d\mu(z_1) d\mu(z_2).$$

Let

$$V(\mathcal{E}) = \inf_\mu I[\mu].$$

Then

$$\text{cap}(\mathcal{E}) = \exp\left[-V(\mathcal{E})\right].$$

DISCUSSION. We have in mind that $\mu(z)$ is a charge distribution on a two-dimensional surface $\mathcal{E}$, so that $I[\mu]$ is the self-energy associated with the distribution μ. In physical equilibrium, this functional is a minimum, and so the potential due to the physical distribution of unit charge on $\mathcal{E}$ is

$$V(E) = \inf_\mu I[\mu].$$

The further definition that $\ln(\text{cap}(\mathcal{E})) - -V(\mathcal{E})$ explains the name of logarithmic capacity for $\text{cap}(\mathcal{E})$.

Finally, we remark that we have emphasized the role of capacity as a point-set measure of the region of inaccurate approximation of Padé approximants to meromorphic functions. Another popular point-set measure is Hausdorff measure, and the results of the previous section generalize to convergence in α-dimensional Hausdorff measure [Wallin, 1974; Lubinsky, 1980]. In this case, the measure is defined by

$$\Lambda_\alpha(\mathcal{E}) = \min\left\{ \sum_{i=1}^\infty \left[\delta(D_i)\right]^\alpha \right\},$$

where the minimum is taken over all possible denumerable families of circular disks D_i which cover the set $\mathcal{E}$, and $\delta(D_i)$ is the diameter of the disk D_i. Two-dimensional Hausdorff measure is similar to area in its properties, and one-dimensional Hausdorff measure is more like length.

The following Boutroux–Cartan lemma is most useful in the context of one-dimensional Hausdorff measure.

THEOREM 6.6.7. *For any* $\eta > 0$, *the polynomial* $p_n(z) = \prod_{i=1}^{n}(z - z_i)$ *satisfies the inequality*

$$|p_n(z)| > (\eta/e)^n, \qquad (6.16)$$

which is valid in the z-plane outside no more than n circles of radii r_i *which obey the inequality*

$$\sum_{i=1}^{n} r_i \leq 2\eta.$$

We do not prove this theorem, and we refer to EPA (Chapter 14) for a more complete account of the role of Hausdorff measure in convergence of Padé approximants. We merely note that (6.16) is the basic type of inequality required in this context. We also note that many of these result, also extend, *mutatis mutandis*, to the rational interpolation problem [Walsh, 1969, 1970; Karlsson, 1976].

We conclude this section by repeating that an inspection of the proofs of the theorems of Section 6.5 shows that they prove convergence in capacity, and we have shown that this is a substantially stronger result than convergence in measure. In particular, an interval has zero measure, but nonzero capacity (and nonzero one-dimensional Hausdorff measure), and so a finite interval, or any set containing a finite interval, is not a permissible exceptional set for the theorems of Section 6.5. Finally, we draw attention to a result of Nuttall's, possibly generalizable in the future, that for certain functions with branch points connected by branch cuts, the Padé approximants converge in capacity in the z-plane except in a region which minimizes the capacity of the cuts in the z^{-1}-plane [Nuttall, 1977; Nuttall and Singh, 1977]. This conjecture should be contrasted with the conjecture of Baker, Gammel, and Wills.

6.7 The Padé Conjecture

A conjecture of Baker, Gammel, and Wills, slightly rephrased, is popularly known as the Padé conjecture. The conjecture, which concerns convergence of diagonal Padé approximants to functions analytic in a disk, gave great impetus to the search for convergence theorems for the diagonal sequence, and also led to confidence in the usefulness of the diagonal sequence.

Before commencing the statement of the conjecture, we present Gammel's counterexample [Baker, 1973b] which shows why the conjecture takes its form and is not stronger.

GAMMEL'S COUNTEREXAMPLE. Let

$$f(z) = \sum_{n=0}^{\infty} c_n z^n = 1 + \sum_{k=1}^{\infty} \alpha_k \left\{ \sum_{n=n_k}^{2n_k} \left(\frac{z}{z_k} \right)^n \right\}$$

$$= \underbrace{1 + \alpha_1 \left(\frac{z}{z_1} \right) + \alpha_1 \left(\frac{z}{z_1} \right)^2}_{k=1} \underbrace{+ \alpha_2 \left(\frac{z}{z_2} \right)^3 + \alpha_2 \left(\frac{z}{z_2} \right)^4 + \cdots + \cdots}_{k=2},$$

where the indices n_k are defined by $n_1 = 1$, $n_{k+1} = 2n_k + 1$. Let $c_n = \alpha_k z_k^{-n}$ if n is such that $n_k \leq n < n_{k+1}$. The coefficients $\{\alpha_k\}$ are specified by

$$\alpha_k = \frac{1}{(2k)!} \min\left(|z_k|^{n_k}, |z_k|^{2n_k} \right).$$

This choice of α_k ensures that

$$|c_n| \leq \frac{1}{n!},$$

so that $f(z)$ is holomorphic (by the comparison test). The function may also be expressed by

$$f(z) = 1 + \sum_{k=1}^{\infty} \alpha_k \frac{\left(\dfrac{z}{z_k} \right)^{n_k} - \left(\dfrac{z}{z_k} \right)^{2n_k+1}}{1 - \dfrac{z}{z_k}}. \tag{7.1}$$

Inspection of (7.1) and the $[L/1]$ sequence calculated by the accuracy through order method shows that

$$[n_k/1] - [n_k + 1/1] = [n_k + 2/1] = \cdots = [2n_k - 1/1],$$

revealing a block of length n_k. This result implies that

$$[n_k/1] = [n_k/n_k]$$

(as is obvious by accuracy-through-order criterion anyway), and so $[n_k/n_k]$ has a pole at $z = z_k$. By selecting the sequence $\{z_k, k = 1, 2, \ldots\}$ to be a set of points dense in the plane, and allowing such repetition as is necessary, we construct an analytic function with a subsequence of diagonal Padé approximants (in fact $[2^k - 1/2^k - 1]$ Padé approximants) which diverges everywhere in the plane. In fact, a host of Gammel counterexamples can be constructed with Padé approximants which diverge in any desired region of the z-plane.

The only redeeming features of Gammel's counterexample are that the area of bad approximation at the poles of the $[n_\lambda/n_\lambda]$ Padé approximants tends to zero rapidly with increasing n, and that Baker has shown that another subsequence of diagonal approximants converges pointwise [Baker, 1973b].

Consideration of the implications of the counterexample shows why the Padé conjecture which follows is widely believed to be both true and as strong a result as possible.

CONJECTURE [Baker et al., 1961]. *Let $f(z)$ be analytic in $|z|<R$ except for m poles at $z_1, z_2, \ldots, z_m$ with $0<|z_1|\leqslant|z_2|\leqslant\cdots\leqslant|z_m|<R$, and except for one point z_0 on the boundary $|z|=R$. Further, given any $\varepsilon>0$, there must exist a neighborhood $|z-z_0|<\delta$ in which $|f(z)-f(z_0)|<\varepsilon$, provided $|z|\leqslant R$, which means that $f(z)$ is continuous at z_0 within the circle. Then at least a subsequence of $[M/M]$ Padé approximants converges uniformly to $f(z)$ on any compact subset of*

$$\mathcal{D}=\{z, |z|\leqslant R, z\neq z_i, i=0,1,2,\ldots,m\}.$$

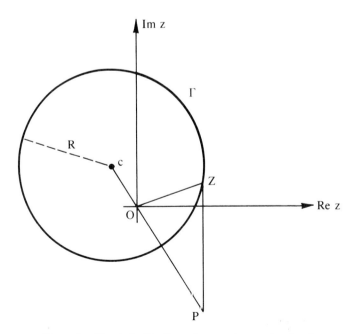

Figure 1.   The circle Γ in the z-plane, center c and radius R.

It is regrettable that no proof yet exists. The importance of this conjecture depends on the homographic invariance property of diagonal Padé approximants. The domain of pointwise convergence of a subsequence may be substantially extended by a development of the Padé conjecture.

QUASITHEOREM [Baker et al., 1961]. *Let $f(z)$ be analytic at $z=0$, and let $\mathcal{D}$ be the union of all circles containing the origin in which $f(z)$ is meromorphic. Then at least a subsequence of $[M/M]$ Padé approximants converges to $f(z)$ pointwise on any compact subset of $\mathcal{D}$ which does not contain the poles of $f(z)$.*

"*Proof.*".   Any point $z \in \mathcal{D}$ lies in the interior of a circle Γ with center c and radius R containing the origin O, as shown in Figure 1. Consider the mapping

$$w = \frac{|R/c|z}{z + c\{|R/c|^2 - 1\}},$$ (7.2)

and let $f(z) = g(w)$. The circle $|w| = 1$ is the image of

$$|z| = |c/R||z + c\{|R/c|^2 - 1\}|.$$ (7.3)

This is a circle for which the origin O is an interior point (if $|c| < R$), and O

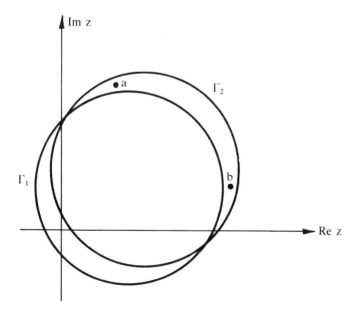

Figure 2.   The complex z-plane, showing the circles Γ_1 and Γ_2.

and P are inverse points provided $z_P = -c\{|R/c|^2 - 1\}$. Γ is given parametrically by $OZ = |c/R| \cdot PZ$. The result "follows" from Theorem 1.5.2.

This result justifies (7.3) and explains the choice (7.2). If $f(z)$ is meromorphic in Γ, then $g(w)$ is meromorphic in $|w| \leq 1$ and the Padé conjecture asserts that a subsequence of the diagonal sequence of Padé approximants converges to $g(w)$ in $|w| \leq 1$ (except at the poles). This quasitheorem is reinterpreted as asserting that the same subsequence of diagonal approximants to $f(z)$ converges at $z \in \Gamma$.

APPLICATION.   Consider the function $f(z) = (z-a)^{3/2}(z-b)^{-3/2}$, where a, b are arbitrary (nonzero) points in the complex plane. The Padé conjecture asserts the convergence, in Γ_1 of Figure 2, of a subsequence. In this example, $z = \infty$ is a regular point, and so convergence at any point outside Γ_2 of a subsequence is asserted.

This example strongly suggests that the poles of the diagonal Padé approximants to $f(z)$ lie on the arc of a circle through O, a, and b. In this case, the mapping $t = z(a-b)/[(z-a)b]$ maps $z=a$ to $t=\infty$ and $z=b$ to $t=-1$. In fact $f(z) = h(t)$ is a Stieltjes series in t, proving that the poles of the diagonal approximants lie on the arc.

An alternative view is that the mapping $u = z^{-1}$ maps the points $z=a$ and $z=b$ to $u=a^{-1}$ and $u=b^{-1}$. The exceptional set for diagonal Padé approximants to $f(z) = F(u)$ in the u-variable minimizes the capacity of the cut, and therefore is the straight line from a^{-1} to b^{-1}. This line is the image of the arc of the circle through $z=0$, $z=a$, and $z=b$ from a to b.

The status of the Padé conjecture and its relation to open problems in rational approximation is surveyed by Walsh [1970].

APPENDIX

A FORTRAN FUNCTION

1. *Specification.* PADE(C,IC,A,IA,B,IB,L,M,X,W,W1,W2,IK,IW) is a FORTRAN IV FUNCTION for calculating Padé approximants.

2. *Method.* The method is based on the solution of the linear equations (1.1.6, 7). A successful FUNCTION call causes the numerator and denominator coefficients of the Padé approximant to be calculated and returns the value of the approximant at some prespecified point X as the value of PADE. If the FUNCTION call is not successful, the approximant has failed a simple empirical degeneracy test which is built into the Gauss–Jordan elimination. The method is designed for data coefficients with full single precision accuracy. Higher precision data coefficients are often warranted: in this case, the code must be modified correspondingly.

3. *Parameters.*

C   a REAL ARRAY of dimension IC. On entry, C contains the given series coefficients and is unchanged on exit.

A   a REAL ARRAY of dimension IA. On exit, $A(1),A(2),\ldots,A(L+1)$ contain the numerator parameters $a_0, a_1,\ldots, a_L$ respectively.

B   a REAL ARRAY of dimension IB. On exit, $B(1),B(2),\ldots,B(M+1)$ contain the numerator parameters $b_0, b_1,\ldots, b_M$ respectively.

L   an INTEGER specifying the degree of the numerator.

M   an INTEGER specifying the degree of the denominator.

W   a two-dimensional DOUBLE PRECISION REAL ARRAY of dimensions (IW,IW) which must be provided for workspace.

W1,W2   two DOUBLE PRECISION REAL ARRAYs of dimension IW which must be provided for workspace.

IK   an INTEGER ARRAY of dimension IW which must be provided for workspace.

IC,IA,IB,IW   INTEGERs specifying various array dimensions, to be set on entry and unchanged on exit.

X   X is the REAL variable of the given power series which must be set with some value on entry. X is unchanged on exit.

4. *Error indicators.* If the integers IC, IA, IB, and IW are inconsistently chosen (we require that $L \geqslant 0$, $M \geqslant 0$, $IW \geqslant M$, $IB > M$, $IA > L$ and $IC > L+M$), or the approximant is degenerate, an error message is printed, and control is returned to the calling program. If the approximant

ENCYCLOPEDIA OF MATHEMATICS and Its Applications, Gian-Carlo Rota (ed.). Vol. 13: George A. Baker, Jr., and Peter R. Graves-Morris, Padé Approximants: Basic Theory, Part I   ISBN 0-201-13512-4

is evaluated at a pole, an error message is printed. The arrays A, B are set correctly, and the artificial value PADE( )=0 is returned.

5. *Example.* The following program calculates the [4/4] Padé approximant of exp(x) to estimate e. ($e=2.71828182\dots$.) Channel 7 is used for input and channel 2 is used for output.

```
      PROGRAM BOOKPROG
      DIMENSION CC(10),AA(5),BB(5),WW(10,10),WX(10),WY(10),IK(10)
      DOUBLE PRECISION WW,WX,WY
      CC(1)=1.0
      DO 1 I=1,8
1     CC(I+1)=CC(I)/I
2     FORMAT(1X,33HTHE GIVEN SERIES COEFFICIENTS ARE/9F13.8)
      WRITE(2,2) (CC(I),I=1,9)
3     FORMAT (2I3,F13.5)
      READ (7,3) L,M,X
      LP1=L+1
      MP1=M+1
4     FORMAT(1X,//11H FORM THE [,I4,1H/,I4,13H] PADE AT X =,F13.8//)
      WRITE (2,4) L,M,X
      Y=PADE(CC,10,AA,5,BB,5,L,M,X,WW,WX,WY,IK,10)
5     FORMAT(1X,30HTHE NUMERATOR COEFFICIENTS ARE//9F13.8)
      WRITE(2,5) (AA(I),I=1,LP1)
6     FORMAT(1X/33H THE DENOMINATOR COEFFICIENTS ARE//9F13.8)
      WRITE(2,6) (BB(I),I=1,MP1)
7     FORMAT(1X/42H THE VALUE OF THE PADE APPROXIMANT AT X IS,F13.8)
      WRITE(2,7) Y
      STOP
      END


      FUNCTION PADE (C,IC,A,IA,B,IB,L,M,X,W,W1,W2,IK,IW)
C
C     THE REAL FUNCTION PADE RETURNS THE VALUE OF THE [L/M] PADE
C     APPROXIMANT EVALUATED AT X. THE PADE NUMERATOR COEFFICIENTS
C     ARE STORED IN THE FIRST L+1 LOCATIONS OF THE ARRAY A.
C     THE DENOMINATOR COEFFICIENTS ARE STORED IN THE FIRST M+1
C     LOCATIONS OF THE ARRAY B.
C     DOUBLE PRECISION ARRAYS W(IW,IW), W1(IW), W2(IW) AND AN
C     INTEGER ARRAY IK(IW) ARE REQUIRED FOR WORKSPACE. THE STRICT
C     INEQUALITIES IW>0 AND IW .GE. M, IB>M, IA>L, IC>L+M ARE
C     NECESSARY TO FULFIL THE STORAGE REQUIREMENTS AT ENTRY.
C
      DIMENSION C(IC),B(IB),A(IA),W(IW,IW),W1(IW),W2(IW),IK(IW)
      DOUBLE PRECISION W,W1,W2
      DOUBLE PRECISION ONE,OUGHT,DET,T
      DATA ONE,OUGHT/0.1D 01,0.0D 00/
C     NEXT STATEMENT APPLIES TO COMPUTERS WITH 12 FIGURE PRECISION
      EPS=0.1E-10
      LP1=L+1
      B(1)=1.0
      IF(L) 40,1,1
1     IF(M) 40,6,2
2     IF(IW-M) 40,3,3
```

290

```
3       IF(IB-M) 40,40,4
4       IF(IA-L) 40,40,5
5       IF(IC-L-M) 40,40,8
6       DO 7 I=1,LP1
7       A(I)=C(I)
        D=1.0
        GO TO 35
8       DO 13 I=1,M
        DO 12 J=1,M
9       IF (L+I-J) 11,10,10
10      I1=L+I-J+1
        W(I,J)=C(I1)
        GO TO 12
11      W(I,J)=OUGHT
12      CONTINUE
        I1=L+I+1
        W1(I)=-C(I1)
13      CONTINUE
C       SOLVE DENOMINATOR EQUATIONS USING GAUSS JORDAN ELIMINATION
C       WITH FULL PIVOTING AND DOUBLE PRECISION.
        DET=ONE
        DO 14 J=1,M
14      IK(J)=0
        DO 30 I=1,M
        T=OUGHT
        DO 19 J=1,M
        IF(IK(J)-1) 15,19,15
15      DO 18 K=1,M
        IF(IK(K)-1) 16,18,30
16      IF(DABS(T)-DABS(W(J,K))) 17,17,18
17      IROW-J
        ICOL=K
        T=W(J,K)
18      CONTINUE
19      CONTINUE
        IK(ICOL)=IK(ICOL)+1
        IF(IROW-ICOL) 20,22,20
20      DET--DET
        DO 21 N=1,M
        T=W(IROW,N)
        W(IROW,N)=W(ICOL,N)
21      W(ICOL,N)=T
        T=W1(IROW)
        W1(IROW)=W1(ICOL)
        W1(ICOL)=T
22      W2(I)=W(ICOL,ICOL)
        IM1=I-1
        IF(I.EQ.1) GO TO 25
        T=DEXP(DLOG(DABS(DET))/DBLE(FLOAT(IM1)))
        IF(DABS(W2(I))-T*DBLE(FLOAT(I))*EPS) 23,25,25
23      DET=OUGHT
```

```
24      FORMAT(1X,5HTHE [,I4,1H/,I4,
       *45H] PADE APPROXIMANT APPARENTLY IS REDUCIBLE OR/
       *49H DOES NOT EXIST. IF THIS IS SO, TRY USING A LOWER/
       *50H ORDER APPROXIMANT. OTHERWISE TRY HIGHER PRECISION/
       *41H THROUGHOUT OR THE NAG LIBRARY ALGORITHM.).
        WRITE(2,24) L,M
        RETURN
25      DET=DET*W2(I)
        W(ICOL,ICOL)=ONE
        DO 26 N=1,M
26      W(ICOL,N)=W(ICOL,N)/W2(I)
        W1(ICOL)=W1(ICOL)/W2(I)
        DO 29 LI=1,M
        IF(LI-ICOL) 27,29,27
27      T=W(LI,ICOL)
        W(LI,ICOL)=OUGHT
        DO 28 N=1,M
28      W(LI,N)=W(LI,N)-W(ICOL,N)*T
        W1(LI)=W1(LI)-W1(ICOL)*T
29      CONTINUE
30      CONTINUE
C       EVALUATE DENOMINATOR
        DO 31 I=1,M
31      B(I+1)=W1(I)
        D=B(M+1)
        DO 32 I=1,M
        I1=M+1-I
32      D=X*D+B(I1)
C       EVALUATE NUMERATOR
        DO 34 I=1,LP1
        K=M+1
        IF(K.GT.I) K=I
        Y=0.0
        DO 33 J=1,K
        I1=I-J+1
33      Y=Y+B(J)*C(I1)
        A(I)=Y
34      CONTINUE
        Y=ABS(D)/M
        IF(Y.LT.EPS) GO TO 42
35      IF(L) 40,36,37
36      PADE=C(1)/D
        RETURN
37      Y=A(L+1)
        DO 38 I=1,L
        I1=L-I+1
38      Y=Y*X+A(I1)
        PADE=Y/D
        RETURN
39      FORMAT(4H IC=,I4,4H IA=,I4,4H IB=,I4,3H L=,I4,3H M-,I4,4H IW=,
       *I4/  43H AND THERE IS AN ERROR AMONG THESE INTEGERS)
```

```
40      WRITE(2,39) IC,IA,IB,L,M,IW
        PADE=0.0
        RETURN
41      FORMAT(1X,E16.8,40H APPARENTLY IS A POLE OF THE APPROXIMANT)
42      WRITE(2,41) X
        PADE=0.0
        RETURN
        END
```

The output from the program BOOKPROG is:

```
THE GIVEN SERIES COEFFICIENTS ARE
   1.00000000   1.00000000   0.50000000   0.16666667   0.04166667 .....
FORM THE [   4/   4] PADE AT X =   1.00000000
THE NUMERATOR COEFFICIENTS ARE
   1.00000000   0.50000000   0.10714286   0.01190476   0.00059524
THE DENOMINATOR COEFFICIENTS ARE
   1.00000000  -0.50000000   0.10714286  -0.01190476   0.00059524
THE VALUE OF THE PADE APPROXIMANT AT X IS   2.71828172
```

Bibliography

References

Abd-Elall, L. F., Delves, L. M., and Reid, J. K. (1970), "A numerical method for locating the zeroes and poles of a meromorphic function," in P. Rabinowitz, (ed.), *Numerical Methods for Nonlinear Algebraic Equations*, Gordon and Breach, London, pp. 47–59.

Abramowitz, M. and Stegun, I. (1964), *Handbook of Mathematical Functions*, Dover.

Aitken, A. C. (1926), "On Bernoulli's numerical solution of algebraic equations," *Proc. Roy. Soc. Edin.*, **46**, 289–305.

Aitken, A. C. (1964), *Determinants and Matrices*, Oliver and Boyd.

Akhiezer, N. I. (1965), *The Classical Moment Problem*, Oliver and Boyd.

Alabiso, C. and Butera, P. (1975), "*N*-variable rational approximants and method of moments," *J. Math. Phys.*, **16**, 840–841.

Alabiso, C., Butera, P., and Prosperi, G. M. (1970), "Resolvent operator, Padé approximation and bound states in potential scattering," *Lett. Nuovo. Cim.*, **3**, 831–839.

Alabiso, C., Butera, P., and Prosperi, G. M. (1971), "Variational principles and Padé approximants. Bound states in potential theory," *Nucl. Phys. B*, **31**, 141–162.

Alabiso, C., Butera, P., and Prosperi, G. M. (1972a), "Variational principles and Padé approximants. Resonances and phase shifts in potential scattering," *Nucl. Phys. B*, **42**, 493–517.

Alabiso, C., Butera, P. and Prosperi, G. M. (1972b), "Variational Padé solution of the Bethe-Salpeter equation," *Nucl. Phys. B*, **46**, 593–614.

Alder, K., Trautmann, D., and Viollier, R. D. (1973), "Anwendung der Padé Approximation in der Kernphysik," *Z. für Naturforschung*, **28**, 321–331.

Allen, G. D., Chui, C. K., Madych, W. R., Narcowich, F. J., and Smith, P. W. (1974), "Padé approximation and Gaussian quadrature," *Bull. Austr. Math. Soc.*, **10**, 263–270.

Allen, G. D., Chui, C. K., Madych, W. R., Narcowich, F. J., and Smith, P. W. (1975), "Padé approximants of Stieltjes series," *J. Approx. Theory*, **14**, 302–316.

Allen, G. D. and Narcowich, F. J. (1975), "On representation and approximation of a class of operator-valued analytic functions," *Bull. Am. Math. Soc.*, **81**, 410–412.

Amos, A. T. (1978), "Padé approximants and Rayleigh Schrödinger perturbation theory," *J. Phys. B*, **11**, 2053–2060.

Anderson, G. (1740), Letter from Anderson in Leyden to Jones, in S. J. Rigaud (ed.), *Correspondence of Scientific Men of the Seventeenth Century*, George Olms, Hildesheim (1965), Vol. I, p. 361.

Argyris, J. H., Vaz, L. E., and William, K. J. (1977), "Higher order methods for transient diffusion analysis," *Computer Methods in Appl. Mech. and Engineering*, **12**, 243–278.

294

Arms, R. J. and Edrei, A. (1970), "The Padé tables and continued fractions generated by totally positive sequences," in *Mathematical Essays*, Ohio University Press, Athens, Ohio, pp. 1–21.

Axelsson, O. (1969), "A class of A-stable methods," *B.I.T.*, **9**, 185–199.

Axelsson, O. (1972), "A note on a class of strongly A-stable methods," *B.I.T.*, **12**, 1–4.

Baker, G. A., Jr. (1960), "An implicit, numerical method for solving the n-dimensional heat equation," *Quarterly of Appl. Math.*, **17**, 440–443.

Baker, G. A., Jr. (1961) "Application of the Padé approximant method to the investigation of some magnetic properties of the Ising model," *Phys. Rev.*, **124**, 768–774.

Baker, G. A., Jr. (1965), "The theory and application of the Padé approximant method," *Adv. in Theoretical Phys.*, **1**, 1–58.

Baker, G. A., Jr. (1969), "Best error bounds for Padé approximants to convergent series of Stieltjes," *J. Math. Phys.*, **10**, 814–820.

Baker, G. A., Jr. (1970), "The Padé approximant and some related generalizations", in G. A. Baker, Jr., and J. L. Gammel (eds.), *The Padé Approximant in Theoretical Physics*, Academic Press, New York, pp. 1–39.

Baker, G. A., Jr. (1972), "Converging bounds for the free energy in certain statistical mechanical problems," *J. Math. Phys.*, **13**, 1862–1864.

Baker, G. A., Jr. (1973a), "Recursive calculation of Pade approximants," in P. R. Graves-Morris (ed.), *Padé Approximants and Their Applications*, Academic Press.

Baker, G. A., Jr. (1973b), "Existence and Convergence of Subsequences of Padé approximants," *J. Math. Anal. Appl.*, **43**, 498–528.

Baker, G. A., Jr. (1974), "Certain invariance properties of the Padé approximant," *Rocky Mtn. J. Math.*, **4**, 141–149.

Baker, G. A., Jr. (1975), "Convergence of Padé approximants using the solution of linear functional equations," *J. Math. Phys.*, **16**, 813–822.

Baker, G. A., Jr. (1976), "A theorem on convergence of Padé approximants," *Studies in Applied Math.*, **55**, 107–117.

Baker, G. A., Jr. (1977), "The application of Padé approximants to critical phenomena," in E. B. Saff and R. H. Varga (eds.), *Padé and Rational Approximation*, Academic Press, New York, pp. 323–337.

Baker, G. A., Jr. and Chisholm, J. S. R., "The validity of perturbation series with zero radius of convergence," *J. Math. Phys.*, **7**, 1900–1902.

Baker, G. A., Jr. and Gammel, J. L. (1961), "The Padé approximant," *J. Math. Anal. Appl.*, **2**, 21–30.

Baker, G. A., Jr. and Gammel, J. L. (1971), "Application of the principle of the minimum maximum modulus to generalized moment problems and some remarks on quantum field theory," *J. Math. Anal. Appl.*, **33**, 197–211.

Baker, G. A., Jr., Gammel, J. L., and Wills, J. G. (1961), "An investigation of the applicability of the Padé approximant method," *J. Math. Anal. Appl.*, **2**, 405–418.

Baker, G. A., Jr., Gilbert, H. E., Eve, J., and Rushbrooke, G. S. (1967), "High temperature expansions for the spin-$\frac{1}{2}$ Heisenberg model," *Phys. Rev.*, **164**, 800–817.

Baker, G. A., Jr. and Graves-Morris, P. R. (1977), "Convergence of rows of the Padé table," *J. Math. Anal. Appl.*, **57**, 323–339.

Baker, G. A., Jr. and Gubernatis, J. E. (1980), "An asymptotic, Padé approximant method for Legendre series," in *Proc. 1979 Int. Christoffel Symposium*, ed. P. L. Butzer, Birkhäuser Verlag, (submitted).

Baker, G. A., Jr. and Hunter, D. L. (1973), "Methods of series analysis II: Generalized and extended methods with application to the Ising model," *Phys. Rev. B* **7**, 3377–3392.

Baker, G. A., Jr. and Kincaid, J. M. (1979), "The continuous spin Ising model and $\lambda : \phi^4 :_d$ field theory," *Phys. Rev. Lett.*, **42**, 1431–1434; *Errata*, **44**, 434 (1980).

Baker, G. A., Jr. and Kincaid, J. M. (1980), "The continuous-spin Ising model, $g_0 : \phi^4 :_d$ field theory and the renormalization group," *J. Stat. Phys.*, in press.

Baker, G. A., Jr., Nickel, B. G., Green, M. S., and Meiron, D. I. (1976), "Ising model critical indices in three dimensions from the Callan–Symanzik equation," *Phys. Rev. Lett.*, **36**, 1351–1354.

Baker, G. A., Jr. and Moussa, P. (1978), "Relation in the Ising model of the Lee-Yang branch point and critical behaviour," *J. Appl. Phys.*, **49**, 1360–1362.

Baker, G. A., Jr. and Oliphant, T. A. (1960), "An implicit, numerical method for solving the two-dimensional heat equation," *Quarterly of Appl. Math.*, **17**, 361–373.

Bareiss, E. H. (1969), "Numerical solution of linear equations with Toeplitz matrices," *Num. Math.*, **13**, 404–424.

Barnsley, M. F. (1973), "The bounding properties of the multipoint Padé approximant to a series of Stieltjes on the real line," *J. Math. Phys.*, **14**, 299–313.

Barnsley, M. F. (1976), "Padé approximant bounds for the difference of two series of Stieltjes," *J. Math. Phys.*, **17**, 559–565.

Barnsley, M. F. and Baker, G. A., Jr. (1976), "Bivariational bounds in a complex Hilbert space, and correction terms," *J. Math. Phys.*, **17**, 1019–1027.

Barnsley, M. F. and Bessis, D. (1979), "Constructive methods based on analytic characterizations and their applications to non-linear elliptic and parabolic differential equations," *J. Math. Phys.*, **20**, 1135–1145.

Barnsley, M. F., Bessis, D., and Moussa, P. (1979), "The Diophantine moment problem and the analytic structure in the activity of the ferromagnetic Ising model," *J. Math. Phys.*, **20**, 535–546.

Barnsley, M. F. and Robinson, P. D. (1974a), "Pade approximant bounds and approximate solutions for Kirkwood Riseman integral equations," *J. Inst. Math. Appl.*, **14**, 251–285.

Barnsley, M. F. and Robinson, P. D. (1974b), "Bivariational Bounds," *Proc. Roy. Soc. London A* **338**, 527–533.

Barnsley, M. F. and Robinson, P. D. (1974c), "Dual variational principles and Padé-type approximants," *J. Inst. Math. Appl.*, **14**, 229–249.

Barnsley, M. F. and Robinson, P. D. (1978), "Rational approximant bounds for a class of two variable Stieltjes functions," *SIAM J. Math. Anal.*, **9**, 272–290.

Basdevant, J.-L. (1969), "Padé approximants," in M. Nikolic (ed.), *Methods in Subnuclear Physics*, Gordon and Breach, New York, pp. 1–40.

Basdevant, J.-L. (1972), "Padé approximants in strong interaction physics," in D. Bessis (ed.), *Cargèse Lectures in Physics*, vol. 5, Gordon and Breach, New York, pp. 431–459.

Basdevant, J.-L. (1973), "Strong interaction physics and the Padé approximation in quantum field theory," in P. R. Graves-Morris (ed.), *Padé Approximants*, Inst. of Physics, Bristol, pp. 77–100.

Basdevant, J.-L., Bessis, D., and Zinn-Justin, J. (1968), "Padé approximants in strong interactions. Two body pion and Kaon systems," *Phys. Lett.*, **27B**, 230–233.

Basdevant, J.-L., Bessis, D., and Zinn-Justin, J. (1969), "Padé approximants in strong interactions. Two body pion and Kaon systems," *Nuovo Cim.*, **60A**, 185–238.

Basdevant, J.-L. and Lee, B. W. (1969a), "Padé approximants in the σ-model with unitary $\pi-\pi$ amplitudes with the current algebra constraints," *Phys. Lett.*, **29B**, 437–441.

Basdevant, J.-L. and Lee, B. W. (1969b), "Padé approximants and bound states: exponential potential," *Nucl. Phys. B*, **13**, 182–188.

Bauer, F. L. (1959), "The quotient-difference and epsilon algorithm," in R. Langer (ed.), *Numerical Approximation*, Univ. of Wisconsin Press, Madison, pp. 361–370.

Bauer, F. L., Rutishauser, H., and Stiefel, E. (1963), "New aspects in numerical quadrature," in N. C. Metropolis, A. H. Traub, J. Todd, and C. B. Tompkins (eds.), *Proc. of 15th Symposium in Applied Mathematics*, Amer. Math. Soc. Press, Vol. 15, pp. 199–218.

Bauhoff, W. (1977), "Padé approximants in the Wick–Cutkosky model," *J. Phys. A*, **10**, 129–134.

Beardon, A. F. (1968a), "The convergence of Padé approximants," *J. Math. Anal. Appl.*, **21**, 344–346.

Beardon, A. F. (1968b), "On the location of poles of Padé approximants," *J. Math. Anal. Appl.*, **21**, 469–474.

Bell, G. I. and Glasstone, S. (1970), "Nuclear Reactor Theory," van Nostrand-Reinhold, N.Y.

Bender, C. and Wu, T. T. (1968), "Analytic structure of energy levels in a field theory model," *Phys. Rev. Lett.*, **21**, 406–409.

Benofy, L. P. and Gammel, J. L. (1977), "Variational principles and matrix Padé approximants," in *Padé and Rational Approximation*, eds. E. B. Saff, and R. H. Varga, Academic Press, N.Y., 339–356.

Benofy, L. P., Gammel, J. L., and Mery, P. (1976), "The off-shell momentum as a variational parameter in calculations of matrix Padé approximants in potential scattering," *Phys. Rev. D*, **13**, 3111.

Bernstein, S. (1928), "Sur les fonctions absolument monotones," *Acta Math.*, **52**, 1–66.

Bessis, D., ed. (1972), *Cargèse Lectures in Physics*, Vol. 5, Gordon and Breach, New York.

Bessis, D. (1973), "Topics in the theory of Padé approximants," ed. P. R. Graves-Morris, Inst. of Phys., Bristol, 19–44.

Bessis, D. (1976), "Construction of variational bounds for the *N*-body eigenstate problem by the method of Padé approximants," in *Padé Approximants Method and its Application to Mechanics*, ed. H. Cabannes, *Springer Lecture Notes in Phys. no. 37*, Berlin, 17–31.

Bessis, D. (1979), "A new method for the combinatorics of the topological expansion," *Comm. Math. Phys.*, **69**, 147–163.

Bessis, D., Drouffe, J. M., and Moussa, P. (1976), "Positivity constraints for the Ising ferromagnetic model," *J. Phys. A*, **9**, 2105–2124.

Bessis, D., Epele, L., and Villani, M. (1974), "Summation of regularized perturbation expansions for singular interactions," *J. Math. Phys.*, **15**, 2071–2078.

Bessis, D., Mery, P., and Turchetti, G. (1974), "Angular momentum analysis of four nucleon Green's function," *Phys. Rev. D*, **10**, 1992–2009.

Bessis, D., Mery, P., and Turchetti, G. (1977), "Variational bounds from matrix Padé approximants in potential scattering," *Phys. Rev. D*, **15**, 2345–2353.

Bessis, D., Moussa, P., and Villani, M. (1975), "Monotonic converging variational approximants to the functional integrals in quantum statistical mechanics," *J. Math. Phys.*, **16**, 2318–2325.

Bessis, D. and Pusterla, M. (1967), "Unitary Padé approximants for the *S*-matrix in strong coupling field theory and application to the calculation of ρ and f^0 Regge trajectories," *Phys. Lett.*, **25B**, 279–281.

Bessis, D. and Pusterla, M. (1968), "Unitary Padé approximants in strong coupling field theory and application to the ρ and f^0 meson trajectories," *Nuovo Cim.*, **54A**, 243–294.

Bessis, D. and Turchetti, G. (1972), "Renormalization of the σ model through Ward identities," in *Cargèse Lectures in Physics*, Vol. 5, ed. D. Bessis, Gordon and Breach, New York, 119–178.

Bessis, D. and Turchetti, G. (1977), "Variational matrix Padé approximants in potential scattering and low energy Lagrangian field theory," *Nucl. Phys. B*, **123**, 173–188.

Bessis, D. and Villani, M. (1975), "Perturbative-variational approximations to the spectral properties of semi-bounded Hilbert space operators, based on the moment problem with finite or diverging moments. Application to quantum mechanical systems," *J. Math. Phys.*, **16**, 462–474.

Biswas, S. N. and Vidhani, T. (1973), "Solution of Schrödinger equation with continued fractions," *J. Phys. A*, **6**, 468–477.

Blanch, G. (1964), "The numerical evaluation of continued fractions," *SIAM Rev.*, **6**, 383–421.

Bombelli, R. (1572), *L'Algebra*, Venezia.

Boas, R. P. (1954), *Entire Functions*, Academic Press, New York.

Brent, R. P., Gustavson, F. G. and Yun, D. Y. Y. (1980), "Fast solution of Toeplitz systems of equations and computation of Padé approximants," *J. Algorithms*, **1**, 259–295.

Brezin, E., Le Guillou, J. C., and Zinn-Justin, J. (1977), "Perturbation theory at large order. I The ϕ^{2^N} interaction," *Phys. Rev. D*, **15**, 1544–1558.

Brezinski, C. (1971), "Accélération de suites à convergence logarithmique," *Comptes Rendus*, **273A**, 727–730.

Brezinski, C. (1972), "Conditions d'application et de convergence de procédés d'extrapolation," *Num. Math.*, **20**, 64–79.

Brezinski, C. (1974), "Review of acceleration methods," *Rendiconti di Mat.*, **7**, 303–316.

Brezinski, C. (1976), "Computation of Padé approximants and continued fractions," *J. Comp. Appl. Math.*, **2**, 113–123.

Brezinski, C. (1977), *Accélération de la Convergence en Analyse Numérique*, Springer Lecture Notes in Mathematics, No. 584, Berlin.

Brezinski, C. (1978a), "Convergence acceleration of some sequences by the ε-algorithm," *Num. Math.*, **29**, 173–177.

Brezinski, C. (1978b), "Padé type approximants for double power series," *J. Indian Math. Soc.*, **42**, 267–282.

Brezinski, C. (1979), "Rational approximation to formal power series," *J. Approx. Theory*, **25**, 295–317.

Brezinski, C. (1980), "A general extrapolation algorithm," *Num. Math.*, **35**, 175–187.

Brezinski, C. and Rieu, A. C. (1974), "Solution of systems of equations using the ε-algorithm, and application to boundary value problems," *Math. Comp.*, **28**, 731–734.

Bulirsch, R. and Stoer, J. (1964), "Fehler abschätzungen und Extrapolation mit rationalen Functionen bei Verfahren vom Richardson-Typus," *Num. Math.*, **6**, 413–427.

Bultheel, A. (1979), "Recurisive algorithms for the Padé table: two approaches," in L. Wuytack (ed.); *Padé Approximation and Its Applications*, Springer Lecture Notes in Mathematics, No. 765, pp. 211–230.

Bultheel, A. (1980a), "Recursive algorithms for the matrix Padé problem," *Math. Comp.*, **35**, 875–892.

Bultheel, A. (1980b), "Division algorithms for continued fractions and the Padé table," *J. Comp. Appld. Math.*, **6**, 259–266.

Bultheel, A. and Wuytack, L., (1981), "Stability of numerical methods for computing Padé approximants", in *Approximation Theory III*, ed. E. W. Cheney, Academic Press, New York.

Bus, J. C. P. and Dekker, T. J. (1975), Two efficient algorithms with guaranteed convergence for finding a zero of a function," *A. C. M. Trans. on Math. Software*, **1**, 330–345.

Butcher, J. C. (1963), "Coefficients for the study of Runge-Kutta integration processes," *J. Australian Math. Soc.*, **3**, 185–201.

Butcher, J. C. (1964), "Implicit Runge-Kutta Processes," *Math. Comp.*, **18**, 50–64.

Carleman, T. (1926), *Les Fonctions Quasi-Analytiques*, Gauthier Villars, Paris; translated by J. L. Gammel, Los Alamos Tech. report 4702, (1971).

Carroll, A., Baker, G. A., Jr., and Gammel, J. L. (1977), "A comment on a paper of Hamer applying Padé approximants to a 1 + 1 dimensional lattice model of Q.C.D., *Nucl. Phys. B*, **129**, 361–364.

Carslaw, H. S. and Jaeger, J. C. (1959), *Conduction of Heat in Solids*, Oxford University Press.

Cartan, H. (1928), "Sur les systèmes de fonctions holomorphes à variétés linéaires et leurs applications," *Ann. Sci. Ecole. Norm. Sup. (3)*, **45**, 255–346.

Caser, S., Piquet, C., and Vermeulen, J. L. (1969), "Padé approximants for a Yukawa potential," *Nucl. Phys. B*, **14**, 119–132.

Cavendish, J. C., Culham, W. E., and Varga, R. S. (1972), "A comparison of Crank-Nicholson and Chebychev rational methods for numerically solving linear parabolic equations," *J. Comp. Phys.*, **10**, 354–368.

Cauchy, M. A.-L. (1821), "Cours d'analyse" de l'Ecole Royale Polytechnique; première partie, L'Imprimerie Royale, Paris.

Char, B. W. (1980), "On Stieltjes continued fraction for the Gamma Function," *Math. Comp.*, **34**, 547–551.

Cheney, E. W. (1966), *Introduction to Approximation Theory*, McGraw-Hill.

Chipman, F. H. (1971), "*A*-stable Runge-Kutta processes," *B.I.T.*, **11**, 384–388.

Chisholm, J. S. R. (1963), "Solution of linear integral equations using Padé approximants," *J. Math. Phys.*, **4**, 1506–1510.

Chisholm, J. S. R. (1966), "Approximation by sequences of Padé approximants in regions of meromorphy," *J. Math. Phys.*, **7**, 39–46.

Chisholm, J. S. R. (1973), "Rational approximants defined from double power series," *Math. Comp.*, **27**, 841–848.

Chisholm, J. S. R. (1977a), "*N*-variable rational approximants," in *Padé and Rational Approximation*, eds. E. B. Saff and R. H. Varga, Academic Press, New York, 23–42.

Chisholm, J. S. R. (1977b), "Multivariate approximants with branch points. I. Diagonal approximants," *Proc. Roy. Soc. Lond. A*, **358**, 351–366.

Chisholm, J. S. R. (1978a), "Multivariate approximants with branch cuts," in *Multivariate Approximation*, ed. D. C. Handscomb, Academic Press, London.

Chisholm, J. S. R. (1978b), "Multivariate approximants with branch points. II. Off-diagonal approximants," *Proc. Roy. Soc. Lond. A*, **362**, 43–56.

Chisholm, J. S. R. and Common, A. K. (1980), "Generalizations of Padé approximation for Chebychev and Fourier series," in *Proc. 1979 Int. Christoffel Symposium*, ed. P. L. Butzer, Birkhäuser Verlag.

Chisholm, J. S. R., Genz, A. C., and Rowlands, G. L. (1972), "Accelerated convergence of sequences of quadrature approximations," *J. Comp. Phys.*, **10**, 284–307.

Chisholm, J. S. R. and Graves-Morris, P. R. (1975), "Generalizations of the theorem of de Montessus to two-variable approximants," *Proc. Roy. Soc. Lond. A*, **342**, 341–372.

Chisholm, J. S. R. and Hughes Jones, R. (1975), "Relative scale covariance of N variable approximants," *Proc. Roy. Soc. Lond. A*, **344**, 465–470.

Chisholm, J. S. R. and McEwan, J. (1974), "Rational approximants defined from power series in N-variables," *Proc. Roy. Soc. Lond. A*, **336**, 421–452.

Chisholm, J. S. R. and Roberts, D. E. (1976), "Rotationally covariant approximants derived from double power series," *Proc. Roy. Soc. Lond. A*, **351**, 585–591.

Christoffel, E. B. (1877), "Sur une classe particulière de fonctions entières et de fractions continues," *Ann. di Mat.*, **8**, 1–10.

Chui, C. K., Shisha, O., and Smith, P. W. (1974), "Padé approximants as limits of best rational approximants," *J. Approx. Theory*, **12**, 201–204.

Chui, C. K., Shisha, O., and Smith, P. W. (1975), "Best local approximation," *J. Approx. Theory*, **15**, 371–381.

Chui, C. K., Smith, P. W., and Ward, J. D. (1978), "Best L_2 local approximation," *J. Approx. Theory*, **22**, 254–261.

Claessens, G. (1976), "A new algorithm for osculatory rational interpolation," *Num. Math.*, **27**, 77–83.

Claessens, G. (1978a), "On the Newton-Padé approximation problem," *J. Approx. Theory*, **22**, 150–160.

Claessens, G. (1978b), "On the structure of the Newton-Padé table," *J. Approx. Theory*, **22**, 304–319.

Claessens, G. (1978c), "Generalized ε-algorithm for rational interpolation," *Num. Math.*, **29**, 227–231.

Claessens, G. and Wuytack, L. (1979), "On the computation of non-normal Padé approximants," *J. Comp. Appl. Math.*, **5**, 283–289.

Clenshaw, C. W. and Lord, K. (1974), "Rational approximations from Chebychev series," in *Studies in Numerical Analysis*, ed. B. K. P. Scaife, Academic Press, London, 95–113.

Cody, W. J., Meinardus, G., and Varga, R. S. (1969), "Chebychev rational approximations to e^{-x} in $[0,\infty)$ and applications in heat conduction equations," *J. Approx. Theory*, **2**, 50–65.

Common, A. K. (1968), "Padé approximants and bounds to series of Stieltjes," *J. Math. Phys.*, **9**, 32–38.

Common, A. K. (1969a), "Properties of Legendre expansions related to Stieltjes series and applications to π–π scattering," *Nuovo Cim.*, **63A**, 863–891.

Common, A. K. (1969b), "Properties of generalisations to Padé approximants," *J. Math. Phys.*, **10**, 1875–1880.

Common, A. K. (1970), "Some consequences of relations of π–π partial wave amplitudes to Stieltjes series," *Nuovo Cim.*, **65A**, 581–596.

Common, A. K. (1979), "Calculation of Yukawa scattering amplitude and impact parameter amplitudes using Legendre–Padé approximants," *J. Phys. A*, **12**, 2563–2572.

Common, A. K. and Graves Morris, P. R. (1974), "Some properties of Chisholm approximants," *J. Inst. Math. Appl.*, **13**, 229–232.

Common, A. K. and Stacey, T. W. (1979a), "The convergence of Legendre-Padé approximants to the Coulomb and other scattering amplitudes," *J. Phys. A*, **11**, 275–289.

Common, A. K. and Stacey, T. W. (1979b), "Legendre–Padé approximants and their application in potential scattering," *J. Phys. A*, **11**, 259–273.

Common, A. K. and Stacey, T. W. (1979c), "Convergent series of Legendre–Padé approximants to the real and imaginary parts of the scattering amplitudes," *J. Phys. A*, **12**, 1399–1417.

Conn, R. W. (1974), "Higher order variational principles and Padé approximants for linear functionals," *Nucl. Sci. Eng.*, **55**, 468–470.

Copley, L. A., Elias, D. K., and Masson, D. (1968), "Broken symmetry model of meson interactions," *Phys. Rev.*, **173**, 1552–1563.

Copley, L. A. and Masson, D. (1967), "Padé approximant calculation of π–π scattering," *Phys. Rev.*, **164**, 2059–2062.

Copson, E. T. (1948), *An Introduction to the Theory of a Complex Variable*, Oxford University Press.

Corcoran, C. T. and Langhoff, P. W. (1977), "Moment theory approximations for non-negative spectral densities," *J. Math. Phys.*, **18**, 651–657.

Cordellier, F. (1979a), "Sur la régularité des procédés δ^2 d'Aitken et W de Lubkin," in L. Wuytack (ed.), *Padé Approximation and Its Applications*, Springer Lecture Notes in Math., No. 765, Berlin.

Cordellier, F. (1979b), Demonstration algébrique de l' extension de l'identité de Wynn aux tables de Padé non-normales," in L. Wuytack (ed.), *Padé Approximation and Its Applications*, Springer Lecture Notes in Math., No. 765, Berlin.

Courant, R. and Hilbert, D. (1953), *Methods of Mathematical Physics*, Vol. I, Interscience.

Crank, J. (1975), *The Mathematics of Diffusion*, Oxford University Press.

Dekker, T. J. (1969), "Finding a zero by means of successive linear interpolation," in *Constructive Aspects of the Fundamental Theorem of Algebra*, eds. B. Dejon and P. Henrici, J. Wiley, 37–48.

Delves, L. M. and Phillips, A. C. (1969), "Present status of the nuclear three body problem," *Rev. Mod. Phys.*, **41**, 497–530.

Dennis, J. J. and Wall, H. S. (1945), "The limit-circle case for a positive definite J-fraction," *Duke Math. J.*, **12**, 255–273.

Dienes, P. (1957), "The Taylor series," Dover, New York.

Dijkstra, D. (1977), "A continued fraction expansion for a generalization of Dawson's integral," *Math. Comp.*, **31**, 503–510.

Dimock, J. (1974), "Asymptotic perturbation expansion in the $P(\phi)_2$ quantum field theory," *Comm. Math. Phys.*, **35**, 347–356.

Dirac, P. A. M. (1958), "Principles of Quantum Mechanics," Oxford University Press.

Domb, C. (1974), "Ising model," in *Phase Transitions and Critical Phenomena*," vol. 3, eds. C. Domb and M. S. Green, Academic Press, London, 357–485.

Domb, C. and Sykes, M. F. (1961), "Use of series expansions for the Ising model susceptibility and excluded value problem," *J. Math. Phys.*, **2**, 63–67.

Donnelly, J. D. P. (1966), "The Padé table," in *Methods of Numerical Approximation*, ed. D. C. Handscomb, Pergamon Press, Oxford, 125–130.

Drew, D. and Murphy, J. A. (1977), "Branch points, M-fractions and rational approximants generated by linear equations," *J. Inst. Math. Appl.*, **19**, 169–185.

Dyson, F. J. (1952), "Divergence of Perturbation Theory in Quantum Electrodynamics," *Phys. Rev.* **85**, 631–632.

Eckmann, J. P., Magnen, J., and Sénéor, R. (1974), "Decay properties and Borel summability for the Schwinger functions in $P(\phi)_2$ theories," *Comm. Math. Phys.*, **39**, 251–271.

Edrei, A. (1939), "Sur les determinants récurrents et les singularités d'une fonction données par son développement de Taylor," *Compositio Math.*, **7**, 20–88.

Edrei, A. (1953), "Proof of a conjecture of Schoenberg on the generating function of a totally positive sequence," *Can. J. Math.*, **5**, 86–94.

Edrei, A. (1975a), "Convergence of complete Padé tables of trigonometric functions," *J. Approx. Theory*, **15**, 278–293.

Edrei, A. (1975b), "The Padé table of functions having a finite number of essential singularities," *Pacific J. Math.*, **56**, 429–453.

Ehle, B. L. (1968), "High order A-stable methods for numerical solution of systems of differential equations," *B.I.T.*, **8**, 276–278.

Ehle, B. L. (1973), "A-stable methods and Padé approximants to the exponential," *SIAM J. Math. Anal.*, **4**, 671–680.

Ehle, B. L. (1976), "On certain order constrained Chebychev rational approximations," *J. Approx. Theory*, **17**, 297–306.

Erdélyi, A. (1956), *Asymptotic Expansions*, Dover, New York.

Euler, L. (1737), "De fractionibus continuis," *Comm. Acad. Sci. Imper. Petropol.*, **9**.

Fair, W. G. (1964), "Padé approximants to the solution of the Ricatti equation," *Math. Comp.*, **18**, 627–634.

Fairweather, G. (1971), "A survey of discrete Galerkin methods for parabolic equations in one space variable," *Math. Colloq. U.C.T.*, **7**, 43–77.

Feldman, J. S. and Osterwalder, K. (1976), "The Wightman axioms and the mass gap for weakly coupled $(\phi^4)_3$ quantum field theories," *Ann. Phys.*, **97**, 80–135.

Ferer, M., Moore, M. A., and Wortis, M. (1971), "Some critical properties of the nearest-neighbor, classical Heisenberg model for the f.c.c. lattice in finite field for temperatures greater than T_c," *Phys. Rev. B*, **4**, 3954–3963.

Ferrar, W. L. (1938), *Convergence*, Oxford University Press.

Fisher, M. E. (1977), "Series expansion approximants for singular functions of many variables," in *Statistical Mechanics and Statistical Methods in Theory and Application*, ed. Uzi Landman, Plenum Press, 3–31.

Fisher, M. E. and Kerr, R. M. (1977), "Partial differential approximants for multicritical singularities," *Phys. Rev. Lett.*, **39**, 667–670.

Fleischer, J. (1972), "Analytic continuation of scattering amplitudes and Padé approximants," *Nucl. Phys. B*, **37**, 59–76.

Fleischer, J. (1973a), "Nonlinear Padé approximants for Legendre series," *J. Math. Phys.*, **14**, 246–248.

Fleischer, J. (1973b), "Generalizations of Padé approximants," in *Padé Approximants*, ed. P. R. Graves-Morris, Inst. of Phys., Bristol, 126–131.

Fleischer, J., Gammel, J. L., and Menzel, M. T. (1973), "Matrix Padé approximants for the $^1S_0$ and $^3P_0$ partial waves in nucleon-nucleon scattering," *Phys. Rev. D*, **8**, 1545–1552.

Fleischer, J. and Tjon, J. A. (1975), "Bethe-Salpeter equation for $J=0$ nucleon-nucleon scattering with one boson exchange," *Nucl. Phys. B*, **84**, 375–396.

Fleischer, J. and Tjon, J. A. (1980), "Bethe-Salpeter equation for elastic nucleon-nucleon scattering," *Phys. Rev. D*, **21**, 87–94.

Fogli, G. L., Pellicoro, M. F., and Villani, M. (1971), "A summation method for a class of series with divergent terms," *Nuovo Cim.*, **6A**, 79–97.

Fogli, G. L., Pellicoro, M. F., and Villani, M. (1972), "An approach to the radiative corrections in Q.E.D. in the framework of the Padé method," *Nuovo Cim.*, **11A**, 153–177.

Ford, W. B. (1960), *Asymptotic Series and Divergent Series*, Chelsea, New York.

Frank, W. M., Land, D. J., and Spector, R. M. (1971), "Singular potentials," *Rev. Mod. Phys.*, **43**, 36–98.

Fratamico, G., Ortolani, F., and Turchetti, G. (1976) "Exact solutions from the variational [1/1] matrix Padé approximant in potential scattering," *Lett. Nuovo Cim.*, **17**, 582–584.

Frobenius, G. (1881), "Ueber Relationen zwischen den Näherungsbrüchen von Potenzreihen," *J. für Reine und Angewandte Math.*, **90**, 1–17.

Gallucci, M. A. and Jones, W. B. (1976), "Rational approximations corresponding to Newton series (Newton-Padé approximants)," *J. Approx. Theory*, **17**, 366–392.

Gammel, J. L. (1973), "Review of two recent generalizations of the Padé approximant," in *Padé Approximants and their Applications*, ed. P. R. Graves-Morris, Academic Press, London, 3–9.

Gammel, J. L. and McDonald, F. A. (1966), "Applications of the Padé approximant to scattering theory," *Phys. Rev.*, **142**, 1245–1254.

Gammel, J. L., Rousseau, C. C., and Saylor, D. P. (1967), "A Generalization of the Padé, Approximant," *J. Math. Anal. Appl.*, **20**, 416–420.

Gargantini, J. and Henrici, P. (1967), "A continued fraction algorithm for the computation of higher transcendental functions in the complex plane," *Math. Comp.*, **21**, 18–29.

Garibotti, C. R. (1972), "Schwinger variational principle and Padé approximants," *Ann. Phys.*, **71**, 486–496.

Garibotti, C. R. and Grinstein, F. F. (1978a), "Summation of partial waves for long range potentials I," *J. Math. Phys.*, **19**, 821–829.

Garibotti, C. R. and Grinstein, F. F. (1978b), "Punctual Padé approximants as a regularization procedure for divergent and oscillatory partial wave expansion of the scattering amplitude," *J. Math. Phys.*, **19**, 2405–2409.

Garibotti, C. R. and Grinstein, F. F. (1979), "Summation of partial waves for long range potentials II," *J. Math. Phys.*, **20**, 141–147.

Garibotti, C. R., Pellicoro, M. F., and Villani, M. (1970), "Padé method in singular potentials," *Nuovo Cim.*, **66A**, 749–766.

Garibotti, C. R. and Villani, M. (1969a), "Continuation in the coupling constant for the total T and K matrices," *Nuovo Cim.*, **59A**, 107–123.

Garibotti, C. R. and Villani, M. (1969b), "Padé approximant and the Jost function," *Nuovo Cim.*, **61A**, 747–754.

Garnett, J. (1973), *Analytic Capacity and Measure*, Springer Lecture Notes in Mathematics, No. 297.

Garside, G. R., Jarratt, P., and Mack, C. (1968), "A new method for solving polynomial equations," *Comp. J.*, **11**, 87–90.

Gauss, C. F. (1813), "Disquisitiones generales circa seriam infinitam

$$1 + \frac{\alpha\beta}{1\cdot\gamma}x + \frac{\alpha(\alpha+1)\beta(\beta+1)}{1\cdot 2\gamma(\gamma+1)}x\cdot x + \cdots,$$

Commentationes Societatis Regiae Scientiorum Goettingensis Recentiores, **2**.

Gauthier, P. M. (1977), "On the possibility of rational approximation," in *Padé and rational approximation*, eds. E. B. Saff and R. S. Varga, Academic Press, New York, pp. 261–264.

Gautschi, W. (1967), "Computational aspects of three term recurrence relations," *SIAM Review*, **9**, 24–82.

Gear, C. W. (1971), *Numerical Initial Value Problems in Ordinary Differential Equations*, Prentice Hall.

Geddes, K. O. (1979), "Symbolic computation of Padé approximants," *A.C.M. Trans. Math. Software*, **5**, 218–233.

Gekeler, E. (1972), "On the solution of systems of equations by the epsilon algorithm of Wynn," *Math. Comp.*, **26**, 427–436.

Genz, A. C. (1972), "An adaptive multidimensional quadrature procedure," *Comp. Phys. Comm.*, **4**, 11–15.

Genz, A. C. (1973), "The applications of the ε-algorithm to quadrature problems," in *Padé Approximants*, ed. P. R. Graves-Morris, Inst. of Physics, Bristol, 105–116.

Genz, A. C. (1974), "Some extrapolation methods for the numerical calculation of multidimensional integrals," in *Software for Numerical Mathematics*, ed. D. J. Evans, Academic Press, New York, 159–172.

Genz, A. C. (1977), "A non-linear method for the convergence of multidimensional sequences," *J. Comp. Appl. Math.*, **3**, 181–184.

Germain-Bonne, B. (1979), "Ensembles des suites et de procédés liés pour l'accélération de la convergence," in L. Wuytack (ed.) *Padé Approximation and Its Applications*, Springer Lecture Notes in Mathematics, No. 765, pp. 116–134.

Gersten, A., Owen, D. A., Gammel, J. L., Mery, P., and Turchetti, G. (1976), "Matrix Padé approximants and the Bethe-Salpeter equation of the nucleon-nucleon interaction," *Phys. Rev. D*, **13**, 1140–1143.

Gilewicz, J. (1978), *Approximants de Padé*, Springer Lecture Notes in Mathematics, No. 667.

Giraud, B. G. (1978), "Lower bounds to bound state eigenvalues," *Phys. Rev. C*, **17**, 800–809.

Giraud, B. G., Khalil, A. B., and Moussa, P. (1976), "Padé approximants for distorted waves," *Phys. Rev. C*, **14**, 1679–1687.

Giraud, B. G. and Turchetti, G. (1978), "Rigorous lower bounds to the energy levels with finite summations," *Lett. Nuovo Cim.*, **21**, 605–608.

Goldberger, M. L. and Watson, K. M. (1964), *Collision Theory*, J. Wiley, New York.

Goldhammer, P. and Feenberg, E. (1956), "Refinement of the Brillouin Wigner perturbation method," *Phys. Rev.*, **101**, 1233–1234.

Gončar, A. A. and López, Guillermo, L. (1978), *Math. U.S.S.R. Sbornik*, **34**, 449–459.

Gordon, R. G. (1968), "Error bounds in equilibrium statistical mechanics," *J. Math. Phys.*, **9**, 655–663.

Götz, U., Rösel, F., Trautmann, D., and Jochin, H. (1976), "Influence of the Mott–Schwinger interaction on the elastic scattering of protons," *Z. Phys. A*, **278**, 139–143.

Graffi, S., Grecchi, V., and Simon, B. (1970), "Borel summability: application to the anharmonic oscillator," *Phys. Lett.*, **32B**, 631–634.

Graffi, S. and Grecchi, V. (1978), "Borel summability and indeterminacy of the Stieltjes moment problem: application to the anharmonic oscillators," *J. Math. Phys.*, **19**, 1002–1006.

Graffi, S., Grecchi, V., and Turchetti, G. (1971), "Summation methods for the perturbation series of the generalized harmonic oscillator," *Nuovo Cim.*, **4B**, 313–340.

Gragg, W. B. (1965), "On extrapolation algorithms for ordinary initial value problems," *SIAM J. Num. Anal.*, **2**, 384–403.

Gragg, W. B. (1968), "Truncation error bounds for g-fractions," *Num. Math.*, **11**, 370–379.

Gragg, W. B. (1970), "Truncation error bounds for π-fractions," *Bull. Am. Math. Soc.*, **76**, 1091–1094.

Gragg, W. B. (1972), "The Padé table and its relation to certain algorithms of numerical analysis," *SIAM Review*, **14**, 1–62.

Gragg, W. B. (1974), "Matrix interpretations and applications of the continued fraction algorithm," *Rocky Mtn. J. Math.*, **4**, 213–225.

Gragg, W. B. (1977), "Laurent, Fourier and Chebychev Padé tables," in *Padé and Rational Approximation*, eds. E. B. Saff and R. H. Varga, 61–70.

Gragg, W. B. and Johnson, G. D. (1974), *The Laurent Padé Table*, Info. Proc. 74, North Holland, Amsterdam, **3**, 632–637.

Graves-Morris, P. R. (1973), "Padé approximants and potential scattering," in *Padé Approximants*, ed. P. R. Graves-Morris, Inst. of Phys., Bristol, 64–76.

Graves-Morris, P. R. (1975), "Convergence of rows of the Padé table," in H. Cabannes (ed.), *Padé Approximation and Its Application to Mechanics*, Springer Lecture Notes in Physics, No. 47, pp. 55–68.

Graves-Morris, P. R. (1977), "Generalisations of the theorem of de Montessus using Canterbury approximants," in E. B. Saff and R. S. Varga (eds.), *Padé and Rational Approximation*, Academic Press, New York, pp. 73–82.

Graves-Morris, P. R. (1978a), "Padé approximants for integral equations?," *J. Inst. Math. Appl.*, **21**, pp. 375–378.

Graves-Morris, P. R. (1978b), "Applications of matrix Padé approximants in potential theory," *Ann. Phys.*, **114**, 290–295.

Graves-Morris, P. R. (1979), "The numerical calculation of Padé approximants," in L. Wuytack (ed.), *Padé Approximation and Its Applications*, Springer Lecture Notes in Mathematics, No. 765, pp. 231–245.

Graves-Morris, P. R. (1980a), "Practical, reliable, rational interpolation," *J. Inst. Math. Appl.*, **25**, 267–286.

Graves-Morris, P. R. (1980b), "A generalized Q.D. algorithm," *J. Comp. Appl. Math.*, **6**, 247–249.

Graves-Morris, P. R. (1981) "The Convergence of Ray Sequences of Padé Approximants of Stieltjes Series" (preprint).

Graves-Morris, P. R. and Hopkins, T. R. (1981), "Reliable rational interpolation," *Num. Math.*, **36**, 111–128.

Graves-Morris, P. R., Hughes Jones, R., and Makinson, G. J. (1974), "The calculation of some rational approximants in two variables," *J. Inst. Math. Appl.*, **13**, 311–320.

Graves-Morris, P. R. and Rennison, J. F. (1974), "Analyticity in the coupling strength," *J. Math. Phys.*, **15**, 230–233.

Graves-Morris, P. R. and Roberts, D. E. (1975), "Calculation of Canterbury approximants," *Comp. Phys. Comm.*, **10**, 234–244; *Erratum*, (1977), **13**, 72.

Graves-Morris, P. R. and Samwell, C. J. (1975), "Canterbury approximants in potential scattering," *J. Phys. G.*, **1**, 805–814.

Gray, H. L., Atchison, T. A., and McWilliams, G. V. (1971), "Higher order G-transformations," *SIAM J. Num. Anal.*, **8**, 365–381.

Grenander, U. and Szegö, G. (1958), *Toeplitz Forms and Their Applications*, Univ. of California Press, Berkeley.

Grinstein, F. F. (1980), "Summation of partial wave expansions in the scattering by short range potentials," *J. Math. Phys.*, **21**, 112–119.

Guttmann, A. J. (1969), "Determination of critical behaviour in lattice statistics from series expansions III," *J. Phys. C*, **2**, 1900–1907.

Guttman, A. J. (1975a), "On the recurrence relation method of series analysis," *J. Phys. A*, **8**, 1081–1088.

Guttmann, A. J. (1975b), "Derivation of 'mimic expansions' from regular perturbation expansions in fluid mechanics," *J. Inst. Math. Appl.*, **15**, 307–315.

Guttmann, A. J. and Joyce, G. S. (1972), "On a new method of series analysis in lattice statistics," *J. Phys. A*, **5**, L81–L84.

Hadamard, J. (1892), "Essai sur l'étude des fonctions données par leur developpement de Taylor," $2^{\text{ième}}$ partie, *J. de Math.*, **4**, 101–186.

Hadamard, J. ("1968"), *Oeuvres*, Vol. 2, CNRS, Paris, pp. 24–60.

Hamburger, H. (1920), "Ueber eine Erweiterung des Stieltjes'schen Momentenproblems," *Math. Annalen*, **81**, 235–319.

Hamburger, H. (1921), "Ueber eine Erweiterung des Stieltjes'schen Momentenproblems," *Math. Annalen*, **82**, 120–164, 168–187.

Hardy, G. H. (1956), "Divergent Series," Oxford University Press.

Haymaker, R. W. (1968), "Application of analyticity properties to the numerical solution of the Bethe-Salpeter equation," *Phys. Rev.*, **165**, 1790–1802.

Haymaker, R. W. and Schlessinger, L. (1970), "Padé approximants as a computational tool for solving the Schrödinger and Bethe-Salpeter equations," in *The Padé Approximant in Theoretical Physics*, eds. G. A. Baker, Jr. and J. L. Gammel, Academic Press, New York, 257–288.

Henrici, P. (1958), "The quotient difference algorithm," *Nat. Bur. Stand.—Appl. Math Series*, **49**, 23–46.

Henrici, P. (1964), *Elements of Numerical Analysis*, Wiley.

Henrici, P. (1974), *Applied and Computational Complex Analysis*, Vol. I, Wiley.

Henrici, P. and Pfluger, P. (1966), "Truncation error estimates for Stieltjes fractions," *Num. Math.*, **9**, 120–138.

Hildebrand, F. B. (1956), "Introduction to Numerical Analysis," McGraw-Hill.

Hille, E. (1959), *Analytic Function Theory*, Vol. I, Ginn and Co.

Hille, E. (1962), *Analytic Function Theory*, Vol. II, Ginn and Co.

Hillion, P. (1977a), "Remarks on rational approximation of multiple power series," *J. Inst. Math. Appl.*, **19**, 281–293.

Hillion, P. (1977b), "Approximating functions with a given singularity," *J. Math. Phys.*, **18**, 465–470.

Hofman, H. M., Starkand, Y., and Kirson, M. W. (1976), *Nucl. Phys.*, A **266**, 138–162.

Holdeman, J. T., Jr. (1969), "A method for the approximation of functions defined by formal series expansion in orthogonal polynomials," *Math. Comp.*, **23**, 275–287.

Householder, A. S. (1970), *The Numerical Treatment of a Single Non-linear Equation*, McGraw-Hill, New York.

Householder, A. S. (1971), "The Padé table, the Frobenius identities and the q.d. algorithm," *Lin. Algebra and Its Appl.*, **4**, 161–174.

Householder, A. S. and Stewart, G. W., III (1969), "Bigradients, Hankel determinants and the Padé table," in B. Dejon and P. Henrici (eds.), *Constructive Aspects of the Fundamental Theorem of Algebra*, Academic Press, New York, pp. 131–150.

Hughes, Jones R. (1976), "General rational approximants in *N*-variables," *J. Approx. Theory*, **16**, 201–233.

Hughes, Jones R. and Makinson, G. J. (1974), "The generation of Chisholm rational approximants to power series in two-variables," *J. Inst. Math. Appl.*, **13**, 299–310.

Hunter, D. L. and Baker, G. A., Jr. (1973), "Methods of series analysis I: comparison of methods used in critical phenomena," *Phys. Rev. B*, **7**, 3346–3376.

Hunter, D. L. and Baker, G. A., Jr. (1979), "Methods of series analysis III: integral approximant methods," *Phys. Rev. B*, **19**, 3808–3821.

Ince, E. L. (1944), *Ordinary Differential Equations*, Dover Publications, New York.

Isaacson, E. and Keller, H. B. (1966), *Analysis of Numerical Methods*, J. Wiley.

Isenberg, C. (1963), "Moment calculations in lattice dynamics I, F.C.C. lattice with nearest neighbor interactions," *Phys. Rev.*, **132**, 2427–2433.

Iserles, A. (1979), "A note on Padé approximations and generalized hypergeometric functions," B.I.T., **19**, 543–545.

Jacobi, C. G. J. (1846), "Über die Darstellung einer Reihe gegebner Werthe durch eine gebrochne rationale Function," *J. für Reine Angewandte Math.*, **30**, 127–156.

Jarratt, P. (1970), "A review of methods for solving non-linear algebraic equations in one variable," in *Numerical Methods for Non-linear Algebraic Equations*, ed. P. Rabinowitz, Gordon and Breach, London, 1–26.

Jones, W. B. (1977), "Multiple point Padé tables," in E. B. Saff and R. H. Varga (eds.), *Padé and Rational Approximation*, Academic Press, New York, pp. 163–172.

Jones, W. B. and Thron, W. J. (1966), "Further properties of *T*-fractions," *Math. Annalen*, **166**, 106–118.

Jones, W. B. and Thron, W. J. (1971), "A posteriori bounds for the truncation error of continued fractions," *SIAM J. Num. Anal.*, **8**, 693–705.

Jones, W. B. and Thron, W. J. (1974), "Numerical stability in evaluating continued fractions," *Math. Comp.*, **28**, 795–810.

Jones, W. B. and Thron, W. J. (1975), "On the convergence of Padé approximants," *SIAM J. Math. Anal.*, **6**, 9–16.

Jones, W. B. and Thron, W. J. (1977), "Two point Padé tables and *T*-fractions," *Bull. Am. Math. Soc.*, **83**, 388–390.

Jones, W. B. and Thron, W. J. (1979), "Sequences of meromorphic functions corresponding to formal Laurent series," *SIAM J. Math. Anal.*, **10**, 1–17.

Jones, W. B. and Thron, W. J. (1980), *Continued Fractions, Analytic Theory and Applications*, Addison-Wesley, Reading, Mass.

Jones, W. B., Thron, W. J., and Waadeland, H. (1980), "A strong Stieltjes moment problem," Trans. Am. Math. Soc., **261**, 503–528.

Joyce, D. C. (1971), "Survey of extrapolation processes in numerical analysis," *SIAM Rev.*, **13**, 435–490.

Joyce, G. S. and Guttmann, A. J. (1973), "A new method of series analysis," in *Padé Approximants and their Applications*, ed. P. R. Graves-Morris, Academic Press, London, 163–167.

Kahaner, D. K. (1972), "Numerical quadrature by the ε-algorithm," *Math. Comp.*, **26**, 689–694.

Kailath, T., Vieira, A., and Morf, M. (1978), "Inverses of Toeplitz operators, innovations and orthogonal polynomials," *SIAM Review*, **20**, 106–119.

Karlin, S. (1968), *Total Positivity I*, Stanford Univ. Press.

Karlsson, J. (1976), "Rational interpolation and best rational interpolation," *J. Math. Anal. Appl.*, **53**, 38–52.

Karlsson, J. and von Sydow, B. (1976), "The convergence of Padé approximants to series of Stieltjes," *Arkiv for Matematik*, **14**, 43–53.

Karlsson, J. and Wallin, H. (1977), "Rational approximation by an interpolation procedure in several variables," in *Padé and Rational Approximation*, eds. E. B. Saff and R. H. Varga, Academic Press, New York, 83–100.

Khalil, A. B. (1977), "The K-matrix in distorted wave theory," *Nuovo Cim.*, **36A**, 354–366.

Killingbeck, J. (1977), "Quantum mechanical perturbation theory," *Rep. Prog. Phys.*, **40**, 963–1031.

Killingbeck, J. (1980), "The harmonic oscillator with λx^m perturbation," *J. Phys. A*, **13**, 49–56.

Kogbetlianz, E. G. (1960), "Generation of elementary functions," in A. Ralston and H. S. Wilf (eds.), *Mathematical Methods for Digital Computers*, Wiley, pp. 7–35.

Kronecker, L. (1881), "Zur Theorie der Elimination einer Variabeln aus zwei algebraischen Gleichungen," *Monatsberichte der Königlich Preussichen Akademie der Wissenschaften zu Berlin*, 535–600.

Laguerre, E. (1879), "Sur la réduction en fractions continues d'une fonction qui satisfait à une équation linéaire du premier ordre à coefficients rationnels," *Bull. Soc. Math de France*, **8**, 21–27.

Laguerre, E. (1885), "Sur la réduction en fractions continues d'une fraction qui satisfait à une équation différentielle linéaire du premier ordre dont les coefficients sont rationnels," *J. de Math, Pures et Appliqués*, **1**, 135–165.

Lambert, J. D. (1974), "Two unconventional classes of methods for stiff systems," in R. A. Willoughby (ed.), *Stiff Differential Equations*, pp. 171–186.

Lambert, J. D. (1975), "Computational methods in ordinary differential equations," J. Wiley.

Lambert, J. D. and Shaw, B., (1965), "On the numerical solution of $y' = f(x, y)$ by a class of formulae based on rational approximation," *Math. Comp.*, **19**, 456–462.

Lambert, J. D. and Shaw, B. (1966), "A generalization of multistep methods for ordinary differential equations," *Num. Math.*, **8**, 250–263.

Lanczos, C. (1952), "Solution of systems of linear equations by minimized iterations," *J. Res. of Nat. Bur. Stan.*, **49**, 33–53.

Lanczos, C. (1957), "Applied Analysis," Pitman Press.

Larkin, F. M. (1967), "Some techniques for rational interpolation," *Comp. J.*, **10**, 178–187.

Larkin, F. M. (1980), "Root finding by fitting rational fractions," *Math. Comp.*, **35**, 803–816.

Lee, T. D. and Yang, C. N. (1952), "Statistical theory of equations of state and phase transitions II. Lattice gas and Ising model," *Phys. Rev.*, **87**, 410–419.

Leighton, W. and Scott, W. T. (1939), "A general continued fraction expansion," *Bull. Am. Math. Soc.*, **45**, 596–605.

Levin, D. (1973), "Development of non-linear transformations for improving convergence of sequences," *Int. J. Comp. Math.*, **B3**, 371–388.

Levin, D. (1976), "General order Padé type rational approximants defined from double power series," *J. Inst. Math. Appl.*, **18**, 1–8.

Loeffel, J. J., Martin, A., Simon, B., and Wightman, A. S., *Phys. Lett.*, **30B**, 656–658.

Longman, I. M. (1966), "The application of rational approximations to the solution of problems in theoretical seismology," *Bull. Seismol. Soc. America*, **56**, 1045–1065.

Longman, I. M. (1972), "Numerical Laplace transform inversion of a function arising in viscoelasticity," *J. Comp. Phys.*, **10**, 224–231.

Longman, I. M. (1973), "Use of Padé table for approximate Laplace transform inversion," in P. R. Graves-Morris (ed.), *Padé Approximants and Their Applications*, Academic Press, London, pp. 131–134.

Longman, I. M. (1974), "Best rational function approximation for Laplace transform inversion," *SIAM J. Math. Anal.*, **5**, 574–580.

Lopez, C. and Yndurain, F. J. (1973), "Model independent calculation of Kp dispersion relations; extrapolation with Padé techniques," *Nucl. Phys. B*, **64**, 315–333.

Lovelace, C. (1964), "Practical theory of three-particle states I Non-relativistic," *Phys. Rev.*, **135B**, 1225–1249.

Lovelace, C. and Masson, D. (1962), "Calculation of Regge poles by continued fractions," *Nuovo Cim.*, **26**, 472–484.

Lubinsky, D. S. (1980), "Exceptional Sets of Padé approximants," Ph.D. thesis, Univ. of Witwatersrand.

Lubkin, S. (1952), "A method of summing infinite series," *J. Res. Nat. Bur. Stand.*, **48**, 228–254.

Luke, Y. L. (1958), "The Padé table and the τ-method," *J. Math and Phys.*, **37**, 110–127.

Luke, Y. L. (1960), "On the economic representations of transcendental functions," *J. Math. and Phys.*, **38**, 279–294.

Luke, Y. L. (1962), "On the approximate inversion of some Laplace transforms," in *Proc. 4th. U.S. Nat. Cong. Appl. Mech.*, pp. 269–276.

Luke, Y. L. (1970), "Evaluation of the gamma function by means of Padé approximations," *SIAM J. Num. Anal.*, **1**, 266–281.

Luke, Y. L. (1975), "On the error in the Padé approximants for a form of incomplete gamma function including the exponential," *SIAM J. Math. Anal.*, **6**, 829–839.

Luke, Y. L. (1977), "On the error in Padé approximations for functions defined by Stieltjes integrals," *Comp. and Math. with Appl.*, **3**, 307–314.

Luke, Y. L. (1978), "Error estimation in numerical inversion of Laplace transform using Padé approximation," *J. Franklin Inst.*, **305**, 259–273.

Luke, Y. L. (1980), "Computations of coefficients in the polynomials of Padé approximations by solving systems of linear equations," *J. Comp. and Appl. Math.*, **6**, 213–218.

Luke, Y. L., Fair, W., and Wimp, J. (1975), "Predictor-corrector formulas based on rational interpolants," *Int. J. Computers and Math. with Appl.*, **1**, 3–12.

Lutterodt, C. L. (1974), "A two-dimensional analogue of Padé approximant theory," *J. Phys. A*, **7**, 1027–1037.

Lyness, J. and Ninham, B. W. (1967), "Numerical quadrature and asymptotic expansions," *Math. Comp.*, **21**, 162–178.

McCabe, J. H. (1974), "A continued fraction expansion, with a truncation estimate, for Dawson's integral," *Math. Comp.*, **28**, 811–816.

McCabe, J. H. (1975), "A formal extension of the Padé table to include two point Padé quotients," *J. Inst. Math. Appl.*, **15**, 363–372.

McCabe, J. H. and Murphy, J. A. (1976), "Continued fractions which correspond to power series at two points," *J. Inst. Math. Appl.*, **17**, 233–247.

McCleod, J. B. (1971), "A note on the ε-algorithm," *Computing*, **7**, 17–24.

McCoy, B. M. and Wu, T. T. (1973), "The two dimensional Ising Model," Harvard Univ. Press, Cambridge.

McEliece, R. J. and Shearer, J. B. (1978), "A property of Euclid's algorithm and its application to Padé approximation," *SIAM J. Appl. Math.*, **34**, 611–615.

Maehly, H. J. (1956), October monthly progress report, Institute for advanced study, Princeton.

Maehly, H. J. (1960), "Rational approximations for transcendental functions," in *Proc. Int. Conf. on Inform. Proc., 1959*, Butterworth, pp. 57–62.

Machly, H. J. and Witzgall, Ch. (1960), "Tschebyscheff—Approximationen in kleinen Intervallen II. Stetigkeitsätze für gebrochen rationale Approximationen," *Num. Math.*, **2**, 293–307.

Magnen, J. and Sénéor, R. (1976), "The infinite volume limit of the ϕ_3^4 model," *Ann. d'Inst. H. Poincaré A*, **24**, 95–159.

Magnus, A. (1962), "Expansion of power series into *P*-functions," *Math. Z.*, **80**, 209–216.

Markov, A. (1884), "Démonstration de certaines inégalités de Tchebychef," *Math. Ann.*, **24**, 172–180.

Marx, I. (1963), "Remark concerning a non-linear sequence to sequence transformation," *J. Math. and Phys.*, **42**, 334–335.

Mason, J. C. (1967), "Chebychev polynomial approximations for the *L*-membrane eigenvalue problem," *SIAM J. Appl. Math.*, **15**, 172–181.

Masson, D. (1967a), "Padé approximant and the partial-wave integral equation," *J. Math. Phys.*, **8**, 512–514.

Masson, D. (1967b), "Analyticity in the potential strength," *J. Math. Phys.*, **8**, 2308–2314.

Meinguet, J. (1970), "On the solubility of the Cauchy interpolation problem," in *Approximation Theory*, ed. A. Talbot, Academic Press, London, 137–163.

Mery, P. (1977), "A variational approach to operator and matrix Padé approximation. Applications to potential scattering and field theory," in *Padé and Rational Approximation*, eds. E. B. Saff and R. H. Varga, Academic Press, New York, 375–388.

Merz, G. (1968), "Padésche Näherungsbrüche und Iterationsverfahren höherer Ordnung," *Comp.*, **3**, 165–183.

Michalik, B. (1970), "Distortion operator method and Padé approximants," *Ann. Phys.*, **57**, 201–213.

Mills, W. H. (1975), "Continued fraction and linear recurrences," *Math. Comp.*, **29**, 173–180.

Milne-Thomson, L. M. (1960), *Calculus of Finite Differences*, Macmillan, London.

de Montessus de Ballore, R. (1902), "Sur les fractions continues algébriques," *Bull. Soc. Math. de France*, **30**, 28–36.

de Montessus de Ballore, R. (1905), "Sur les fractions continues algébriques," *Rend. di Palermo*, **19**, 1–73.

Muir, T. (1960), *A Treatise on the Theory of Determinants*, Dover, New York.

Murphy, J. A. (1971), "Certain rational fraction approximations to $(1+x^2)^{-1/2}$," *J. Inst. Math. Appl.*, **7**, 138–150.

Murphy, J. A. and O'Donohoe, M. R. (1977), "A class of algorithms for rational approximation of functions formally defined by power series," *J. Appl. Math. and Phys. (ZAMP)*, **28**, 1121–1131.

Narcowich, F. J. and Allen, G. D. (1975), "Convergence of diagonal operator-valued Padé approximants to Dyson expansion," *Comm. Math. Phys.*, **45**, 153–157.

Nelson, E. (1973), "Construction of quantum fields from Markoff fields," *J. Funct. Anal.*, **12**, 97–112.

Newton, R. G. (1966), *Scattering Theory of Waves and Particles*, McGraw-Hill.

Ninham, B. W. and Lyness, J. (1969), "Asymptotic expansions for the error functional," *Math. Comp.*, **23**, 71–83.

Nørsett, S. P. (1974), "One step methods of Hermite type for numerical integration of stiff systems," *BIT*, **14**, 63–77.

Nuttall, J. (1966), "Padé approximants and bounds on the Bethe-Salpeter amplitude," *Phys. Lett.*, **23**, 492.

Nuttall, J. (1967), "Convergence of Padé approximants for the Bethe-Salpeter amplitude," *Phys. Rev.*, **157**, 1312–1316.

Nuttall, J. (1970a), "Connection of Padé approximants with stationary variational principles and the convergence of certain Padé approximants," in *The Padé Approximant in Theoretical Physics*, eds. G. A. Baker, Jr. and J. L. Gammel, Academic Press, New York 219–230.

Nuttall, J. (1970b), "Convergence of Padé approximants of meromorphic functions," *J. Math. Anal. Appl.*, **31**, 147–153.

Nuttall, J. (1973), "Variational principles and Padé approximants," in P. R. Graves-Morris (ed.), *Padé Approximants and Their Applications*, Academic Press, New York, pp. 29–40.

Nuttall, J. (1977), "The convergence of Padé approximants to functions with branch points," in E. B. Saff and R. H. Varga (eds.), *Padé and Rational Approximation*, Academic Press, New York, pp. 101–109.

Nuttall, J. and Singh, S. R. (1977), "Orthogonal polynomials and Padé approximants associated with a system of arcs," *J. Approx. Theory*, **21**, 1–42.

Ostrowski, A. (1925), "Uber Folgen analytischer Funktionen und einige Verschärfungen des Picarden Satzes," *Math. Zeit.*, **24**, 215–258.

Padé, H. (1892), "Sur la représentation approchée d'une fonction par des fractions rationelles," *Ann. de l'Ecole Normale Sup. 3ième Série*, **9**, Suppl., 3–93.

Padé, H. (1894), "Sur les séries entières convergents on divergents, et les fractions continues rationelles," *Acta Math.*, **18**, 97–112.

Padé, H. (1899), "Memoire sur les développements en fractions continues de la fonction exponentielle pouvant servir d'introduction à la théorie des fractions continues algébriques," *Ann. Sci. Ecole Norm. Sup.*, **16**, 395–426.

Padé, H. (1900), "Sur la distribution des réduites anormales d'une fonction," *Comptes Rendus*, **130**, 102–104.

Padé, H. (1901), "Sur l'expression générale de la fonction rationelle approchée de $(1+x)^m$," *Comptes Rendus*, **132**, 754–756.

Padé, H. (1907), "Recherches sur la convergence de développements en fractions continues d'une certaine catégorie de fonctions," *Ann. Sci. Ecole Norm. Sup.*, **24**.

Paydon, J. F. and Wall, H. S. (1942), "The continued fraction as a sequence of linear transformations," *Duke Math. J.*, **9**, 360–372.

Pekeris, C. L. (1958), "Ground state of two electron atoms," *Phys. Rev.*, **112**, 1649–1658.

Pekeris, C. L. (1959), "Binding energy of Helium," *Phys. Rev.*, **115**, 1216–1221.

Peres, A. (1963), "Mechanical model for quantum field theory," *J. Math. Phys.*, **4**, 332–333.

Perron, O. (1957), *Die Lehre von den Kettenbrüchen*, B. G. Teubner, Stuttgart.

Pfeuty, P., Jasnow, D., and Fisher, M. E. (1974), "Cross over scaling functions for exchange anisotropy," *Phys. Rev. B*, **10**, 2088–2112.

Pindor, M. (1976), "A simplified algorithm for calculating the Padé table derived from the Baker and Longman schemes," *J. Comp. Appl. Math.*, **2**, 255–258.

Pindor, M. (1979a), "Padé approximants and rational functions as tools for finding poles and zeros of analytical functions measured experimentally," in L. Wuytack (ed.), *Padé Approximation and Its Applications*, Academic Press, New York, pp. 338–347.

Pindor, M. (1979b), "Matrix variational Padé approximants and multistep square well potentials," *Lett. Math. Phys.*, **3**, 223–228.

Pindor, M. (1980), "Unambiguous results from variational matrix Padé approximants," CERN report.

Pommerenke, Ch. (1973), "Padé approximants and convergence in capacity," *J. Math. Anal. Appl.*, **41**, 775–780.

Pringsheim, A. (1910), "Uber Konvergenz und Funktionen-theoretischen Charakter gewisser limitärperiodischer Kettenbrüche," *Sitzungsberichte der Math.-Physikalische Klasse der Kgl. Bayerischen Akademie der Wissenschaften zu München*, **6**, 1–52.

Reed, M. and Simon, B. (1979), "Scattering Theory," vol. 3 in *Methods of Modern Math. Physics*, Academic Press, New York.

Rehr, J. J., Joyce, G. S., and Guttmann, A. J. (1980), "A recurrence technique for confluent singularity analysis of power series," *J. Phys. A*, **13**, 1587–1602.

Reid, W. T. (1972), *Riccati equations*, Academic Press, London.

Richardson, L. F. (1927), "The deferred approach to the limit, I. Single lattice," *Phil. Trans. Roy. Soc. Lond.*, **A226**, 299–349.

Riesz, F. and Nagy, B. v. Sz. (1955), *Functional Analysis*, Ungar.

Riesz, M. (1922a), "Sur le problème des moments," *Arkiv för Matematik, Astronomi och Fysik*, **16** (12), 1–23.

Riesz, M. (1922b), "Sur le problème des moments," *Arkiv för Matematik, Astronomi och Fysik*, **16** (19), 1–21.

Riesz, M. (1923), "Sur le problème des moments," *Arkiv för Matematik, Astronomi och Fysik*, **17**, (16), 1–52.

Riordan, J. (1966), *Combinatorial Identities*, Wiley.

Rissanen, J. (1973), "Algorithms for triangular decomposition of block Hankel and Toeplitz matrices with applications to factoring positive matrix polynomials," *Math. Comp.*, **27**, 147–155.

Rivlin, T. J. (1969), *An Introduction to the Theory of Functions*, Blaisdell.

Roberts, D. E. (1977), "An analysis of double power series using rotationally covariant approximants," *J. Comp. Appl. Math.*, **3**, 257–262.

Roberts, D. E., Griffiths, H. P., and Wood, D. W. (1975), "The analysis of double power series using Canterbury approximants," *J. Phys. A*, **8**, 1365–1372.

Robinson, P. D. and Barnsley, M. F. (1979), "Pointwise bivariational bounds on solutions of Fredholm integral equations," *SIAM J. Num. Anal.*, **16**, 135–144.

Rogers, L. J. (1907), "On the representation of certain asymptotic series as convergent continued fractions," *Proc. Lond. Math. Soc.*, **4**, 72–89.

Roth, A. (1938), "Approximationseigenschaften und Strahlengrenzwerte unendlich vieler linearer Gleichungen," *Comm. Math. Helv.*, **11**, 77–125.

Roth, R. (1965), "The qualifying examination," *Math. Mag.*, **38**, 166–167.

Rudin, W. (1976), "Principles of Mathematical Analysis," McGraw-Hill.

Ruijgrok, Th. W. (1968), "A new formulation of the theory of strong coupling," *Nucl. Phys. B*, **8**, 591–608.

Ruijgrok, Th. W. (1972), "A numerical study of analyticity in the coupling constant," *Nucl. Phys. B*, **39**, 616–642.

Runge, C. (1885), "Zur Theorie der eindeutigen analytischen Funktionen," *Acta Math.*, **6**, 229–244.

Rutishauser, H. (1954), "Der Quotienten-Differenzen-Algorithmus," *Zeit. für Angewandte Math.*, **5**, 233–251.

Rutishauser, H. (1957), *Der Quotienten-Differenzen-Algorithmus*, Birkhäuser Verlag, Basel.

Saff, E. B. (1972), "An extension of Montessus de Ballore's theorem on the convergence of interpolating rational functions," *J. Approx. Theory*, **6**, 63–67.

Saff, E. B., Schönhage, A., and Varga, R. S. (1978), "Geometrical convergence to e^{-z} by rational functions with real poles", *Num. Math.*, **25**, 307–322.

Saff, E. B. and Varga, R. S. (1975), "On the zeros and poles of Padé approximants to e^z," *Num. Math.*, **25**, 1–14.

Saff, E. B. and Varga, R. S. (1976a), "Zero-free parabolic regions for sequences of polynomials," *SIAM J. Math. Anal.*, **7**, 344–357.

Saff, E. B. and Varga, R. S. (1976b), "On the sharpness of theorems concerning zero free regions for certain sequences of polynomials," *Num. Math.*, **26**, 345–354.

Saff, E. B. and Varga, R. S. (1977), "On the zeros and poles of Padé approximants to e^z. II," in E. B. Saff and R. S. Varga (eds.), *Padé and Rational Approximation*, Academic Press, New York, pp. 195–214.

Saff, E. B. and Varga, R. S. (1978), "Non-uniqueness of best rational approximations to real functions on real intervals," *J. Approx. Theory*, **23**, 78–85.

Saff, E. B., Varga, R. S., and Ni, W.-C. (1976), "Geometric convergence of rational approximations to e^{-z} in infinite sectors," *Num. Math.*, **26**, 211–225.

Salzer, H. E. (1962), "Note on osculatory rational interpolation," *Math. Comp.*, **16**, 486–491.

Schlessinger, L. and Schwartz, C. (1966), "Analyticity as a computational tool," *Phys. Rev. Lett.*, **16**, 1173–1174.

Schoenberg, I. J. (1951), "On the Pólya frequency functions I: The totally positive functions and their Laplace transforms," *J. d'Anal. Math.*, **1**, 331–374.

Schwartz, C. (1966), "Information content of the Born series," *J. Comp. Phys.*, **1**, 21–28.

Schwartz, C. and Zemach, C. (1966), "Theory and calculation with the Bethe-Salpeter equation," *Phys. Rev.*, **141**, 1454–1467.

Schweber, S. S. (1961), *Introduction to Relativistic Quantum Field Theory*, Harper and Row, New York.

Schofield, D. F. (1972), "Continued fraction method for perturbation theory," *Phys. Rev. Lett.*, **29**, 811–814.

Scott, W. T. and Wall, H. S. (1940a), "Continued fraction expansions for arbitrary power series," *Ann. Math.*, **41**, 328–349.

Scott, W. T. and Wall, H. S. (1940b), "A convergence theorem for continued fractions," *Trans. Am. Math. Soc.*, **47**, 155–172.

Shafer, R. E. (1974), "On quadratic approximation," *SIAM J. Num. Anal.*, **11**, 447–460.

Shanks, D. (1955), "Non-linear transformations of divergent and slowly convergent sequences," *J. Math. Phys.*, **34**, 1–42.

Sheludyak, Yu. E. and Rabinovich, V. A. (1979), "On the values of the critical indices of the three dimensional Ising model," *High Temp.*, **17**, 40–43.

Short, L. (1978), "The practical evaluation of multivariate approximants with branch points," *Proc. Roy. Soc. London A*, **362**, 57–69.

Sidi, A. (1979), "Convergence properties of some non-linear sequence transformations," *Math. Comp.*, **33**, 315–326.

Sidi, A. (1980a), "Some aspects of two-point Padé approximants," *J. Comp. Appld. Math.*, **6**, 9–17.

Sidi, A. (1980b), "Analysis of the convergence of the T-transformation for power series," *Math. Comp.*, **35**, 833–850.

Siemieniuch, J. L. (1976), "Properties of certain rational approximations to e^{-z}," *B.I.T.*, **16**, 172–191.

Simon, B. (1970), "Coupling constant analyticity for the anharmonic oscillator," *Ann. Phys.*, **58**, 76–136.

Simon, B. (1972), "The anharmonic oscillator—a singular perturbation theory," in D. Bessis (ed.), *Cargèse Lecture Notes in Physics*, Vol. 5, Gordon and Breach, pp. 383–414.

Smith, D. A. and Ford, W. F. (1979), "Acceleration of linear and logarithmic convergence," *SIAM J. Num. Anal.*, **16**, 223–240.

Smith, I. M., Siemieniuch, J. L., and Gladwell, I. (1977), "Evaluating Nørsett methods for integrating differential equations in time," *Int. J. Num. Anal. Meth. in Geomechanics*, **1**, 57–74.

Smithies, F. (1958), *Integral Equations*, Cambridge University Press.

Starkand, Y. (1976), "Subroutine for calculation of matrix Padé approximants," *Comm. Comp. Phys.*, **11**, 325–330.

Stern, M. S. and Warburton, A. E. A. (1972), "Finite difference solution of the partial wave Schrödinger equation," *J. Phys. A*, **5**, 112–124.

Stieltjes, T. J. (1884), "Quelques recherches sur les quadratures dites méchaniques," *Ann. Sci. Ecole. Norm. Sup.*, **1**, 409–426.

Stieltjes, T. J. (1889), "Sur la réduction en fraction continue d'une série précédant suivant les puissances descendents d'une variable," *Ann. Fac. Sci. Toulouse*, **3H**, 1–17.

Stieltjes, T. J. (1894), "Recherches sur les fractions continues," *Ann. Fac. Sci. Toulouse*, **8J**, 1–122; **9A**, 1–47.

Stoer, J. (1961), "Uber zwei Algorithmen zur Interpolation mit rationalen Funktionen," *Num. Math.*, **3**, 285–304.

Stone, M. H. (1932), *Linear Transformations in Hilbert Space*, Am. Math. Soc., Providence.

Takehasi and Mori, (1971), "Estimation of errors in the numerical quadrature of analytic functions," *Applicable Analysis*, **1**, 201–229.

Talbot, A. (1979), "The accurate numerical inversion of Laplace transforms," *J. Inst. Math. Appl.*, **23**, 97–120.

Tani, S. (1965), "Padé approximants in potential scattering," *Phys. Rev.*, **139**, B1011–1020.

Tani, S. (1966a), "Complete continuity of the kernel in generalized potential scattering I. Short range interaction without strong singularity," *Ann. Phys.*, **37**, 411–450.

Tani, S. (1966b), "Complete continuity of the kernel in generalized potential scattering II. Generalized Fourier series expansion," *Ann. Phys.*, **37**, 451–406.

Taylor, J. M. (1978), "The condition of Gram matrices and related problems," *Proc. Roy. Soc. Edin.*, **80A**, 45–56.

Tchebycheff, P. (1858), "Sur les fractions continues," *J. de Math.*, **8**, 289–323.

Tchebycheff, P. (1874), "Sur les valeurs limites des intégrales," *J. de Math. Pures et Appl.*, **19**, 157–160.

Thacher, H. C. and Tukey, J. (1960), "Rational interpolation made easy by a recursive algorithm," unpublished manuscript.

Thacher, H. C. (1974), "Numerical application of the generalized Euler transformation," in J. L. Rosenfeld (ed.), *Information Processing 74*, North Holland, Amsterdam, pp. 627–631.

Thiele, T. N. (1909), "Interpolationsrechnung," Teubner.

Thompson, C. J., Guttman, A. J., and Ninham, B. W. (1969), "Determination of critical behaviour in lattice statistics from series expansions, II," *J. Phys. C*, **2**, 1889–1899.

Thron, W. J. (1948), "Some properties of continued fractions $1 + d_0 z + K(z/(1 + d_n z))$," *Bull. Am. Math. Soc.*, **54**, 206–218.

Thron, W. J. (1961), "Convergence regions for continued fractions and other infinite processes," *Am. Math. Monthly*, **68**, 734–750.

Thron, W. J. (1974), "A survey of recent convergence results for continued fractions," *Rocky Mtn. J. Math.*, **4**, 273–282.

Thron, W. J. (1977), "Two point Padé tables, *T*-fractions and sequences of Schur," in E. B. Saff and R. S. Varga (eds.), *Padé and Rational Approximation*, Academic Press, New York, pp. 215–225.

Titchmarsh, E. C. (1939), *Theory of Functions*, Oxford University Press.

Tjon, J. A. (1970), "The Padé approximant in three body calculations," *Phys. Rev. D*, **1**, 2109–2112.

Tjon, J. A. (1973), "Application of Padé approximants in the three body problem," in *Padé Approximants*, ed. P. R. Graves-Morris, Inst. of Phys., Bristol, 241–252.

Tjon, J. A. (1977), "Operator Padé approximants and three body scattering," in *Padé and Rational Approximation*, eds. E. B. Saff and R. S. Varga, Academic Press, New York, 389–396.

Traub, J. F. (1964), *Iterative Methods for the Solution of Equations*, Prentice-Hall, Englewood Cliffs, N.J.

Trench, W. (1964), "An algorithm for the inversion of finite Toeplitz matrices," *SIAM J. Appl. Math.*, **12**, 515–522.

Trench, W. (1965), "An algorithm for the inversion of finite Hankel matrices," *SIAM J. Appl. Math.*, **13**, 1102–1107.

Tricomi, F. G. (1957), *Integral Equations*, Interscience.

Trudi, N. (1862), *Teoria de' Determinati e Loro Applicazioni*, Libreria Scientifica e Industriale de B. Pellerano, Napoli.

Turchetti, G. (1976), "Variational principles and matrix approximants in potential theory," *Lett. Nuovo Cim.*, **15**, 129–133.

Turchetti, G. (1978), "Variational matrix Padé approximants in two body scattering," *Fortsch. Physik*, **26**, 1–28.

Van Rossum, H. (1971), "On the poles of Padé approximants to e^z," *Nieuw Archief voor Wiskunde*, **29**, 37–45.

Van Vleck, E. B. (1903), "On an extension of the 1894 memoir of Stieltjes," *Trans. Am. Math. Soc.*, **4**, 297–332.

Van Vleck, E. B. (1904), "On the convergence of algebraic continued fractions whose coefficients have limiting values," *Trans. Am. Math. Soc.*, **5**, 253–262.

Vargu, R. S. (1961), "On higher order stable implicit methods for solving parabolic partial differential equations," *J. Math. Phys.*, **40**, 220–231.

Varga, R. S. (1962), *Matrix Iterative Analysis*, Prentice-Hall, Englewood Cliffs.

Villani, M. (1972), "A summation method for perturbative series with divergent terms," in D. Bessis (ed.), *Cargèse Lecture Notes in Physics*, Vol. 5, Gordon and Breach, New York, pp. 461–474.

Viskovatov, B. (1803–1806), "De la méthode générale pour réduire toutes sortes des quantités en fractions continues," *Memoires de L'Academie Impériale des Sciences de St. Petersburg*, **1**, 226–247.

Vorobyev, Yu. V. (1965), *Method of Moments in Applied Mathematics*, Gordon and Breach, New York.

Waadeland, H. (1979), "General *T*-fractions corresponding to functions satisfying certain boundedness conditions," *J. Approx. Theory*, **26**, 317–328.

Wall, H. S. (1931), "On the Padé approximants associated with a positive definite power series," *Trans. Am. Math. Soc.*, **33**, 511–532.

Wall, H. S. (1932a), "On the relationship among the diagonal files of a Padé table," *Bull. Am. Math. Soc.*, **38**, 752–760.

Wall, H. S. (1932b), "General theorems on the convergence of sequences of Padé approximants," *Trans. Am. Math. Soc.*, **34**, 409–416.

Wall, H. S. (1945), "Note on a certain continued fraction," *Bull. Am. Math. Soc.*, **51**, 930–934.

Wall, H. S. (1948), *The Analytic Theory of Continued Fractions*, Van Nostrand, Princeton, N.J.

Wallin, H. (1972), "On the convergence theory of Padé approximants," in P. L. Butzer, J.-P. Kahane, and B. Sz-Nagy (eds.), *Linear Operators and Approximation* (Proc. Conf. at Oberwolfach), Birkhäuser, Basel, pp. 461–469.

Wallin, H. (1974), "The convergence of Padé approximants and the size of the power series coefficients," *Appl. Anal.*, **4**, 235–251.

Walsh, J. L. (1964a), "Convergence of sequences of rational functions of best approximation. I," *Math. Ann.*, **155**, 252–264.

Walsh, J. L. (1964b), "Padé approximants as limits of rational functions of best approximation," *J. Math. and Mech.*, **13**, 305–312.

Walsh, J. L. (1965a), "Convergence of sequences of rational functions of best approximation II," *Trans. Am. Math. Soc.*, **116**, 227–237.

Walsh, J. L. (1965b), "The convergence of sequences of rational functions of best approximation with some free poles," in *Approximation of Functions*, ed. H. L. Garabedian, Elsevier, Amsterdam, 1–16.

Walsh, J. L. (1967), "On the convergence of sequences of rational functions," *SIAM J. Num. Anal.*, **4**, 211–221.

Walsh, J. L. (1969), "Interpolation and approximation by rational functions in the complex domain," *Am. Math. Soc. Colloq. Pub.*, Vol. 20, Providence, R.I.

Walsh, J. L. (1970), "Approximation by rational functions: open problems," *J. Approx. Theory*, **3**, 236–242.

Warburton, A. E. A. and Stern, M. S. (1968), "Two body off-shell potential scattering," *Nuovo Cim.*, **60A**, 131–159.

Warner, D. D. (1976), "An extension of Saff's theorem on the convergence of interpolating rational functions," *J. Approx. Theory*, **18**, 108–118.

Watson, P. J. S. (1973), "Algorithms for differentiation and integration," in P. R. Graves-Morris (ed.), *Padé Approximants and Their Applications*, Academic Press, London, pp. 93–98.

Watson, P. J. S. (1974), "Two variable rational approximants–a new method," *J. Phys. A*, **7**, L167–L170.

Werner, H. (1979), "A reliable method for rational interpolation," in L. Wuytack (ed.), *Padé Approximation and Its Applications*, Springer Lecture Notes in Mathematics, No. 765, pp. 257–277.

Werner, H. (1980), "Ein algorithmus zur rationalen Interpolation," in *Numerische Methoden der Approximationstheorie*, **5**, eds. L. Collatz, G. Meinardus and H. Werner, Birkhäuser, 319–337.

Werner, H. and Schaback, R. (1972), "Praktische Mathematik II," Springer Verlag, Berlin.

Werner, H. and Wuytack, L. (1978), "Non-linear quadrature rules in the presence of a singularity," *Comp. & Math. with Applcns.*, **4**, 237–245 (1981).

Wetterling, W. (1963), "Ein Interpolationsverfahren zur Lösung der linearen Gleichungssysteme, die bei der rationalen Tchebyscheff-Approximation auftreten," *Archiv. Rat. Mech. und Anal.*, **12**, 403–408.

Wheeler, J. C. and Gordon, R. G. (1970), "Bounds for averages using moment constraints," in G. A. Baker, Jr., and J. L. Gammel (eds.), *The Padé Approximant in Theoretical Physics*, Academic Press, New York, pp. 99–128.

Widder, D. V. (1972), *The Laplace Transform*, Princeton University Press.

Wilkinson, J. H. (1963), *Rounding Error in Algebraic Processes*, Notes in Applied Science, No. 32, H.M.S.O., London.

Wilkinson, J. H. (1965), *The Algebraic Eigenvalue Problem*, Oxford University Press.

Wilson, R. (1927), "Divergent continued fractions and polar singularities," *Proc. Lond. Math. Soc.*, **26**, 159–168.

Wilson, R. (1928a), "Divergent continued fractions and polar singularities II. Boundary pole multiple," *Proc. Lond. Math. Soc.*, **27**, 497–512.

Wilson, R. (1928b), "Divergent continued fractions and polar singularities III. Several boundary poles," *Proc. Lond. Math. Soc.*, **28**, 128–144.

Wilson, R. (1930), "Divergent continued fractions and non-polar singularities," *Proc. Lond. Math. Soc.*, **30**, 38–57.

Wood, D. W. and Griffiths, H. P. (1974), "Chisholm approximants and critical phenomena," *J. Phys. A*, **7**, L101–L104.

Wright, K. (1970), "Some relationships between implicit Runge-Kutta, collocation and Lanczos τ-methods and their stability properties," *BIT*, **10**, 217–227.

Wuytack, L. (1973), "An algorithm for rational interpolation similar to the q.d.-algorithm," *Num. Math.*, **20** 418–424.

Wuytack, L. (1974a), "Numerical integration by using non-linear techniques," *J. Comp. Appl. Math.*, **1**, 261–272.

Wuytack, L. (1974b), "On some aspects of the rational interpolation problem," *SIAM J. Num. Anal.*, **11**, 52–60.

Wuytack, L. (1975), "On the osculatory rational interpolation problem," *Math. Comp.*, **29**, 837–843.

Wuytack, L. (1979), "Commented bibliography," in L. Wuytack (ed.), *Padé Approximation and Its Applications*, Springer Lecture Notes in Mathematics, No. 765, pp. 375–392.

Wynn, P. (1956), "On a device for calculating the $e_m(S_n)$ transformation," *Math. Tables and A.C.*, **10**, 91–96.

Wynn, P. (1960), "The rational approximation of functions which are formally defined by power series expansions," *Math. Comp.*, **14**, 147–186.

Wynn, P. (1961a), "The epsilon algorithm and operational formulas of numerical analysis," *Math. Comp.*, **15**, 151–158.

Wynn, P. (1961b), "L'ε-algoritmo e la tavola di Padé," *Rendi. Mat. Roma*, **20**, 403–408.

Wynn, P. (1962a), "The numerical efficiency of certain continued fraction expansions," *Proc. Kon. Ned. Akad. Wetensch. A*, **65**, 127–154.

Wynn, P. (1962b), "Acceleration techniques for iterated vector and matrix problems," *Math. Comp.*, **16**, 301–322.

Wynn, P. (1963), "Continued fractions whose coefficients obey a non-commutative law of multiplication," *Arch. Rat. Mech. Anal.*, **12**, 273–312.

Wynn, P. (1964), "General purpose vector epsilon algorithm procedures," *Num. Math.*, **6**, 22–36.

Wynn, P. (1966a), "Upon systems of recursions which obtain among quotients of the Padé table," *Num. Math.*, **8**, 264–269.

Wynn, P. (1966b), "On the convergence and stability of the ε-algorithm," *SIAM J. Num. Anal.*, **3**, 91–122.

Wynn, P. (1967), "A general system of orthogonal polynomials," *Quart. J. Math.*, **18**, 81–96.

Wynn, P. (1968), "Upon the Padé table derived from a Stieltjes series," *SIAM J. Num. Anal.*, **5**, 805–834.

Wynn, P. (1973), "On the zeros of certain confluent hypergeometric functions," *Proc. Am. Math. Soc.*, **40**, 173–182.

Wynn, P. (1974), "Some recent developments in the theories of continued fractions and the Padé table," *Rocky Mtn. J. Math.*, **4**, 297–323.

Wynn, P. (1977), "The transformation of series by the use of Padé quotients and more general approximants," in E. B. Saff and R. S. Varga (eds.), *Padé and Rational Approximation*, Academic Press, New York, pp. 121–146.

Young, R. C., Biedenharn, L. C., and Feenberg, E. (1957), "Continued fraction approximants to the Brillouin-Wigner Perturbation series," *Phys. Rev.*, **106**, 1151–1155.

Zinn-Justin, J. (1970), "Strong interaction dynamics with Padé approximants," *Phys. Repts.*, **1C**, 56–102.

Zinn-Justin, J. (1971), "Convergence of Padé approximants in the general case," in A. Visconti (ed.), *Colloquium on Advanced Computing Methods in Theoretical Physics*, CNRS, Marseille.

Zinn-Justin, J. (1973), "Recent developments in the theory of Padé approximants," in A. Visconti (ed.), *International Colloquium on Advanced Computing Methods in Theoretical Physics*, Vol. 2, CNRS, Marseille, p. C-XIII-1.

Zohar, S. (1974), "The solution of a Toeplitz set of linear equations," *J. Assoc. Comp. Mech.*, **21**, 272–276.

Bibliography of General Padé-Approximant Reviews

Baker, G. A., Jr. (1965), "The theory and application of the Padé approximant method," *Advances in Theoretical Physics*, **1**, 1–58.

Baker, G. A., Jr. (1975), *The Essentials of Padé Approximants*, Academic Press, New York.

Baker, G. A., Jr. and Gammel, J. L., eds. (1970), *The Padé Approximant in Theoretical Physics*, Academic Press, London.

Basdevant, J.-L. (1972), "The Padé approximant and its physical applications," *Fortschr. Phys.*, **20**, 282–331.

Bausset, M., ed. (1973), *Accélération des Convergences*, Centre Universitaire de Toulon.

Bessis, D., ed. (1972), *Cargèse Lectures in Physics*, Vol. **5**, Gordon and Breach, New York.

Bessis, D., Gilewicz, J., and Mery, P., eds. (1975), *Proceedings of a Workshop on Padé Approximants*, C.N.R.S., Marseille.

Brezinski, C. (1977), *Accélération de la Convergence en Analyse Numérique*, Springer Lecture Notes in Mathematics, No. 584.

Brezinski, C. (1980), "Padé type approximation and general orthogonal polynomials," Birkhäuser Verlag.

Cabannes, H., ed. (1976), *Padé Approximant Method and its Application to Mechanics* (Proceedings of the Euromech Colloquium, Toulon, 1975), Springer Lecture Notes in Physics, No. 47.

Chui, C. K. (1976), "Recent results on Padé approximants and related problems," in G. G. Lorentz, C. K. Chui and L. L. Schumaker (eds.), *Approximation Theory*, Academic Press, New York, pp. 79–116.

Claessens, G. (1976), "Some aspects of the rational Hermite interpolation table and its applications," Ph. D. Thesis, University of Antwerp.

Gilewicz, J. (1978), *Approximants de Padé*, Springer Lecture Notes in Mathematics, No. 667.

Gragg, W. B. (1972), "The Padé table and its relation to certain algorithms of numerical analysis," *SIAM Review*, **14**, 1–62.

Graves-Morris, P. R., ed. (1973), *Padé Approximants and Their Applications*, Academic Press, London.

Graves-Morris, P. R., ed. (1973), *Padé Approximants*, Inst. of Phys. Press, London.

Jones, W. B. and Thron, W. J. (1974), "Proceedings of the international conference on Padé approximants, continued fractions and related topics," *Rocky Mtn. J. Math.*, **4**, 135–397.

Padé, H. (1892), "Sur la representation approchée d'une fonction par des fractions rationelles," *Annales de l'Ecole Normale*, **9**, suppl. 3, 93.

Saff, E. B. and Varga, R. S., eds. (1977), *Padé and Rational Interpolation*, Academic Press, New York.

Warner, D. D. (1974), "Hermite interpolation with rational functions," Ph. D. Thesis, University of California.

Wimp, J. (1981), "Sequence Transformations and Their Applications," Academic Press, New York.

Wuytack, L. (1976), "Applications of Padé approximation in numerical analysis," in Schaback, R. and Scherer, K. (eds.), *Approximation Theory*, Springer Lecture Notes in Mathematics, Vol. 556, pp. 453–461.

Wuytack, L., ed. (1979), *Padé Approximation and Its Applications*, Springer Lecture Notes in Mathematics, Vol. 765.

Zinn-Justin, J. (1970), "Strong interaction dynamics with Padé approximants," *Phys. Reports*, **1C**, 56–102.

Selected bibliography on Continued Fractions and Method of Moments

Akhiezer, N. I. (1965), *The Classical Moment Problem*, Oliver and Boyd, London.

Jones, W. B. and Thron, W. J. (1980), "Continued fractions. Analytic theory and applications," in G.-C. Rota (ed.), *Encyclopedia of Mathematics and Its Applications*, Addison-Wesley, Reading, Mass.

Khovanskii, A. N. (1963), *Application of Continued Fractions and Their Generalizations to Problems in Approximation Theory*, Noordhoff.

Perron, O. (1957), *Die Lehre von der Kettenbrüchen*, Vol. 2, Teubner B. G., Stuttgart.

Shohat, J. A. and Tamarkin, J. D. (1963), The Problem of Moments, American Math. Soc. Publ., Vol. 1, Providence, R.I.

Vorobyev, Yu. V. (1965), *Method of Moments in Applied Mathematics*, Gordon and Breach, New York.

Wall, H. S. (1929), "On the Padé approximants associated with the continued fraction and series of Stieltjes," *Trans. Am. Math. Soc.*, **31**, 91–115.

Wall, H. S. (1948), *The Analytic Theory of Continued Fractions*, Van Nostrand, Princeton.

Index for Part I and Part II